Springer Series in Statistics

Probability and its Applications

A Series of the Applied Probability Trust

Springer Series in Statistics

Anderson: Continuous-Time Markov Chains: An Applications-Oriented Approach.
Andrews/Herzberg: Data: A Collection of Problems from Many Fields for the Student and Research Worker.
Anscombe: Computing in Statistical Science through APL.
Berger: Statistical Decision Theory and Bayesian Analysis, 2nd edition.
Bolfarine/Zacks: Prediction Theory for Finite Populations.
Brémaud: Point Processes and Queues: Martingale Dynamics.
Brockwell/Davis: Time Series: Theory and Methods, 2nd edition.
Choi: ARMA Model Identification
Daley/Vere-Jones: An Introduction to the Theory of Point Processes.
Dzhaparidze: Parameter Estimation and Hypothesis Testing in Spectral Analysis of Stationary Time Series.
Farrell: Multivariate Calculation.
Fienberg/Hoaglin/Kruskal/Tanur (Eds.): A Statistical Model: Frederick Mosteller's Contributions to Statistics, Science, and Public Policy.
Goodman/Kruskal: Measures of Association for Cross Classifications.
Grandell: Aspects of Risk Theory.
Hall: The Bootstrap and Edgeworth Expansion.
Härdle: Smoothing Techniques: With Implementation in S.
Hartigan: Bayes Theory.
Heyer: Theory of Statistical Experiments.
Jolliffe: Principal Component Analysis.
Kotz/Johnson (Eds.): Breakthroughs in Statistics Volume I.
Kotz/Johnson (Eds.): Breakthroughs in Statistics Volume II.
Kres: Statistical Tables for Multivariate Analysis.
Leadbetter/Lindgren/Rootzén: Extremes and Related Properties of Random Sequences and Processes.
Le Cam: Asymptotic Methods in Statistical Decision Theory.
Le Cam/Yang: Asymptotics in Statistics: Some Basic Concepts.
Manoukian: Modern Concepts and Theorems of Mathematical Statistics.
Miller, Jr.: Simultaneous Statistical Inference, 2nd edition.
Mosteller/Wallace: Applied Bayesian and Classical Inference: The Case of *The Federalist Papers.*
Pollard: Convergence of Stochastic Processes.
Pratt/Gibbons: Concepts of Nonparametric Theory.
Read/Cressie: Goodness-of-Fit Statistics for Discrete Multivariate Data.
Reiss: Approximate Distributions of Order Statistics: With Applications to Nonparametric Statistics.
Ross: Nonlinear Estimation.
Sachs: Applied Statistics: A Handbook of Techniques, 2nd edition.
Salsburg: The Use of Restricted Significance Tests in Clinical Trials.
Särndal/Swensson/Wretman: Model Assisted Survey Sampling.
Seneta: Non-Negative Matrices and Markov Chains.

(continued after index)

Petar Todorovic

An Introduction to Stochastic Processes and Their Applications

With 15 Illustrations

Springer-Verlag
New York Berlin Heidelberg London Paris
Tokyo Hong Kong Barcelona Budapest

Petar Todorovic
Department of Statistics and
Applied Probability
University of California—Santa Barbara
Santa Barbara, CA 93106
USA

Series Editors:
J. Gani
Department of Statistics
University of California
Santa Barbara, CA 93106
USA

C.C. Heyde
Department of Statistics
Institute of Advanced Studies
The Australian National University
GPO Box 4, Canberra ACT 2601
Australia

Mathematics Subject Classification (1991): 60G07, 60G12, 60J25

Library of Congress Cataloging-in-Publication Data
Todorovic, P. (Petar)
An introduction to stochastic processes and their applications / by P. Todorovic.
p. cm. — (Springer series in statistics)
Includes bibliographical references and index.
ISBN-13:978-1-4613-9744-1 e-ISBN-13:978-1-4613-9742-7
DOI: 10.1007/978-1-4613-9742-7
1. Stochastic processes. I. Series.
QA274.T64 1992
519.2—dc20 91-46692

Printed on acid-free paper.

Production managed by Henry Krell; manufacturing supervised by Robert Paella.
Typeset by Asco Trade Typesetting Ltd., Hong Kong.

9 8 7 6 5 4 3 2 1

ISBN-13:978-1-4613-9744-1

To my wife Živadinka

Preface

This text on stochastic processes and their applications is based on a set of lectures given during the past several years at the University of California, Santa Barbara (UCSB). It is an introductory graduate course designed for classroom purposes. Its objective is to provide graduate students of statistics with an overview of some basic methods and techniques in the theory of stochastic processes. The only prerequisites are some rudiments of measure and integration theory and an intermediate course in probability theory. There are more than 50 examples and applications and 243 problems and complements which appear at the end of each chapter.

The book consists of 10 chapters. Basic concepts and definitions are provided in Chapter 1. This chapter also contains a number of motivating examples and applications illustrating the practical use of the concepts. The last five sections are devoted to topics such as separability, continuity, and measurability of random processes, which are discussed in some detail.

The concept of a simple point process on R_+ is introduced in Chapter 2. Using the coupling inequality and Le Cam's lemma, it is shown that if its counting function is stochastically continuous and has independent increments, the point process is Poisson. When the counting function is Markovian, the sequence of arrival times is also a Markov process. Some related topics such as independent thinning and marked point processes are also discussed. In the final section, an application of these results to flood modeling is presented.

Chapter 3 provides a short introduction to the theory of one-dimensional Brownian motion. Principal topics here are hitting times, extremes, the reflection principle, properties of sample paths and the law of the iterated logarithm. The chapter ends with a discussion of the Langevin equation, the Ornstein–Uhlenbeck process and stochastic integration.

Chapter 4 deals with the theory of Gaussian processes. It begins with a brief account of the relevant matrix theory and the definition of a multivariate Gaussian distribution and its characteristic function, which is worked out in detail. We show that a system of random variables is Gaussian if and only if every linear combination of these variables is normal. We also discuss the Markov–Gaussian process and prove Doob's theorem, which asserts that the only stationary Gaussian processes are the Ornstein–Uhlenbeck processes.

Chapter 5 contains a brief introduction to the Hilbert space L_2, which has some particular features not shared by other L_p spaces. Here the emphasis is on those topics essential in subsequent sections. They include the Riesz–Fisher theorem, the structure of L_2 spaces, the concept of orthogonal projection and orthogonal basis, separability, and linear and projection operators.

Chapter 6 deals with the theory of second order (or L_2) processes, which are characterized up to Hilbert space isometry by their covariance functions. The focus here is on the covariance function and its properties. It is natural to have criteria for continuity, etc., expressed in terms of the covariance function. Expansion of the covariance function in terms of its eigenvalues and eigenfunctions, as well as the Karhunen–Loève expansion are discussed in some detail.

The first part of Chapter 7 is concerned with the spectral analysis of (wide sense) stationary processes. The gist of this section is the "spectral representation" of a stationary process, which establishes an isometric isomorphism between the closed linear manifold spanned by the random variables of the process and a certain L_2 space of complex functions. With the groundwork laid, the problem of estimation (and its special cases filtering and prediction) can now be investigated. The method for solving the prediction problem described here is due to Yaglom. Its starting point is the spectral representation of the process. However, the results obtained are most useful for rational spectral densities. Finally, the Wold decomposition is also considered in some detail.

Chapter 8, an introduction to Markov processes, consists of three parts. The first lists some basic features of homogeneous Markov processes: it is shown that the existence of a stationary measure is a necessary and sufficient condition for the process to be strictly stationary. The second part treats a class of homogeneous Markov processes with countable state space. The focus here is on the transition probability and its properties. If sample paths of the Markov process are right continuous, then its transition probability is not only uniformly continuous but also differentiable. This is used to derive Kolmogorov's backward and forward differential equations. In this section we also introduce the concept of the "strong Markov" property and discuss the structure of Markov chains. The last part is concerned with homogeneous diffusion. We briefly describe Itô's approach, which shows that a diffusion process is governed by a first-order stochastic differential equation which depends on a standard Brownian motion process.

Chapter 9 provides an introduction to the application of semigroup theory to Markov processes, whereas Chapter 10 discusses some rudiments of the theory of discrete parameter martingales.

I would like to point out that after Chapter 1 (or at least the first half of it) one can move directly to most of the other chapters. Chapter 5, however, is a necessary prerequisite for reading Chapters 6, 7, and 9. The course has been tested over years on graduate students of statistics at the University of California, Santa Barbara, and contains material suitable for an introductory as well as a more advanced course in stochastic processes.

For encouragement, support, and valuable advice, I am glad to thank Dr. Joe Gani. I am also grateful to the referees including William Griffith and Gennady Samorodnitsky for their comments on the first draft of this book. My special thanks to Chris Heyde for his extraordinarily careful reading of the whole manuscript and for correcting numerous errors and misprints. Finally, I acknowledge with warm thanks my indebtedness to colleagues and students at the UCSB Department of Statistics and Applied Probability.

Petar Todorovic

Contents

Preface vii

CHAPTER 1
Basic Concepts and Definitions 1

1.1. Definition of a Stochastic Process 1
1.2. Sample Functions 2
1.3. Equivalent Stochastic Processes 5
1.4. Kolmogorov Construction 7
1.5. Principal Classes of Random Processes 8
1.6. Some Applications 12
1.7. Separability 18
1.8. Some Examples 21
1.9. Continuity Concepts 22
1.10. More on Separability and Continuity 25
1.11. Measurable Random Processes 27
Problems and Complements 30

CHAPTER 2
The Poisson Process and Its Ramifications 34

2.1. Introduction 34
2.2. Simple Point Process on R_+ 35
2.3. Some Auxiliary Results 37
2.4. Definition of a Poisson Process 40
2.5. Arrival Times $\{\tau_k\}$ 43
2.6. Markov Property of $N(t)$ and Its Implications 46
2.7. Doubly Stochastic Poisson Process 50
2.8. Thinning of a Point Process 51
2.9. Marked Point Processes 53
2.10. Modeling of Floods 56
Problems and Complements 58

CHAPTER 3
Elements of Brownian Motion 62

3.1. Definitions and Preliminaries 62
3.2. Hitting Times 65
3.3. Extremes of $\xi(t)$ 67
3.4. Some Properties of the Brownian Paths 71
3.5. Law of the Iterated Logarithm 74
3.6. Some Extensions 79
3.7. The Ornstein–Uhlenbeck Process 81
3.8. Stochastic Integration 85
Problems and Complements 88

CHAPTER 4
Gaussian Processes 92

4.1. Review of Elements of Matrix Analysis 92
4.2. Gaussian Systems 93
4.3. Some Characterizations of the Normal Distribution 96
4.4. The Gaussian Process 97
4.5. Markov Gaussian Process 99
4.6. Stationary Gaussian Process 102
Problems and Complements 103

CHAPTER 5
L_2 Space 106

5.1. Definitions and Preliminaries 106
5.2. Convergence in Quadratic Mean 110
5.3. Remarks on the Structure of L_2 113
5.4. Orthogonal Projection 115
5.5. Orthogonal Basis 118
5.6. Existence of a Complete Orthonormal Sequence in L_2 121
5.7. Linear Operators in a Hilbert Space 122
5.8. Projection Operators 125
Problems and Complements 126

CHAPTER 6
Second-Order Processes 129

6.1. Covariance Function $C(s, t)$ 129
6.2. Quadratic Mean Continuity and Differentiability 132
6.3. Eigenvalues and Eigenfunctions of $C(s, t)$ 136
6.4. Karhunen–Loève Expansion 139
6.5. Stationary Stochastic Processes 143
6.6. Remarks on the Ergodicity Property 145
Problems and Complements 148

CHAPTER 7
Spectral Analysis of Stationary Processes 150

7.1. Preliminaries 150
7.2. Proof of the Bochner–Khinchin and Herglotz Theorems 153
7.3. Random Measures 157

7.4. Process with Orthogonal Increments 160
7.5. Spectral Representation 162
7.6. Ramifications of Spectral Representation 166
7.7. Estimation, Prediction, and Filtering 169
7.8. An Application 172
7.9. Linear Transformations 174
7.10. Linear Prediction, General Remarks 176
7.11. The Wold Decomposition 179
7.12. Discrete Parameter Processes 182
7.13. Linear Prediction 185
7.14. Evaluation of the Spectral Characteristic $\varphi(\lambda, h)$ 188
7.15. General Form of Rational Spectral Density 192
Problems and Complements 196

CHAPTER 8
Markov Processes I 200

8.1. Introduction 200
8.2. Invariant Measures 203
8.3. Countable State Space 205
8.4. Birth and Death Process 211
8.5. Sample Function Properties 214
8.6. Strong Markov Processes 216
8.7. Structure of a Markov Chain 218
8.8. Homogeneous Diffusion 223
Problems and Complements 227

CHAPTER 9
Markov Processes II: Application of Semigroup Theory 232

9.1. Introduction and Preliminaries 232
9.2. Generator of a Semigroup 234
9.3. The Resolvent 238
9.4. Uniqueness Theorem 241
9.5. The Hille–Yosida Theorem 243
9.6. Examples 247
9.7. Some Refinements and Extensions 252
Problems and Complements 256

CHAPTER 10
Discrete Parameter Martingales 258

10.1. Conditional Expectation 258
10.2. Discrete Parameter Martingales 260
10.3. Examples 263
10.4. The Upcrossing Inequality 266
10.5. Convergence of Submartingales 268
10.6. Uniformly Integrable Martingales 272
Problems and Complements 276

Bibliography 279

Index 284

CHAPTER 1

Basic Concepts and Definitions

1.1. Definition of a Stochastic Process

Generally speaking, a stochastic or random process (in this book both terms will be used in an equivalent sense) is a family of random variables defined on a common probability space, indexed by the elements of an ordered set T, which is called the parameter set. Most often, T is taken to be an interval of time and the random variable indexed by an element $t \in T$ is said to describe the state of the process at time t. Random processes considered here are specified by the following definition:

Definition 1.1. A stochastic process is a family of (real- or complex-valued) random variables $\{X(t); t \in T\}$, defined on a common probability space $\{\Omega, \mathscr{B}, P\}$, where the parameter set T is a subset of the real line R.

In the following we call a stochastic process real if the random variables $X(t)$ are real-valued for all $t \in T$. If the random variables $X(t)$ are all complex-valued, the random process is called complex.

The parameter set T is often called the domain of definition of the stochastic process $X(t)$. If $T = N_+ = \{0, 1, \ldots\}$ the process is said to be a discrete parameter stochastic process. If T is an interval of the real line, the process is said to have a continuous parameter. Most often, T is either N_+ or $R_+ = [0, \infty)$.

Let $\{X(t); t \in T\}$ be a real-valued random process and $\{t_1, \ldots, t_n\} \subset T$, where $t_1 < t_2 < \cdots < t_n$, then

$$F_{t_1, \ldots, t_n}(x_1, \ldots, x_n) = P\{X(t_1) \leq x_1, \ldots, X(t_n) \leq x_n\} \tag{1.1.1}$$

is a finite-dimensional marginal distribution function of the process

$\{X(t); t \in T\}$. One of the basic characteristics of a real random process are its marginal distribution functions.

Definition 1.2. We shall say that the random process $\{X(t); t \in T\}$ is given if all its marginal distribution functions

$$\{F_{t_1,\ldots,t_n}(\cdot,\ldots,\cdot)\} \tag{1.1.2}$$

obtained for all finite subsets $\{t_1,\ldots,t_n\} \subset T$ are given.

Marginal distribution functions (1.2) must satisfy the following consistency conditions:

(i)
$$F_{t_{k_1},\ldots,t_{k_n}}(x_{k_1},\ldots,x_{k_n}) = F_{t_1,\ldots,t_n}(x_1,\ldots,x_n) \tag{1.1.3}$$

for any permutation $k_1,\ldots,k_n$ of $1,\ldots,n$.

(ii) For any $1 \leq k < n$ and $x_1,\ldots,x_k \in R$,

$$F_{t_1,\ldots,t_k}(x_1,\ldots,x_k) = F_{t_1,\ldots,t_n}(x_1,\ldots,x_k,\infty,\ldots,\infty).$$

Many future statements will hold for both real- and complex-valued random processes. For this and other reasons, it is convenient to introduce the concept of a state space $\{S, \mathscr{L}\}$ of a process. Here, S is the set of values of the process and $\mathscr{L}$ is a σ-algebra of subsets of S. For instance, if the process is real, S will be the real line or a part of it and $\mathscr{L}$ will be the σ-field of Borel subsets of S.

1.2. Sample Functions

From Definition 1.1.1, it follows that for each $t \in T$ fixed, $X(t) = X(t,\cdot)$ is a random variable defined on Ω. On the other hand, for any $\omega \in \Omega$ fixed, $X(\cdot,\omega)$ is a function defined on the parameter set T. Any such function is called a realization, trajectory, sample path, or sample function of the stochastic process $X(t)$. In Figure 1.1, a realization of the process $X(t)$ is depicted when $T = [0, T_0]$.

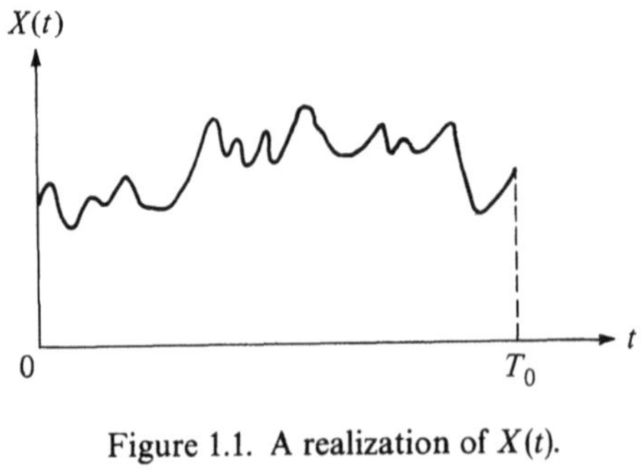

Figure 1.1. A realization of $X(t)$.

A stochastic process (say, real) $\{X(t); t \in T\}$ is said to be directly given if the sample space $\Omega = R^T$, where R^T is the set of all functions on the parameter set T with values in R, i.e., any $\omega \in \Omega$ is a mapping

$$\omega; T \to R.$$

The stochastic process $\{X(t); t \in T\}$ is then defined as a family of coordinate mappings on R^T to R. In other words, for any $t \in T$, $X(t, \cdot)$ is a random variable on R^T defined as follows:

$$\text{for any } \omega(\cdot) \in R^T, \qquad X(t, \omega) = \omega(t).$$

EXAMPLE 1.2.1. Let X and Y be independent random variables on a probability space $\{\Omega, \mathscr{B}, P\}$ with

$$H_X(x) = P\{X \leq x\} \quad \text{and} \quad H_Y(y) = P\{Y \leq y\}.$$

Let $\{\xi(t); t \geq 0\}$ be a stochastic process defined as

$$\xi(t) = tX + Y.$$

In Figure 1.2 are some sample functions of $\xi(t)$.

Clearly,

$$F_t(x) = P\{tX + Y \leq x\} = \int_{-\infty}^{\infty} P\left\{X \leq \frac{x-y}{t}\right\} dH_Y(y)$$
$$= \int_{-\infty}^{\infty} H_X\left(\frac{x-y}{t}\right) dH_Y(y).$$

For any $0 < t_1 < t_2 < \cdots < t_n$, we have

$$F_{t_1,\ldots,t_n}(x_1, \ldots, x_n) = P\{\xi(t_1) \leq x_1, \ldots, \xi(t_n) \leq x_n\}$$
$$= \int_{-\infty}^{\infty} P\left\{X \leq \frac{x_1 - y}{t_1}, \ldots, X \leq \frac{x_n - y}{t_n}\right\} dH_Y(y)$$
$$= \int_{-\infty}^{\infty} H_X\left(\min_{1 \leq i \leq n} \frac{x_i - y}{t_i}\right) dH_Y(y).$$

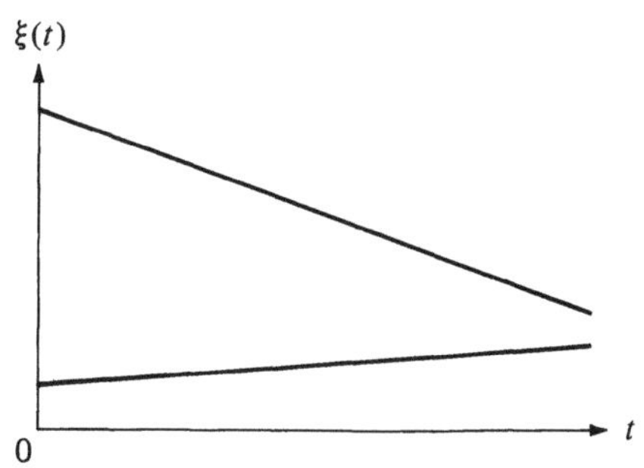

Figure 1.2. Realizations of $\xi(t)$.

It is also easy to verify that

$$E\{\xi(t)\xi(s)\} = stE\{X^2\} + E\{Y^2\},$$

assuming that $E\{X\} = E\{Y\} = 0$ and that $X, Y \in L_2\{\Omega, \mathscr{B}, P\}$.

Before concluding this section, a few remarks on the structure of the σ-field $\mathscr{B}$, in the case when Ω is the space of functions R^T, seem appropriate. Let B_1, $B_2, \ldots, B_n$ be n arbitrary Borel subsets of the real line, and consider the following subset of R^T:

$$\{\omega; X(t_1) \in B_1, \ldots, X(t_n) \in B_n\},$$

where $\{t_1, \ldots, t_n\} \subset T$ and $X(t, \cdot)$ is coordinate mapping on R^T to R. This subset is called an n-dimensional Borel cylinder. Then the σ-algebra $\mathscr{B}$ is defined as the least σ-algebra of subsets of R^T containing all the finite-dimensional Borel cylinders.

The problem with this σ-field is that the sample space $\Omega = R^T$ is too big and the σ-algebra $\mathscr{B}$ is too small, in the sense that many important subsets of Ω are not members of $\mathscr{B}$. Therefore, we cannot compute their probabilities because the probability measure P is defined on $\mathscr{B}$. For instance, the subset of all continuous functions of R^T is not an event (i.e., a member of $\mathscr{B}$). The subset

$$\left\{\omega; \sup_{t \in T} X(t) \leq x\right\} = \bigcap_{t \in T} \{X(t) \leq x\}$$

does not belong in general to $\mathscr{B}$. As a matter of fact, any subset of R^T that is an intersection or union of uncountably many events from $\mathscr{B}$ may not necessarily be an element of $\mathscr{B}$. Such problems do not arise when T is a countable set.

These difficulties are alleviated by the separability concept introduced by Doob which shall be discussed in a later section. The following example is quite instructive.

Example 1.2.2. Let $\{\xi(t); t \geq 0\}$ be a real stochastic process on a probability space $\{\Omega, \mathscr{B}, P\}$ given by

$$\xi(t) = t + X,$$

where $X \sim N(0, 1)$ (i.e., X has a standard normal distribution). The sample functions of the process $\xi(t)$ are straight lines on $T = [0, \infty)$. Let $D \subset T$ be a countable set, say $D = \{t_1, t_2, \ldots\}$, and consider the random event

$$A = \{\xi(t) = 0 \qquad \text{for at least one } t \in D\}.$$

That $A \in \mathscr{B}$ can be easily seen if we write it as

$$A = \bigcup_{t_i \in D} \{\xi(t_i) = 0\} = \bigcup_{t_i \in D} \{X = -t_i\}.$$

Because each $\{X = -t_i\} \in \mathscr{B}$, it follows that $A \in \mathscr{B}$. It is also easy to see that $P(A) = 0$ because $P\{X = -t_i\} = 0$ for each $t_i \in D$.

On the other hand, if we choose D to be $[0, 1]$, then the subset $B \subset \Omega$ defined as

$$B = \{\xi(t) = 0 \text{ for at least one } t \in [0, 1]\}$$

$$= \bigcup_{t \in [0,1]} \{X = -t\}$$

is not necessarily an event because it is an uncountable union of random events. But,

$$\bigcup_{t \in [0,1]} \{X = -t\} = \{X \in [-1, 0]\} \in \mathscr{B}.$$

We also see that

$$P(B) = (2\pi)^{-1} \int_{-1}^{0} e^{-x^2/2}\, dx > 0.$$

Therefore, sometimes an uncountable union of null events can be an event with positive probability. However, a countable union of null events is always a null event.

1.3. Equivalent Stochastic Processes

Let $\{X(t); t \in T\}$ and $\{Y(t); t \in T\}$ be two stochastic processes defined on a common probability space $\{\Omega, \mathscr{B}, P\}$ and taking values in the same state space $\{S, \mathscr{L}\}$.

Definition 1.3.1. If, for each $n = 1, 2, \ldots$

$$P\{X(t_1) \in B_1, \ldots, X(t_n) \in B_n\} = P\{Y(t_1) \in B_1, \ldots, Y(t_n) \in B_n\} \quad (1.3.1)$$

for all $\{t_1, \ldots, t_n\} \subset T$ and $\{B_1, \ldots, B_n\} \subset \mathscr{L}$, the random processes $X(t)$ and $Y(t)$ are said to be stochastically equivalent in the wide sense.

Definition 1.3.2. If, for every $t \in T$,

$$P\{X(t) = Y(t)\} = 1, \quad (1.3.2)$$

the random processes are said to be stochastically equivalent, or just equivalent.

Let us show that (1.3.2) implies (1.3.1): Due to (1.3.2),

$$P\{X(t_1) \in B_1, \ldots, X(t_n) \in B_n\}$$
$$= P\{X(t_1) \in B_1, \ldots, X(t_n) \in B_n, X(t_1) = Y(t_1), \ldots, X(t_n) = Y(t_n)\}$$

$$= P\{Y(t_1) \in B_1, \ldots, Y(t_n) \in B_n, X(t_1) = Y(t_1), \ldots, X(t_n) = Y(t_n)\}$$
$$= P\{Y(t_1) \in B_1, \ldots, Y(t_n) \in B_n\}.$$

Definition 1.3.3. Let $\{\xi(t); t \in T\}$ be a stochastic process on $\{\Omega, \mathscr{B}, P\}$ with state space $\{S, \mathscr{L}\}$. Any other stochastic process $\{\xi(t); t \in T\}$ on the same probability space, equivalent to $\xi(t)$, is called a "version" of $\xi(t)$.

Definition 1.3.4. The stochastic process $\{X(t); t \in T\}$ and $\{Y(t); t \in T\}$ are said to be "indistinguishable" if almost all their sample paths coincide on T. In other words, if

$$P\{X(t) = Y(t) \qquad \forall t \in T\} = 1. \tag{1.3.3}$$

This is the strongest form of equivalence, which clearly implies the other two. The following example shows that two equivalent processes may have quite different sample paths. In other words, (1.3.2) holds but not (1.3.3).

EXAMPLE 1.3.1. Let $\Omega = [0, 1]$, $\mathscr{B}$ the σ-field of Borel subsets of $[0, 1]$, P the Lebesgue measure, and $T = [0, 1]$. Consider two stochastic processes $\{X(t); t \in T\}$ and $\{Y(t); t \in T\}$ on $\{\Omega, \mathscr{B}, P\}$ defined as follows:

$$X(t, \omega) = 0 \quad \text{on } T \text{ for each } \omega \in \Omega,$$

$$Y(t, \omega) = 0 \quad \text{on } T \text{ except at the point } t = \omega \text{ when } Y(\omega, \omega) = 1.$$

For any $t \in T$ fixed,

$$\{\omega; X(t, \omega) \neq Y(t, \omega)\} = \{\omega; \omega = t\} = \{t\}.$$

Because $P\{t\} = 0$, it follows that

$$P\{X(t) = Y(t)\} = 1 \quad \text{for each } t \in T.$$

In other words, these two processes are stochastically equivalent and yet their sample paths are different. In addition,

$$P\{\omega; X(t, \omega) = Y(t, \omega)\ \forall t \in T\} = 0,$$

$$\sup_{t \in T} X(t) = 0, \quad \sup_{t \in T} Y(t) = 1.$$

Remark 1.3.1. At this point, it would be of some interest to elucidate the principal distinction between Definitions 1.3.2 and 1.3.4. The point is that in Definition 1.3.2 the null set Λ_t on which $X(t)$ and $Y(t)$ may differ depends on t. As we have seen in Example 1.2.2, the union

$$\bigcup_{t \in T} \Lambda_t$$

of uncountably many null events may not be a null event. On the other hand, in Definition 1.3.4 there is just one null event Λ such that

$$X(t, \omega) = Y(t, \omega) \quad \text{on } T$$

for every $\omega \in \Lambda^c$.

Under what conditions are Definitions 1.3.2 and 1.3.4 equivalent? This is clearly the case if T is a countable set. For continuous time, we need some conventions. Let $\{\xi(t); t \in T\}$ be a stochastic process with state space $\{S, \mathscr{L}\}$, which is a metric space. The stochastic process is said to be Cadlag if each of its trajectory is a right continuous function and has limits on the left.

Proposition 1.3.1. *Let $\{X(t); t \in T\}$ and $\{Y(t); t \in T\}$ be stochastically equivalent and both right continuous. Then $X(t)$ and $Y(t)$ are indistinguishable.*

Proof. Let Q be the set of rational numbers. For each $r \in Q \cap T$, $P\{X(r) \neq Y(r)\} = 0$. Consequently, $P(G) = 0$, where

$$G = \bigcup_{r \in Q \cap T} \{X(r) \neq Y(r)\}.$$

However, by right continuity,

$$\{X(t) \neq Y(t)\} \subset G$$

for any $t \in T$. Therefore,

$$\bigcup_{t \in T} \{X(t) \neq Y(t)\} \subset G,$$

which shows that

$$P\{X(t) = Y(t) \quad \forall t \in T\} = 1. \qquad \square$$

1.4. Kolmogorov Construction

Let $\{\xi(t); t \in T\}$ be a stochastic process on $\{\Omega, \mathscr{B}, P\}$ with state space $\{R, \mathscr{R}\}$, where R is the real line and $\mathscr{R}$ the σ-algebra of Borel subsets of R. The stochastic process determines a consistent family of marginal distribution functions by

$$F_{t_1, \ldots, t_n}(x_1, \ldots, x_n) = P\{\xi(t_1) \leq x_1, \ldots, \xi(t_n) \leq x_n\}.$$

Is the converse true? In other words, given a consistent family of distribution functions, does there exist a stochastic process for which these distributions are its marginal distributions? The following theorem due to Kolmogorov (which is not going to be proved here) provides an affirmative answer to this question.

Theorem 1.4.1. *Assume that*

$$\{F_{t_1, \ldots, t_n}(x_1, \ldots, x_n)\}, \{t_1, \ldots, t_n\} \subset T, \qquad n = 1, 2, \ldots \tag{1.4.1}$$

is a consistent family of distribution functions, then there exists a stochastic process with (1.4.1) *as its marginal distribution functions.*

As the probability space on which this process is defined we may use

$$\{R^T, \mathscr{B}, P\}$$

(see Section 1.2 for definitions) and the stochastic process is specified by

$$X(t, \omega) = \omega(t) \quad \text{for each } \omega \in R^T \quad \text{and } t \in T.$$

The method in the Kolmogorov theorem used to construct a stochastic process starting from a consistent family of distribution functions leads to a rather large set of sample functions, namely, R^T. Often it is desirable to construct a process whose sample functions are subject to some regularity conditions. For instance, we may want them to be Borel measurable, or continuous, and so on. Denote this particular subset of R^T by Ω_0. In order that there exists a process $\{\xi_0(t); t \in T\}$ stochastically equivalent to $\{\xi(t); t \in T\}$, which can be realized in Ω_0, it is necessary and sufficient that $P^*(\Omega_0) = 1$, where P^* is the outer measure induced by P as follows: For any $M \subset R^T$,

$$P^*(M) = \inf\{P(C); C \supset M\},$$

where $C \subset R^T$ is a cylinder set. In such a case, the system $\{\Omega_0, \mathscr{B}_0, P^*\}$ is the new probability space with $\mathscr{B}_0 = \mathscr{B} \cap \Omega_0$. We are not going to prove this result here.

1.5. Principal Classes of Random Processes

In this section, we isolate and define several important classes of random processes which have been extensively studied and used in various applications.

According to Definition 1.1.1, every random process is a family of r.v.'s defined on a common probability space. The theory of stochastic processes is concerned with different features of such a family, many of which can be studied by means of their marginal distributions. Of particular interest are those processes endowed with certain stochastic structure. More specifically, those whose marginal distributions are assigned certain properties.

In the sequel, we identify the following five classes of random processes, each of them having a particular stochastic structure.

(i) *Processes with independent increments.* Let

$$\{\xi(t); t \in T\} \tag{1.5.1}$$

be a real-valued stochastic process on a probability space $\{\Omega, \mathscr{B}, P\}$, where $T \subset R$ is an interval.

Definition 1.5.1. The process (1.5.1) is said to have "independent increments" if for any finite subset $\{t_0, t_1, \ldots, t_n\} \subset T$ with $t_0 < t_1 < \cdots < t_n$, the increments

$$\xi(t_0),\ \xi(t_1) - \xi(t_0),\ \ldots,\ \xi(t_n) - \xi(t_{n-1}) \tag{1.5.2}$$

are independent r.v.'s.

From the definition, it readily follows that all the marginal distributions are completely determined by the distribution of $\xi(t)$ for all $t \in T$ and by $\xi(t_2) - \xi(t_1)$, $t_1, t_2 \in T$, $t_1 < t_2$. Two most important processes with independent increments are the Poisson and Wiener (or Brownian motion) process. They will be discussed in some detail in the following chapters.

(ii) *Markov processes.* Consider a random process on $\{\Omega, \mathscr{B}, P\}$,

$$\{\xi(t); t \in T\} \tag{1.5.3}$$

where $T = [0, \infty)$, with values in an R.

Definition 1.5.2. Process (1.5.3) is called Markov with a state space $\{R, \mathscr{R}\}$ if, for any $0 \leq t_1 < t_2 < \cdots < t_n$ and $B \in \mathscr{R}$,

$$P\{\xi(t_n) \in B | \xi(t_1), \ldots, \xi(t_{n-1})\} = P\{\xi(t_n) \in B | \xi(t_{n-1})\} \quad \text{(a.s.).} \tag{1.5.4}$$

Property (1.5.4) of the stochastic process is called the Markov property. To elucidate somewhat its true meaning let us fix $t \in (0, \infty)$ and consider the families of r.v.'s

$$\{\xi(s); s \leq t\} \quad \text{and} \quad \{\xi(u); u \geq t\}. \tag{1.5.5}$$

If we take the instant t to be present, then the first family of r.v.'s represents the "past and present" of the process, whereas the second family is its "future." Now, for any B_1, $B_2 \in \mathscr{R}$ and $s < t < u$, one can easily verify that (1.5.4) implies that the following holds almost surely:

$$P\{\xi(s) \in B_1, \xi(u) \in B_2 | \xi(t)\} = P\{\xi(s) \in B_1 | \xi(t)\} P\{\xi(u) \in B_2 | \xi(t)\}. \tag{1.5.6}$$

Thus, the Markov property, roughly speaking, means that given the present, the future and the past of a Markov process are independent.

Definition 1.5.3. The distribution π on $\mathscr{R}$ defined by

$$\pi(B) = P\{\xi(0) \in B\} \tag{1.5.7}$$

is called the "initial distribution" of the process.

Definition 1.5.4. A version

$$P(s, t, x, B) \tag{1.5.8}$$

of $P\{\xi(t) \in B | \xi(s) = x\}$ having the properties

a. for every $s < t$ and $x \in R$ fixed, $P(s, t, x, \cdot)$ is a probability measure on $\mathscr{R}$,
b. for every $s < t$ and $B \in \mathscr{R}$ fixed, $P(s, t, \cdot, B)$ is an $\mathscr{R}$-measurable function,

is called the transition probability function (or simply the transition probability) of the Markov process.

From the definition, it follows that

$$P(t, t, x, B) = \begin{cases} 1 & \text{if } x \in B \\ 0 & \text{if } x \notin B \end{cases} \tag{1.5.9}$$

for all $t \geq 0$. In addition, the transition probability satisfies the so-called "Chapman–Kolmogorov equation." For any $0 \leq s < t < u$ and $B \in \mathscr{R}$,

$$P(s, u, x, B) = \int_R P(s, t, x, dy)P(t, u, y, B). \tag{1.5.10}$$

The initial distribution π and the transition probability function $P(s, t, x, B)$ determine completely and uniquely the probability measure P of the process. To show this, let $\{B_k\}_1^n \subset \mathscr{R}$ and $0 < t_1 < \cdots < t_n$ be arbitrary. Then taking into account (1.5.4) we clearly have

$$\begin{aligned} P\{\xi(t_1) \in B_1, \ldots, \xi(t_n) \in B_n\} \\ &= \int_R \int_{B_1} \cdots \int_{B_n} P\{\xi(0) \in dx_0, \xi(t_1) \in dx_1, \ldots, \xi(t_n) \in dx_n\} \\ &= \int_R \pi(dx_0) \int_{B_1} P(0, t_1, x_0, dx_1) \cdots \int_{B_n} P(t_{n-1}, t_n, x_{n-1}, dx_n). \end{aligned} \tag{1.5.11}$$

This defines P on the class of measurable cylinders. The rest follows from Theorem 1.4.1.

Definition 1.5.5. A Markov process is said to be "homogeneous" or to have "stationary transition probability" if, for any $0 \leq s < t$,

$$P(s, t, x, B) = P(0, t - s, x, B). \tag{1.5.12}$$

To simplify notation we shall write

$$P(x, t, B) = P(0, t, x, B). \tag{1.5.13}$$

In this case, the Chapman–Kolmogorov equation becomes

$$P(x, s + t, B) = \int_R P(x, t, dy)P(y, s, B). \tag{1.5.14}$$

From (1.5.14), there follows an interesting fact: It is enough to know $P(y, s, B)$ for all $s \leq \varepsilon$, where $\varepsilon > 0$ is arbitrarily small, because, for all $u > \varepsilon$, it is determined by Equation (1.5.14). In other words, the local behavior of the process (in the neighborhood of zero) determines its global behavior.

(iii) *Strictly stationary processes.* In the theory of stochastic processes, an important class consists of those processes whose marginal distributions are invariant under time-shift.

Definition 1.5.6. A real-valued stochastic process $\{\xi(t); t \in T\}$ on a probability space $\{\Omega, \mathscr{B}, P\}$ is said to be "strictly stationary" if for any $\{t_1, \ldots, t_n\} \subset T$, $\{t_1 + h, \ldots, t_n + h\} \subset T$, and any $\{B_k\}_1^n \subset \mathscr{R}$, $n = 1, 2, \ldots$,

$$P\{\xi(t_1 + h) \in B_1, \ldots, \xi(t_n + h) \in B_n\} = P\{\xi(t_1) \in B_1, \ldots, \xi(t_n) \in B_n\}. \quad (1.5.15)$$

In applications of strictly stationary stochastic processes, the parameter set T is usually R_+ or R. If $E|\xi(t)| < \infty$, then from (1.5.15) it follows that

$$E\{\xi(t)\} = E\{\xi(t + h)\} = m \quad \text{(constant)}.$$

Similarly, if the second moment of the process exists, then it is also equal to a constant.

Finally, set $T = R_+$ and let $s, t \in T$. Consider

$$R(s, t) = E\{\xi(s)\xi(t)\}. \quad (1.5.16)$$

Clearly, $R(s, t) = R(t, s)$. Assume for the sake of definiteness that $s < t$. Then due to (1.5.15)

$$E\{\xi(s)\xi(t)\} = E\{\xi(s + h)\xi(t + h)\}. \quad (1.5.17)$$

When we set $h = -s$ in (1.5.17) we obtain

$$R(s, t) = R(0, t - s).$$

The function (1.5.16) is often called the covariance function.

Definition 1.5.7. A real or complex r.v. Z on $\{\Omega, \mathscr{B}, P\}$ is said to be "second order" if $E|Z|^2 < \infty$. The family of all second-order r.v.'s on $\{\Omega, \mathscr{B}, P\}$ is denoted by

$$L_2 = L_2\{\Omega, \mathscr{B}, P\}.$$

Definition 1.5.8. A stochastic process $\{\xi(t); t \in T\}$ on $\{\Omega, \mathscr{B}, P\}$ is second order if

$$\xi(t) \in L_2\{\Omega, \mathscr{B}, P\} \quad \forall t \in T.$$

(iv) *Wide sense stationary processes.* There exists a large class of problems in the theory of stochastic processes and their applications, whose solution requires only knowledge of the first two moments of a process. In such a case, problems are considerably simplified if these moments are invariant under time-shift.

Definition 1.5.9. A second-order stochastic process $\{\xi(t); t \in T\}$ on a probability space $\{\Omega, \mathscr{B}, P\}$ is said to be "wide sense" stationary if, for all $t \in T$,

$$E\{\xi(t)\} = m \quad \text{(constant)}$$

and (1.5.18)

$$R(t, t + \tau) = R(0, \tau).$$

It is clear that the second-order strictly stationary stochastic process is a fortiori wide sense stationary. The time invariance of covariance function with respect to time-shifts implies that the methods of harmonic analysis may play a useful role in the theory of wide sense stationary processes. This will be discussed in some details later.

(v) *Martingales*. Let $\{\xi(t); t \in T\}$ be a real stochastic process on $\{\Omega, \mathscr{B}, P\}$ such that $E\{|\xi(t)|\} < \infty$ for all $t \in T$.

Definition 1.5.10. The stochastic process $\xi(t)$ is said to be a martingale if, for every $t_1 < t_2 < \cdots < t_n$ in T,

$$E\{\xi(t_n)|\xi(t_1), \ldots, \xi(t_{n-1})\} = \xi(t_{n-1}) \quad \text{(a.s.).} \tag{1.5.19}$$

If

$$E\{\xi(t_n)|\xi(t_1), \ldots, \xi(t_{n-1})\} \le \xi(t_{n-1}) \quad \text{(a.s.),} \tag{1.5.20}$$

the process $\xi(t)$ is supermartingale. Finally, if

$$E\{\xi(t_n)|\xi(t_1), \ldots, \xi(t_{n-1})\} \ge \xi(t_{n-1}) \quad \text{(a.s.),} \tag{1.5.21}$$

the process $\xi(t)$ is said to be a submartingale.

1.6. Some Applications

There is a wide variety of applications of stochastic processes. For some of them all that is needed is calculation of one or more parameters of a stochastic process which is of direct interest in the given problem. Some other applications require very sophisticated theoretical methods of great complexity. This book deals with elements and methods of the theory of stochastic processes on an intermediate level, with strong emphasis on their applications. Successful use of the theory depends a great deal on successful modeling. For this reason, it is worthwhile to make a few remarks about it.

Roughly speaking, mathematical modeling is the formulation of a mathematical system designed to "simulate" behavior of certain aspects of a physical or biological phenomenon. For instance, Newton's law of motion of a body falling freely under gravity may serve as an example of a mathematical model.

A mathematical model represents our conceptual image of the phenomenon that we observe and as such it reflects our quantitative understanding of a situation. Its formulation is always based on a certain number of assumptions concerning the fundamental nature of the phenomenon we investigate. If these assumptions are of a general nature, we may finish with a model of great mathematical complexity. On the other hand, too many simplifying assumptions may mean considerable restrictions on the model's ability to provide a sufficiently detailed and accurate description of the system being modeled.

The formulation then of a mathematical model must be a compromise between these two extremes. For this reason one usually talks about "a" mathematical model rather than "the" model. In the rest of this section, a number of examples illustrating mathematical modeling of various systems will be discussed in some details.

1.6.1. Model of a Germination Process

From an agricultural viewpoint, germination is a process which begins when a dry seed is planted in wet soil and ends when the seedling emerges above the ground. The duration of this process is a r.v. T such that $0 < T \leq \infty$. The stochastic nature of the quantity T is a result of the fact that the water uptake (by imbibition) is a purely physical process based on diffusion of water through a porous medium (the seed coat). In addition, the soil matrix potential, the seed soil contact area, and the concentration of soil moisture around the seed are additional factors contributing to random variations of T.

Consider the following problem: Suppose, at time $t = 0$, N seeds of the same plant species are planted and denote by $T_1, T_2, \ldots, T_N$, their respective germination times. How many seeds will germinate in $(0, t]$? Denote by $X(t)$ this number; it is then quite clear that $X(t)$ is a stochastic process defined as follows:

$$X(t) = \sum_{i=1}^{N} I_{\{T_i \leq t\}} \quad \text{for } t > 0 \text{ and } X(0) = 0. \tag{1.6.1}$$

We now make several assumptions. Our physical intuition is not violated by assuming that:

a. $\{T_i\}_1^N$ is an i.i.d. sequence of r.v.'s with a common distribution function

$$H(t) = P\{T \leq t\} \quad \text{with } H(0) = 0 \text{ and } H(\infty) = 1 - p. \tag{1.6.2}$$

Here, $0 \leq p < 1$ is the probability that a seed may fail to germinate. From (1.6.1) and (1.6.2), we have

$$P\{X(t) = k\} = \binom{N}{k}(H(t))^k\{1 - H(t)\}^{N-k}, \tag{1.6.3}$$

Thus, if $H(t)$ is known, the probability (1.6.3) is completely determined. Now make the following assumption:

b. The average proportion of germinations in $(s, s + \Delta s)$ is approximately $\lambda \cdot \Delta s$, where $\lambda > 0$ is a constant.

Next, partition $(0, t]$ into n subintervals of equal length $\Delta t = t/n$. Then due to assumption b, the average number of germinations in $(0, \Delta t]$ is approximately

$$N(1 - p)\lambda\Delta t.$$

Thus, the average number of nongerminating seeds at time Δt is $N(1-p)(1-\lambda\Delta t)$. The average number of germinating seeds in $(\Delta t, 2\Delta t]$ is $N(1-p)(1-\lambda\Delta t)\lambda\Delta t$, so that the average number of nongerminating seeds at time $2\Delta t$ is

$$N(1-p)(1-\lambda\Delta t)^2,$$

and so on. Continuing in this fashion, we conclude that the number of nongerminating seeds at time t is approximately

$$N(1-p)(1-\lambda\Delta t)^n.$$

From this, we deduce that for n sufficiently large

$$P\{T \le t\} \approx \{1-(1-\lambda\Delta t)^n\}(1-p).$$

By letting $\Delta t \to 0$, we obtain that

$$H(t) = (1-p)(1-e^{-\lambda t}). \tag{1.6.4}$$

From this and (1.6.3), the required probability is now easy to obtain.

1.6.2. Modeling of Soil Erosion Effect

Soil erosion is a result of perpetual activities of various geophysical forces acting on the earth's surface. The ensuing degradation of farmland may have profound impact on future food supplies. Because no experimental data exist, there is no choice but to resort to mathematical models in assessing future crop losses due to erosion. Here we present a simplified version of such a model.

Surface erosion occurs by the migration of individual soil particles in response to forces such as wind, raindrop impact, surface runoff, and so on. Erosion reduces plant-available minerals and nutrients which are adsorbed in eroded particles. It also continuously decreases the root zone depth, which, in turn, reduces the water-holding capacity of the soil. This leads to decline of crop yields due to newly induced droughts that become more frequent and severe.

Even under optimal conditions, the production of a particular crop fluctuates from one year to another in a random fashion. The mechanism responsible for these variations is quite complex. It involves climatic and meteorological factors (such as air temperature, wind, solar radiation, rainfall) soil characteristics, and other conditions. Due to these reasons, the effect of soil erosion on crop yield is not directly observable but becomes gradually evident over a long period of time.

Having outlined the mechanism of the soil erosion process, we now embark on the problem of constructing a model to assess its effect on crop production. To this end consider a crop-producing area A and denote by $\{Y_n\}_1^\infty$ the sequence of annual yields of a given crop. If the area A is not affected by erosion and the same agricultural practice is followed year after year, we may assume that $\{Y_i\}_1^\infty$ is an i.i.d. sequence of strictly positive r.v.'s on $\{\Omega, \mathscr{B}, P\}$, with a common distribution function

$$Q(y) = P\{Y \leq y\} \quad \text{such that } E\{Y^i\} < \infty, \quad i = 1, 2. \tag{1.6.5}$$

If, on the other hand, A is subject to soil erosion, the resulting sequence of annual yields $\{X_n\}_1^\infty$ consists of r.v.'s which are neither independent nor identically distributed. To determine this stochastic process, suppose that the soil erosion reduces annual yield by a certain percentage each year and denote by R_i the percentage in the ith year. Then the loss suffered in the first year is

$$Y_1 \cdot \frac{R_1}{100} = Y_1 \cdot U_1.$$

Thus, $X_1 = Y_1 - Y_1 \cdot U_1 = Y_1(1 - U_1)$. The loss suffered in the second year is

$$Y_2(1 - U_1) \cdot \frac{R_2}{100} = Y_2(1 - U_1) \cdot U_2$$

so that $X_2 = Y_2(l - U_1) - Y_2(1 - U_1)U_2 = Y_2(1 - U_1)(1 - U_2)$, and so on. Therefore, the crop yield in year n is

$$X_n = Y_n \prod_{i=1}^{n} Z_i, \qquad Z_i = 1 - U_i. \tag{1.6.6}$$

Our physical intuition is not violated by assuming that $\{Z_i\}_1^\infty$ is an i.i.d. sequence of r.v.'s independent of $\{Y_n\}_1^\infty$, with common support in $(0, 1]$. It seems appropriate to call $L_n = \prod_1^n Z_i$ "the loss-rate function." Notice that $\{L_i\}_1^\infty$ is a Markov chain. It is interesting to note that the quantity $L_n = \prod_1^n Z_i$ first appeared in a paper by Kolmogorov (1941) which is concerned with the modeling of a rock crunching process.

1.6.3. Brownian Motion

The kinetic theory envisages a fluid as composed of a very large number of molecules moving with widely varying velocities in all directions and colliding with each other. In the early days of the theory, the most frequently asked questions were: Are molecules real? Do they actually exist and can we demonstrate their existence? The attempts to answer them led to an exhaustive study of an interesting phenomenon and shed much light on other kinetic properties. The phenomenon takes its name after the English botanist R. Brown who in 1827 noticed the irregular but ceaseless motion of the small particles, e.g., pollen suspended in a liquid. At first, the motion was thought to be of organic origin. After the advent of kinetic theory, it became clear that the only reasonable explanation for it lay in the assumption that the particles were subject to the continual bombardment by the molecules of the surrounding medium.

Suppose at time $t = 0$, when our observation begins, the pollen was at the point $x = 0$. Denote by $\xi(t)$ its position at time $t > 0$ [here $\xi(t)$ denotes one coordinate of the particle]. The chaotic nature of this motion clearly implies that $\xi(t)$ is a r.v. for all $t > 0$. Thus, $\{\xi(t); t \geq 0\}$ is a continuous parameter stochastic process.

Examining the nature of the phenomenon, it seems reasonable to assume that the distribution of $\xi(t + s) - \xi(t)$ does not depend on t if the temperature of the liquid remains constant. It also seems reasonable to assume that the change in position of the particle during the time interval $[t, t + s]$ is independent of anything that happened up to time t. This implies that the process $\xi(t)$ has independent increments. Finally, the trajectories of the process should be continuous functions.

We shall show that under these assumptions the stochastic process $\xi(t)$ is Gaussian (i.e., all the marginal distributions of the process are normal) with

$$E\{\xi(t)\} = 0, \qquad \mathrm{Var}\{\xi(t)\} = t.$$

From this, we have for $s < t$

$$E\{\xi(s)\xi(t)\} = E\{\xi(s)[\xi(t) - \xi(s)] + \xi^2(s)\} = s.$$

In other words,

$$E\{\xi(s)\xi(t)\} = \min(s, t).$$

1.6.4. Modeling of Dispersion in a Porous Medium

Consider a column packed with a porous material saturated with a single incompressible fluid (say, fresh water) under convection (see Figure 1.3). Suppose that at time $t = 0$ a set of tagged (dynamically neutral) particles are introduced into the flow (across the section AB of the column). The transport of these particles in the direction of the flow is called "longitudinal dispersion in a porous medium."

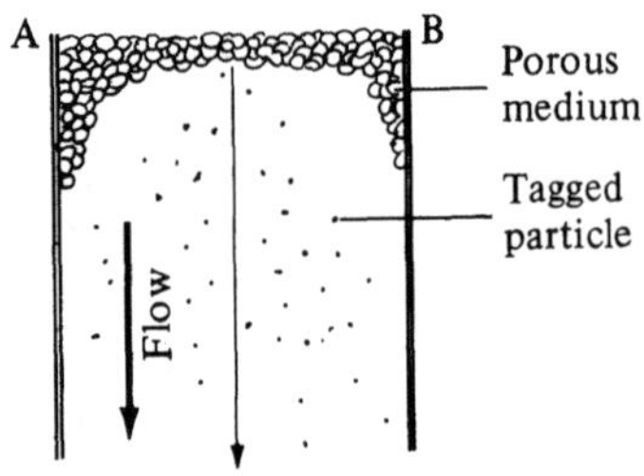

Figure 1.3. Dispersion in a porous medium.

We can construct a model of the dispersion under the following conditions:

(i) the flow is steady and uniform;
(ii) the porous structure is isotropic and homogeneous porous medium;
(iii) tagged particles have identical "transport characteristics" and move independently of one another;
(iv) there is no mass transfer between the solid phase and the fluid.

A tagged particle in the flow through porous medium undergoes a particular kind of random walk. It progresses through pores and voids of the porous structure in a series of steps of random length, with a rest period of random duration between two consecutive steps.

Denote by $X(t)$ the distance of particle from the section AB at time t; then

$$X(0) = 0 \leq X(t_1) \leq X(t_2) \quad \text{if } 0 < t_1 < t_2.$$

It is intuitively clear that $\{X(t); t \geq 0\}$ is a time homogeneous Markov process. Under conditions (i)–(iv), the marginal distribution

$$F_t(x) = P\{X(t) \leq x\} \tag{1.6.7}$$

provides the longitudinal concentration function of tagged particles in the column at any $t \geq 0$. Calculation of (1.6.7) is based on the fact that $X(t)$ can be approximated by a regular Markov jump process.

1.6.5. Queues

A queueing system can be described as follows. Customers arrive at a service station (for instance, a post office, bank, etc.) with a certain number of servers. An arriving customer may have to wait until his turn comes or one of the servers becomes available. He leaves the station upon the completition of the service.

To formulate a mathematical model of a queueing system, we must specify conditions under which the system operates. For instance;

1. in what manner do customers enter the system?
2. in what order are they served?
3. how long are their service times?

Concerning the first question, we shall assume that the arrival time of the nth customer in the system is a r.v. τ_n (we assume that $\tau_0 \equiv 0 < \tau_1 < \tau_2 < \cdots$). Consequently, the number of customers who have entered the system by the time t is a random process $\{N(t); t \geq 0\}$ such that $N(0) \equiv 0$

$$N(t)\colon 0, 1, 2, \ldots \quad \text{for } t > 0 \qquad \text{and} \qquad N(t_1) \leq N(t_2) \quad \text{if } t_1 < t_2. \tag{1.6.8}$$

Clearly, for any $n = 0, 1, \ldots,$

$$\tau_n = \inf\{t; N(t) = n\}, \qquad N(t) = \max\{n; \tau_n \leq t\}. \tag{1.6.9}$$

An arriving customer joins the end of a single line of people waiting to be served. The service is on a "first come first served" basis. When one of servers becomes free, he turns his attention to the customer at the head of the waiting line. This answers the second question.

Finally, concerning the third question we, shall assume that the service times form an i.i.d. sequence of r.v.'s $\{U_n\}_1^\infty$, independent of $\{\tau_n\}_1^\infty$.

Here we shall consider queueing systems with a single server. Under certain conditions such a queue is completely described by the process $\{N(t); t \geq 0\}$ and $\{U_n\}_1^\infty$. The state of the system at any time $t \geq 0$ is specified by the random process

$$\{S(t); t \geq 0\}, \qquad S(0) \equiv 0, \tag{1.6.10}$$

which is the number of waiting customers, including one in the process of being served, at time t.

What kind of information about $S(t)$ is of interest to us? For instance, we would like to know something about the asymptotic behavior of $S(t)$ when $t \to \infty$. When is $S(t)$ asymptotically stationary? When is $S(t)$ a Markov chain or when does it contain an imbedded Markov chain?

The answers to these and other similar questions depend on the process $N(t)$ and $\{U_n\}_1^\infty$. Set $T_1 = \tau_1$ and $T_n = \tau_n - \tau_{n-1}$ for $n \geq 1$. Clearly, $\{T_k\}_1^\infty$ is the sequence of interarrival times. We shall assume that $\{T_k\}_1^\infty$ is also an i.i.d. sequence of r.v.'s Set

$$F_T(t) = P\{T \leq t\}, \qquad F_U(t) = P\{U \leq t\}. \tag{1.6.11}$$

Under these assumptions, a queueing system is completely specified by these distributions. For this reason, it seems convenient to describe a queueing system in terms of the distributions (1.6.11). The most common notational scheme for this purpose is by a triple $A/B/s$, where A specifies F_T, B specifies F_U, and s is the number of servers.

For instance, $M/M/1$ means that

$$F_T(t) = 1 - e^{-\lambda t}, \qquad F_U(t) = 1 - e^{-\alpha t}, \qquad s = 1.$$

On the other hand, $G/G/1$ means that F_T and F_U are some general distributions and $s = 1$, and so on.

1.7. Separability

Let $\{\xi(t); t \in T\}$ be a real stochastic process on a complete probability space $\{\Omega, \mathscr{B}, P\}$. From the definition of a stochastic process, it follows that the mapping

$$\xi(t, \cdot): \Omega \to R$$

is $\mathscr{B}$-measurable for every $t \in T$. In other words,

$$\{\omega; \xi(t, \omega) \in B\} \in \mathscr{B}$$

for each Borel subset $B \subset R$. However,

$$\{\omega; \xi(t, \omega) \in B, t \in T\} = \bigcap_{t \in T} \{\omega; \xi(t, \omega) \in B\}$$

need not be in $\mathscr{B}$ in general, unless T is countable. This then implies that functions like

$$\sup_{t \in T} \xi(t) \qquad \text{and} \qquad \inf_{t \in T} \xi(t)$$

may not be $\mathscr{B}$-measurable either because, for instance,

$$\left\{\sup_{t \in T} \xi(t) \leq x\right\} = \bigcap_{t \in T} \{\xi(t) \leq x\}.$$

Therefore, a large number of important functionals of a continuous parameter stochastic process (all those which depend on an uncountable number of coordinates) may not be random variables. A remedy for this situation is the separability concept introduced by Doob which will be defined next.

Definition 1.7.1. The stochastic process $\{\xi(t); t \in T\}$, where $T \subset R$ is an interval, is said to be separable if there exists a countable dense subset $D \subset T$ and a null event $\Lambda \subset \Omega$ such that

$$\{\omega; \xi(t, \omega) \in C, t \in I \cap D\} - \{\omega; \xi(t, \omega) \in C, t \in I \cap T\} \subset \Lambda \tag{1.7.1}$$

for any closed set C and any open interval I.

The countable set D is called the separant. Note that

$$\{\omega; \xi(t, \omega) \in C, t \in I \cap D\} \supset \{\omega; \xi(t, \omega) \in C, t \in I \cap T\}.$$

From (1.7.1) it follows that

$$\Lambda^c \cap \{\omega; \xi(t, \omega) \in C, t \in I \cap D\} = \Lambda^c \cap \{\omega; \xi(t, \omega) \in C, t \in I \cap T\}, \tag{1.7.2}$$

which clearly implies that the right-hand side of (1.7.2) is an event because the left-hand side is so. The next proposition shows that (roughly speaking) every trajectory of the process $\{\xi(t); t \in T\}$ from Λ^c is as regular on T as its restriction on D.

Proposition 1.7.1. *For each $\omega \in \Lambda^c$ and open interval $I \subset T$,*

$$\overline{\xi(I \cap T, \omega)} = \overline{\xi(I \cap D, \omega)}. \tag{1.7.3}$$

Proof. Here $\overline{\xi(I \cap T, \omega)}$ denotes the closure in R of the set of values assumed by the trajectory $\xi(\cdot, \omega)$ as t varies in $I \cap T$. The other set in (1.7.3) has similar interpretation.

To prove the proposition consider a closed subset $C \subset R$, we then have

$$\xi(I \cap D, \omega) \subset C \Leftrightarrow \overline{\xi(I \cap D, \omega)} \subset C.$$

On the other hand, from (1.7.2) we deduce that

$$\forall \omega \in \Lambda^c, \qquad \xi(l \cap D, \omega) \subset C \Leftrightarrow \xi(l \cap T, \omega) \subset C.$$

Taking $C = \overline{\xi(l \cap D, \omega)}$, it follows from this that

$$\xi(l \cap T, \omega) \subset \overline{\xi(l \cap D, \omega)}$$

from which we conclude that (1.7.3) holds. This proves the proposition. □

From (1.7.3) we obtain that for each $\omega \in \Lambda^c$ and $t \in T$,

$$\xi(t, \omega) \in \overline{\xi(l \cap D, \omega)} \tag{1.7.4}$$

for every open interval l containing t. This, on the other hand, implies that because, by definition, D is dense in T, for every $t \in T$ there exists a sequence $\{u_k\}_1^\infty \subset D$ such that $u_k \to t$. Then, for each $\omega \in \Lambda^c$,

$$\lim_{k \to \infty} \xi(u_k, \omega) = \xi(t, \omega) \tag{1.7.5}$$

(the sequence $\{u_k\}$ may depend on ω). This can be seen as follows. Let $t \in l$, then

$$\lim_{k \to \infty} \xi(u_k, \omega) \in \overline{\xi(l \cap D, \omega)}.$$

Let us show that for every $\omega \in \Lambda^c$ and open set l

$$\sup_{l} \xi(t, \omega) = \sup_{l \cap D} \xi(t, \omega). \tag{1.7.6}$$

From its definition, it follows that the left-hand side of (1.7.6) is the upper bound of the set $\overline{\xi(l, \omega)}$ which belongs to it. Similarly, the right-hand side of (1.7.6) is the upper bound of $\overline{\xi(l \cap D, \omega)}$, and must belong to this set. From this and (1.7.3), the assertion follows. In the same fashion, one can show that

$$\inf_{l} \xi(t, \omega) = \inf_{l \cap D} \xi(t, \omega).$$

Do separable processes exist? The following proposition due to Doob provides an affirmative answer to this question.

Proposition 1.7.2. *Every real-valued stochastic process $\{\xi(t); t \in T\}$ defined on a complete probability space $\{\Omega, \mathscr{B}, P\}$ has a separable version.*

We are not going to prove this proposition here. This is an important result which implies the following: For any real random process $\{\xi(t); t \in T\}$ on a complete probability space $\{\Omega, \mathscr{B}, P\}$, there is a real random process $\{\tilde{\xi}(t); t \in T\}$ defined on the same probability space which is separable and stochastically equivalent to $\xi(t)$. Note that $\tilde{\xi}(t)$ and $\xi(t)$ may have different trajectories. This, however, is not the case if $\tilde{\xi}(t)$ and $\xi(t)$ are both Cadlag (see Proposition 1.3.1).

Remark 1.7.1. If $\{\xi(t); t \in T\}$ is separable, then, as we have seen from (1.7.5), for any $\omega \in \Lambda^c$ and $t \in T$

$$\xi(t, \omega) = \lim_{n \to \infty} \xi(u_k, \omega),$$

where $\{u_k\}_1^\infty \subset D$ (depends on ω) is such that $u_k \to t$. In other words, the values of the sample functions on T are limits of sequences $\{\xi(u_k, \cdot)\}_1^\infty$.

1.8. Some Examples

In this section we present two examples. In the first one we consider a separable version of a nonseparable stochastic process.

Example 1.8.1. Let $\{\Omega, \mathscr{B}, P\}$ be a probability space with $\Omega = [0, 1]$, $\mathscr{B}$ the σ-algebra of Borel subsets of $[0, 1]$, and P the Lebesgue measure. Let $\{\xi(t); t \in [0, 1]\}$ be a stochastic process on the probability space defined as follows:

$$\xi(t, \omega) = \begin{cases} 1 & \text{if } t = \omega \\ 0 & \text{if } t \neq \omega. \end{cases}$$

From this definition, it follows that

$$\{\omega; \xi(t, \omega) = 0\} = \{\omega; \omega \neq t\}.$$

Hence, for any subset $\Gamma \subset T = [0, 1]$,

$$\begin{aligned} \{\omega; \xi(t, \omega) = 0, t \in \Gamma\} &= \bigcap_{t \in \Gamma} \{\omega; \xi(t, \omega) = 0\} \\ &= \bigcap_{t \in \Gamma} \{\omega; \omega \neq t\} = \left(\bigcup_{t \in \Gamma} \{\omega; \omega = t\}\right)^c \\ &= \{\omega; \omega \in \Gamma\}^c = [0, 1] - \Gamma. \end{aligned}$$

From this, it follows that if $I \subset T$ is any open interval and D the set of all rationals,

$$P\{\omega; \xi(t, \omega) = 0, t \in I\} = 1 - P(I),$$

$$P\{\omega; \xi(t, \omega) = 0, t \in I \cap D\} = 1,$$

which clearly shows that the process $\xi(t)$ is not separable.

Now, let $\{\tilde{\xi}(t); t \in [0, 1]\}$ be a stochastic process on the same probability space specified as follows:

$$\tilde{\xi}(t, \omega) = 0 \quad \text{for all } (t, \omega) \in [0, 1] \times \Omega.$$

Then, we clearly have

$$\{\omega; \xi(t,\omega) \neq \tilde{\xi}(t,\omega)\} = \{\omega; t = \omega\} = \{t\},$$

which is a null event. Thus, $\tilde{\xi}(t)$ is an equivalent version of $\xi(t)$. It is also easy to check that $\tilde{\xi}(t)$ is separable.

EXAMPLE 1.8.2. Let $\{\xi(t); t \in T\}$ be a real stochastic process on the probability space $\{R^T, \mathscr{B}, P\}$ (see Section 1.2 for definitions). Denote by

$$C_T = \{x(t); t \in T\}$$

the set of all continuous functions with values in R. Clearly, C_T, which is a subset of R^T, is not an element of $\mathscr{B}$. As matter of fact,

$$\{\omega; \xi(\cdot,\omega) \in C_T\} = \bigcap_{n=1}^{\infty} \bigcup_{k=1}^{\infty} \bigcap_{|t-s|\leq 1/k} \left\{\omega; |\xi(t,\omega) - \xi(s,\omega)| \leq \frac{1}{n}\right\} \notin \mathscr{B}.$$

However, if the process is separable with a set $D \subset T$ as separant and T is closed,

$$\{\omega; \xi(\cdot,\omega) \in C_T\} = \bigcap_{n=1}^{\infty} \bigcup_{k=1}^{\infty} \bigcap_{\substack{s,t \in D \\ |s-t|\leq 1/k}} \left\{\omega; |\xi(t,\omega) - \xi(s,\omega)| \leq \frac{1}{n}\right\} \tag{1.8.1}$$

and as such is an element of $\mathscr{B}$. In order that the realizations of a separable process are continuous with probability 1, it is necessary and sufficient that the probability of the random event (1.8.1) is 1.

1.9. Continuity Concepts

In this section we define three types of continuity of a stochastic process. The first one is stochastic continuity (or continuity in probability). Let $\{\xi(t); t \in T\}$ be a real-valued stochastic process on $\{\Omega, \mathscr{B}, P\}$.

Definition 1.9.1. The process $\{\xi(t); t \in T\}$ is said to be stochastically continuous at a point $t_0 \in T$ if, for any $\varepsilon > 0$,

$$\lim_{t \to t_0} P\{|\xi(t) - \xi(t_0)| > \varepsilon\} = 0. \tag{1.9.1}$$

If (1.9.1) holds in every point $t_0 \in T$, we say that the process is stochastically continuous on T.

Remark 1.9.1. From the definition, we see that stochastic continuity is a regularity condition on bivariate marginal distributions of the process. As matter of fact, we have

$$P\{|\xi(t) - \xi(t_0)| > \varepsilon\} = 1 - \int_{-\infty}^{\infty} P\{\xi(t_0) \in dx\} \int_{x-\varepsilon}^{x+\varepsilon} P\{\xi(t) \in dy | \xi(t_0) = x\}.$$

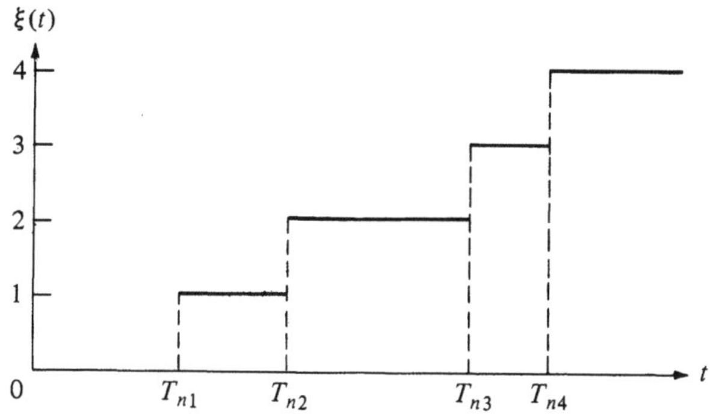

Figure 1.4. A realization of $\xi(t)$.

EXAMPLE 1.9.1. Let $\{T_k\}_1^\infty$ be an i.i.d. sequence of non-negative r.v.'s on a probability space $\{\Omega, \mathscr{B}, P\}$ with $H(t) = P\{T_i \le t\}$ continuous. Set

$$\xi(t) = \sum_{i=1}^{n} I_{\{T_i \le t\}}, \quad t \ge 0. \tag{1.9.2}$$

Clearly,

$$0 \le \xi(t_1) \le \xi(t_2)$$

for all $0 \le t_1 \le t_2$. The realizations of $\xi(t)$ are nondecreasing step functions with unit jumps at points

$$T_{n1} < T_{n2} < \cdots < T_{nn},$$

where $\{T_{ni}\}_1^n$ are order statistics for $\{T_i\}_i^n$. In Figure 1.4 is a trajectory of $\xi(t)$.

Although every sample function of $\xi(t)$ has discontinuities, the random process $\xi(t)$ is stochastically continuous on $[0, \infty)$ because due to Markov's inequality and continuity of $H(\cdot)$, it follows that

$$P\{|\xi(t \pm h) - \xi(t)| > \varepsilon\} \le \frac{n}{\varepsilon}|H(t \pm h) - H(t)| \to 0.$$

as $h \to 0+$ for each $t \ge 0$.

Remark 1.9.2. Condition (1.9.1) is equivalent to

$$\xi(t) \xrightarrow{P} \xi(t_0) \quad \text{as } t \to t_0. \tag{1.9.3}$$

From Example 1.9.1, we see that a process may be stochastically continuous even if each realization of the process is discontinuous. This is so when the probability that a discontinuity will occur in an instant $t \in T$ is zero. So, when is a stochastic process stochastically discontinuous at a point $t_0 \in T$? The answer is only if t_0 is a fixed discontinuity point of the process. What is a fixed discontinuity point?

Definition 1.9.2. Suppose that $\{\xi(t); t \in T\}$ is real-valued and separable, and denote by N_{t_0} the set of sample functions which are discontinuous at $t_0 \in T$. If $P(N_{t_0}) > 0$, we say that $t_0 \in T$ is a fixed discontinuity point of $\xi(t)$.

Definition 1.9.3. Suppose that $\{\xi(t); t \in T\}$ is real-valued and separable. The stochastic process is continuous (a.s.) at a point $t_0 \in T$ if the set of realizations discontinuous at t_0 is negligible. If the process is (a.s.) continuous at every $t \in T$, we say that the process is (a.s.) continuous on T.

Remark 1.9.3. It is apparent that every separable process continuous (a.s.) at a point t is also continuous in probability at t [in this case $P(N_t) = 0$], but the converse is false, in general.

Definition 1.9.4. A stochastic process $\{\xi(t); t \in T\}$ is said to have (a.s.) continuous trajectories if the set of those trajectories which have discontinuities on T is negligible.

If $\{\xi(t); t \in T\}$ is (a.s.) continuous at each $t \in T$, it does not necessarily imply that it has (a.s.) continuous trajectories. Indeed, the set of all those trajectories of the process without discontinuities on T is

$$N^c = \bigcap_{t \in T} N_t^c$$

and this event may not have probability 1.

EXAMPLE 1.9.2. Let $\{\Omega, \mathscr{B}, P\}$ be a probability space with $\Omega = [0, 1]$, $\mathscr{B}$ the σ-algebra of Borel subsets of $[0, 1]$, and P the Lebesgue measure. Let $\{\xi(t); t \in [0, 1]\}$ be defined as follows:

$$\xi(t, \omega) = \begin{cases} 0 & \text{if } t < \omega \\ 1 & \text{if } t \geq \omega \end{cases}$$

(see Figure 1.5). Let $\Gamma = [0, s]$; then

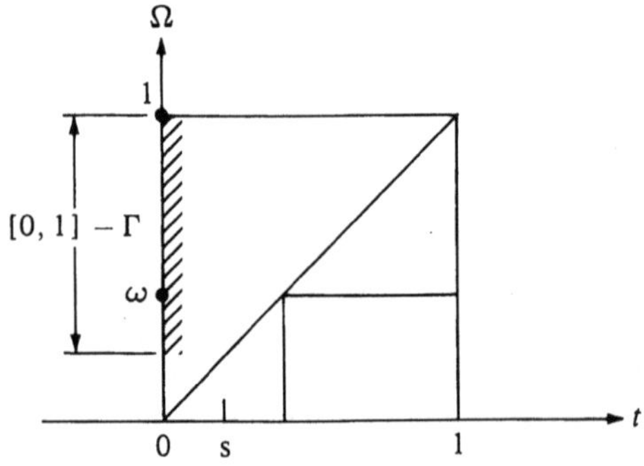

Figure 1.5. Graphical depiction of $\xi(t, \omega)$.

$$\{\omega; \xi(t,\omega) = 0, t \in \Gamma\} = [0,1] - \Gamma,$$

so that

$$P\{\omega; \xi(t,\omega) = 0, t \in \Gamma\} = 1 - s.$$

On the other hand, if D is the set of all rationals

$$P\{\omega; \xi(t,\omega) = 0, t \in D \cap \Gamma\} = 1 - s.$$

This shows that the process is separable. In addition, for $t \in T$, $N_t = \{\omega; \omega = t\}$ so that $P(N_t) = 0$. However,

$$N = \bigcup_{t \in T} N_t = \Omega \quad \text{so that } P(N) = 1.$$

The following theorem by Kolmogorov gives a sufficient condition for (a.s.) continuity of sample functions of a separable stochastic process.

Proposition 1.9.1. *Let $\{\xi(t); t \in T\}$ be a real-valued separable process and T a compact interval. If there exist constants $\alpha > 0$, $\beta > 0$, $C > 0$ such that*

$$E|\xi(t+h) - \xi(t)|^\alpha \leq Ch^{1+\beta}, \tag{1.9.4}$$

then almost all sample functions of $\xi(t)$ are continuous on T.

1.10. More on Separability and Continuity

Given a real-valued process, there will be, generally speaking, several equivalent versions, some of which are separable. This permits us to choose a separable version whose sample paths have some nice properties.

Let $\{\xi(t); t \in T\}$ be real-valued and separable with $D \subset T$ as a separant. Then as we have seen, there exists a negligible subset $\Lambda \subset \Omega$ such that, for each $\omega \in \Lambda^c$, the value of $\xi(\cdot, \omega)$ at every $t \in T$ is the limit

$$\lim_{n\to\infty} \xi(t_n, \omega),$$

where $\{t_n\} \subset D$ and $t_n \to t$ as $n \to \infty$. The sequence $\{t_n\}_1^\infty$, in general, depends on ω.

Consider a real-valued stochastic process $\{\xi(t); t \in T\}$ on a complete probability space $\{\Omega, \mathscr{B}, P\}$, without discontinuity points of the second kind. In other words,

$$\xi(t - 0, \omega) \quad \text{and} \quad \xi(t + 0, \omega)$$

exist for all $t \in T$ and $\omega \in \Omega$. The next simple proposition shows that in such a case we can always choose a version $\tilde{\xi}(t)$ of $\xi(t)$ which is continuous from the right if $\xi(t)$ is stochastically continuous.

Proposition 1.10.1. *Let $\{\xi(t); t \in T\}$ be real-valued stochastically continuous random process on a complete probability space $\{\Omega, \mathscr{B}, P\}$ and $T \subset R$ a compact* *interval. If the process does not have discontinuities of the second kind, there exists a separable version $\tilde{\xi}(t)$ equivalent to $\xi(t)$ whose sample functions are continuous from the right.*

PROOF. Choose a separable version and denote by B the random event that the limit

$$\lim_{n\to\infty} \xi\left(t+\frac{1}{n}\right)$$

exists for every $t \in T$. Let us show that $P(B) = 1$. To this end, let $t \in T$ be fixed and consider

$$B_t = \left\{\omega;\ \lim_{n\to\infty} \xi\left(t+\frac{1}{n}, \omega\right) \text{exists}\right\}.$$

Because the process is separable, this limit exists for all $\omega \in \Lambda^c$, where Λ is a null set. Therefore, $P(B_t) = 1$ for each $t \in T$. On the other hand, due to the separability assumption,

$$B = \bigcap_{t \in D\cap T} B_t \Rightarrow P(B) = 1.$$

Next, set

$$\tilde{\xi}(t,\omega) = \lim_{n\to\infty} \xi\left(t+\frac{1}{n},\omega\right)$$

if $\omega \in B$, and $\tilde{\xi}(t,\omega) = \xi(t,\omega)$ if $\omega \in B^c$; then

$$\{\omega; \tilde{\xi}(t) \neq \xi(t)\} = \bigcup_{r=1}^{\infty} \left\{\omega; |\tilde{\xi}(t) - \xi(t)| > \frac{1}{r}\right\} \cap B.$$

Hence,

$$P\{\omega; \tilde{\xi}(t) \neq \xi(t)\} = \lim_{r\to\infty} P\left(\left\{|\tilde{\xi}(t) - \xi(t)| > \frac{1}{r}\right\} \cap B\right).$$

On the other hand,

$$\begin{aligned} P\left\{|\tilde{\xi}(t) - \xi(t)| > \frac{1}{r}\right\} &= P\left(\bigcup_{n=1}^{\infty}\bigcap_{k=n}^{\infty}\left\{\left|\xi\left(t+\frac{1}{k}\right) - \xi(t)\right| > \frac{1}{r}\right\}\right) \\ &= \lim_{n\to\infty} P\left(\bigcap_{k=n}^{\infty}\left\{\left|\xi\left(t+\frac{1}{k}\right) - \xi(t)\right| > \frac{1}{r}\right\}\right) \\ &\leq \lim_{n\to\infty} P\left\{\left|\xi\left(t+\frac{1}{n}\right) - \xi(t)\right| > \frac{1}{r}\right\} = 0 \end{aligned}$$

due to stochastic continuity. Therefore,

$$P\{\omega; \tilde{\xi}(t) \neq \xi(t)\} = 0, \quad \forall t \in T.$$

Finally, because $\forall\omega \in B$ and $t \in T$

$$\tilde{\xi}(t) = \xi(t + 0),$$

it follows that $\{\tilde{\xi}(t); t \in T\}$ is continuous from the right. This proves the assertion. □

The following result will be needed in the sequel.

Proposition 1.10.2. *Let $\{\xi(t); t \in T\}$ be a real-valued stochastic process and $T \subset R$ a compact interval. A necessary and sufficient condition for $\xi(t)$ to be continuous in probability on T is that*

$$\sup_{|t-s|<h} P\{|\xi(t) - \xi(s)| > \varepsilon\} \to 0 \tag{1.10.1}$$

as $h \to 0$ for any $\varepsilon > 0$.

PROOF. Sufficiency of (1.10.1) is clear. To prove its necessity, assume that $\xi(t)$ is stochastically continuous on T. Then for any fixed ε, $\delta > 0$ and $u \in T$, there exists $h > 0$ so that

$$\sup_{|t-u|<h} P\{|\xi(t) - \xi(u)| > \varepsilon\} < \delta.$$

Let $\{u_i\}_1^n \subset T$, with $u_1 < u_2 < \cdots < u_n$, be a sequence such that $T \subset \bigcup_1^n (u_i - h, u_i + h)$. If I is an open interval contained in one of $(u_i - h, u_i + h)$, then, for any $s, t \in I$,

$$\{|\xi(t) - \xi(s)| > 2\varepsilon\} \subset \{|\xi(t) - \xi(u)| > \varepsilon\} \cup \{|\xi(s) - \xi(u)| > \varepsilon\}.$$

This implies the inequality

$$P\{|\xi(t) - \xi(s)| > 2\varepsilon\} \leq 2\delta,$$

which holds as long as $|t - s| < h$. The proof now follows by letting $\delta \to 0$.

□

1.11. Measurable Random Processes

Let R be the real line and $\mathscr{R}$ the σ-algebra of Borel subsets of R. In the following we shall often use the notation

$$\mathscr{R}_C = \mathscr{R} \cap C = \{B \cap C; B \in \mathscr{R}\}. \tag{1.11.1}$$

Clearly, $\mathscr{R}_C$ is a σ-algebra, such that $\mathscr{R}_C \subset \mathscr{R}$ if $C \in \mathscr{R}$.

Let $\{\xi(t); t \in T\}$ be real-valued stochastic process on a probability space $\{\Omega, \mathscr{B}, P\}$ and $T \subset R$ an interval. From Definition 1.1.1, we know that $\xi(t, \cdot)$ is a $\mathscr{B}$-measurable function on Ω for every $t \in T$. From this, however, it does not follow that the mapping

$$\xi(\cdot,\cdot)\colon T \times \Omega \to R \tag{1.11.2}$$

is $\mathscr{R}_T \times \mathscr{B}$-measurable. We now give the following definition:

Definition 1.11.1. A real-valued stochastic process $\{\xi(t); t \in T\}$ is said to be measurable if the mapping (1.11.2) is $\mathscr{R}_T \times \mathscr{B}$-measurable, i.e., if $\{(\omega, t); \xi(t,\omega) \in B\} \in \mathscr{R}_T \times \mathscr{B}$, for every $B \in \mathscr{R}$.

Remark 1.11.1. If $\{\xi(t); t \in T\}$ is a measurable process. Then from Fubini's theorem, it follows that for almost all $\omega \in \Omega$ the sample functions $\xi(\cdot,\omega)$ are $\mathscr{R}_T$-measurable.

To motivate the consideration of the measurability concept, suppose that for a measurable process $\xi(t)$, we have

$$\iint_{T\times\Omega} |\xi(t,\omega)|\, dt\, P(d\omega) < \infty. \tag{1.11.3}$$

Again invoking Fubini's theorem, we obtain

$$\iint_{T\times\Omega} |\xi(t,\omega)|\, dt\, P(d\omega) = \int_T E\{|\xi(t)|\}\, dt = \int_\Omega P(d\omega) \int_T |\xi(t,\omega)|\, dt.$$

Thus,

$$\int_T E\{|\xi(t)|\}\, dt < \infty$$

and

$$X(\omega) = \int_T |\xi(t,\omega)|\, dt < \infty \quad \text{(a.s.)},$$

where $X(\cdot)$ is a r.v. on $\{\Omega, \mathscr{B}, P\}$.

Next we give an example of a measurable process.

Example 1.11.1. Let $\{\xi_k\}_0^\infty$ be a sequence of r.v.'s on $\{\Omega, \mathscr{B}, P\}$. Let $\{\xi(t); t \in [a,b]\}$ be a stochastic process defined as follows: Set $a < t_1 < t_2 < \cdots < t_n < b$ and write

$$\xi(t) = \xi_0, \quad t \in [a, t_1); \qquad \xi(t) = \xi_n, \quad t \in [t_n, b]$$

and

$$\xi(t) = \xi_i, \quad t \in [t_i, t_{i+1}), \quad i = 1, 2, \ldots, n-1.$$

To show that the stochastic process is measurable, let $\Gamma \in \mathscr{R}$ and consider

$$\{(t,\omega); \xi(t,\omega) \in \Gamma\} = [a,t_1) \times \{\omega; \xi_0 \in \Gamma\} \cup [t_1,t_2) \times \{\omega; \xi_1 \in \Gamma\} \cup \cdots \cup [t_n, b] \times \{\omega; \xi_n \in B\}.$$

Clearly, then ($T = [a,b]$)

$$\{(t,\omega); \xi(t,\omega) \in \Gamma\} \in \mathscr{R}_T \times \mathscr{B},$$

so that $\xi(t)$ is measurable.

Next, we give some sufficient conditions which ensure measurability of a random process.

Proposition 1.11.1. *Let $\{\xi(t); t \in T\}$ be a real-valued stochastic process and $T \subset R$ a compact interval. If $\xi(t)$ is continuous in probability, there exists a version $\tilde{\xi}(t)$ which is separable and measurable.*

PROOF. Due to continuity in probability of $\xi(t)$, we have (see 1.10.1)

$$\sup_{|t-s|<h} P\{|\xi(t) - \xi(s)| > \varepsilon\} \to 0 \quad \text{as } h \to 0. \tag{1.11.4}$$

Now set $T = [a,b]$ and choose

$$a = t_0^n < t_1^n < \cdots < t_{k_n}^n = b$$

such that

$$P\{|\xi(u) - \xi(v)| > \varepsilon\} < \frac{1}{n^2} \tag{1.11.5}$$

for all $u, v \in [t_{j-1}^n, t_j^n]$. Suppose now that for every n, $\{t_j^n\} \subset \{t_j^{n+1}\}$ and let L, consisting of all the t_j^n, be dense in T. Define

$$\xi_n(t,\omega) = \xi(t_j^n,\omega) \quad \text{for } t_j^n \le t < t_{j+1}^n.$$

From Example 1.11.1 we clearly see that $\xi_n(t,\omega)$ is a measurable process for every $n = 1, 2, \ldots$. Next, from (1.11.5) we infer that

$$P\{|\xi(t) - \xi_n(t)| > \varepsilon \text{ i.o.}\} = 0,$$

so that

$$\xi_n(t) \to \xi(t) \quad \text{(a.s.)}, \quad \forall t \in T.$$

Further, because $\{\xi_n(t); t \in T\}$ is a sequence of measurable processes,

$$\tilde{\xi}(t) = \limsup_{n\to\infty} \xi_n(t)$$

is a measurable process such that $\tilde{\xi}(t) \stackrel{d}{=} \xi(t)$.

Finally, from the definition of $\tilde{\xi}(t)$ and $\xi_n(t)$, it is apparent that $\tilde{\xi}(t) = \xi_n(t)$ for all $t \in L$. On the other hand, for every $t \in T$ and $\omega \in \Omega$, $\tilde{\xi}(t)$ is the lim sup of $\{\xi(t_{i_n}^n)\}$ for a sequence $\{t_{i_n}^n\} \subset L$ which increases to t. From this and Definition 1.7.1, it follows that $\tilde{\xi}(t)$ is separable. □

Problems and Complements

1.1. Let $\{\Omega, \mathscr{B}, P\}$ be a probability space. Let ρ be a function on $\mathscr{B} \times \mathscr{B}$ defined by

$$\rho(A, B) = P(A \triangle B),$$

where $A \triangle B = (A - B) \cup (B - A)$. Show that, for any $C \in \mathscr{B}$,

$$\rho(A, B) \leq \rho(A, C) + \rho(B, C).$$

1.2. Show that for any two random events A and B

$$|P(A) - P(B)| \leq \rho(A, B).$$

1.3. Let X_1 and X_2 be independent real-valued r.v.'s on a probability space $\{\Omega, \mathscr{B}, P\}$ with the distribution functions (d.f.'s) $H_1(\cdot)$ and $H_2(\cdot)$, respectively. Let $\{\xi(t); t \geq 0\}$ be a stochastic process defined by

$$\xi(t) = tX_1 + X_2.$$

Calculate $P(A)$, where A is the set of all nondecreasing sample functions of the process.

1.4. Let $\{\xi(t); t \geq 0\}$ be a stochastic process defined by

$$\xi(t) = X + \alpha t, \quad \alpha > 1,$$

where X is a r.v. with the Cauchy distribution. Let $D \subset [0, \infty)$ be finite or countably infinite. Determine:
a. $P\{\xi(t) = 0 \text{ for at least one } t \in D\}$,
b. $P\{\xi(t) = 0 \text{ for at least one } t \in (1, 2]\}$.

1.5. Let X and Y be r.v.'s on $\{\Omega, \mathscr{B}, P\}$, where $Y \sim N(0, 1)$ (standard normal). Let $\{\xi(t); t \geq 0\}$ be a stochastic process defined by

$$\xi(t) = X + t(Y + t).$$

Let A be the set of sample functions of $\xi(t)$ nondecreasing on $[0, \infty)$. Show that A is an event and determine $P(A)$.

1.6. In Problem 1.5 denote by B the set of sample functions which are nonincreasing in $[0, 1]$. Show that B is an event and determine $P(B)$.

1.7. Let X_1 and X_2 be independent r.v.'s on $\{\Omega, \mathscr{B}, P\}$ with common standard normal d.f. Let $\{\xi(t); t \geq 0\}$ be a stochastic process defined by

$$\xi(t) = (X_1 + X_2)t.$$

Determine $F_{t_1, \ldots, t_n}(x_1, \ldots, x_n)$. If A is the set of all non-negative sample functions, show that A is an event and determine $P(A)$.

1.8. Let $\{\xi(t); t \geq 0\}$ be a stochastic process defined by

$$\xi(t) = X \cos(t + U),$$

where X and U are independent r.v.'s U is uniform in $[-\pi, \pi]$, and $E\{X\} = 0$. Determine its covariance function.

1.9. A stochastic process $\{\xi(t); t \in T\}$ is called normal (or Gaussian) if all its marginal distributions are Gaussian. Set

$$\bar{x}(t) = E\{\xi(t)\}, \qquad C(t_1, t_2) = E\{\xi(t_1)\xi(t_2)\} - \bar{x}(t_1)\bar{x}(t_2),$$

then

$$\xi(t) \sim N(\bar{x}(t), C(t,t))$$

for all $t \in T$. Assume that

$$\bar{x}(t) = 3, \qquad C(t_1, t_2) = 4\exp(-0.2|t_2 - t_1|).$$

Show that

$$P\{\xi(t) \leq 2\} \approx .309.$$

1.10. Let $\{\xi(t); t \geq 0\}$ be random process with $\xi(0) = 0$ (a.s.) and marginal probability densities ($t_0 = x_0 = 0$)

$$f_{t_1,\ldots,t_n}(x_1,\ldots,x_n) = \prod_{i=1}^{n} [2\pi(t_i - t_{i-1})]^{-1/2} \exp\left(-\frac{1}{2}\frac{(x_i - x_{i-1})^2}{(t_i - t_{i-1})}\right)$$

($0 < t_1 < \cdots < t_n$). Determine

$$E\{\xi(t)\}, \qquad E\{\xi(t)\}^2, \qquad E\{\xi(s)\xi(t)\} = C(s,t).$$

1.11. Let X and U be independent r.v.'s on $\{\Omega, \mathscr{B}, P\}$, U is uniform in $[0, 2\pi]$, and the probability density of X is defined by

$$f_X(x) = \begin{cases} 2x^3 e^{(-1/2x^4)} & x \geq 0 \\ 0, & x < 0. \end{cases}$$

Let $\{\xi(t); t \geq 0\}$ be defined by

$$\xi(t) = X^2 \cos(2\pi t + U).$$

Show that $\xi(t)$ is a Gaussian process [i.e., every random vector $(\xi(t_1), \ldots, \xi(t_n))$ is normally distributed; see Chapter 4].

1.12. Let $\{\xi(t); t \in T\}$ and $\{X(t); t \in T\}$ be real stochastic processes on $\{\Omega, \mathscr{B}, P\}$. If they are stochastically equivalent, show that they have identical marginal distributions. Under what conditions will they have the same sample functions?

1.13. Let $\{X_i\}_1^\infty$ and $\{Y_i\}_1^\infty$ be r.v.'s on $\{\Omega, \mathscr{B}, P\}$ such that $E\{X_i\} = E\{Y_i\} = 0$, $\operatorname{Var}\{X_i\} = \operatorname{Var}\{Y_i\} = \sigma_i^2 < \infty$, $E\{X_i X_j\} = 0$, $E\{Y_i Y_j\} = 0$, and $E\{X_i Y_j\} = 0$ for all $i \neq j$. Let $\{\xi(t); t \geq 0\}$ be defined by

$$\xi(t) = \sum_{j=1}^{n} \{X_j \cos \lambda_j t + Y_j \sin \lambda_j t\}.$$

Is the process wide sense stationary? Determine its covariance function.

1.14. Let $\{X_n\}_0^\infty$ be an i.i.d. sequence of r.v.'s with $E\{X_i\} = 0$ and $\operatorname{Var}\{X_i\} = \sigma^2 < \infty$. Let $\{N(t); t \geq 0\}$ be a homogeneous Poisson process with $E\{N(t)\} = \lambda t$ and independent of $\{X_n\}_0^\infty$. Is the stochastic process $\{\xi(t); t \geq 0\}$ defined by

$$\xi(t) = X_{N(t)}$$

strictly stationary? [A non-negative integer-valued random process with independent increments $\{N(t); t \geq 0\}$ is homogeneous Poisson process if $0 \leq N(t_1) \leq N(t_2)$ for all $0 \leq t_1 \leq t_2$ and

$$P\{N(t_2) - N(t_1) = k\} = \exp[-\lambda(t_2 - t_1)] \frac{[\lambda(t_2 - t_1)]^k}{k!};$$

see Chapter 2].

1.15. Let Z be a Bernoulli r.v. such that

$$P\{Z = 1\} = p, \qquad P\{Z = -1\} = q, \qquad p + q = 1.$$

Let $\{N(t); t \geq 0\}$ be a homogeneous Poisson process with parameter λ, independent of Z. Consider the stochastic process $\{\xi(t); t \geq 0\}$ defined by

$$\xi(t) = (-1)^{1/2(1-Z)+N(t)}.$$

Determine

$$P\{\xi(t) = 1\}, \qquad E\{\xi(t_1)\xi(t_2)\}.$$

1.16. Let $\{\xi(t); -\infty < t < \infty\}$ be a random telegraph process defined as follows: $\xi(t)$ assumes only two values, -1 and $+1$, with equal probabilities

$$P\{\xi(t) = 1\} = \tfrac{1}{2}, \qquad P\{\xi(t) = -1\} = \tfrac{1}{2}.$$

The sequence of transition times $\{\tau_k\}_{-\infty}^{\infty}$ forms a homogeneous Poisson process with parameter $\lambda > 0$. In other words, the number of transitions (from -1 to $+1$ and vice versa) in $(u, u + t]$ is a homogeneous Poisson process $N(t)$. Show that the process is wide sense stationary and determine its covariance function.

1.17. Let $\{\xi(t); t \in \mathscr{R}\}$ be a sample continuous wide sense stationary stochastic process. Determine

$$\operatorname{Var}\left\{\int_{-u}^{u} \xi(s)\, ds\right\}.$$

1.18. Let $\{\xi(t); t \geq 0\}$ be a process with independent increments. Assume that

$$\varphi_1(t, \lambda) = E\{e^{i\lambda\xi(t)}\} \quad \text{and} \quad \varphi_2(t_1, t_2, \lambda) = E\{e^{i\lambda[\xi(t_2) - \xi(t_1)]}\}$$

are given. Determine the characteristic function of $(\xi(t_1), \ldots, \xi(t_n))$ in terms of φ_1 and φ_2.

1.19. A stochastic process with independent increments is said to be homogeneous if $\xi(t + s) - \xi(s) \stackrel{d}{=} \xi(t)$ for all $s, t \geq 0$. In other words, $\xi(t + s) - \xi(s)$ and $\xi(t)$ have the same distribution. Show that the characteristic function $\varphi_1(t, \lambda)$ is infinitely divisible if continuous at $t = 0$.

1.20. (Continuation) Let a stochastic process with independent increments $\{\xi(t); t \geq 0\}$ be homogeneous. If $\varphi(t; \lambda)$ is continuous at $t = 0$ for all λ, the process is stochastically continuous.

1.21. Let $\{\xi(t); t \in T\}$ be a real stochastic process on a complete probability space $\{\Omega, \mathscr{B}, P\}$, $D \subset T$ countable and everywhere dense, and $\Lambda \subset \Omega$ negligible. If, for any $\omega \notin \Lambda$ and $t \in T$, there exists $\{t_n\}_1^{\infty} \subset D$, $t_n \to t$, such that $\xi(t_n, \omega) \to \xi(t, \omega)$, show that the process is separable.

1.22. If $\{\xi(t); t \in T\}$ is separable and $f: R \to R$ continuous, show that $\{f(\xi(t)); t \in T\}$ is separable.

1.23. If $\{\xi(t); t \in T\}$ is such that $\xi(\cdot, \omega)$ is continuous on T for all $\omega \in \Lambda^c$, where $\Lambda \subset \Omega$ is negligible, show that the process is separable.

1.24. Let $\{\xi(t); t \in T\}$ be stochastically continuous on T and $f: T \to R$. Then $X(t) = \xi(t) + f(t)$ is stochastically continuous in those and only those points of T where $f(t)$ is continuous.

1.25. Let $\{\xi(t); t \in T\}$ be a family of i.i.d. r.v.'s with common probability density $f(\cdot)$. Show that $\xi(t)$ cannot be stochastically continuous at any point $t \in T$.

1.26. Let $\{\xi(t); t \in T\}$ be stochastically continuous at every $t \in T$. Then $\Psi(\xi(t))$ is also stochastically continuous if $\Psi: R \to R$ is continuous.

1.27. In the previous problem, show that $\varphi(t) = E\{\Psi(\xi(t))\}$ is a continuous function.

1.28. Let $\{\xi(t); t \in T\}$ be a stochastically continuous process on $\{\Omega, \mathscr{B}, P\}$. Let $\{X(t); t \in T\}$ be another process on $\{\Omega, \mathscr{B}, P\}$ equivalent to $\xi(t)$. Show that $X(t)$ is also stochastically continuous.

1.29. A stochastic process $\{\xi(t); t \in T\}$ is said to be "bounded in probability" (or stochastically bounded) if

$$\lim_{C \to +\infty} \sup_{t \in T} P\{|\xi(t)| > C\} = 0. \tag{$*$}$$

If $\xi(t)$ is stochastically continuous on $T = [a, b]$, then $(*)$ holds.

1.30. Let $\{\xi(t); t \in T\}$ be a process with independent increments. If $\operatorname{Var}\{\xi(t)\}$ is a continuous function in t, the process is stochastically continuous.

1.31. Let $\{\xi(t); t \in [0, 1]\}$ be a standard Brownian motion [i.e., $\xi(t)$ has independent increments and for any s, $t + s \in [0, 1]$ with $t > 0$

$$P\{\xi(t + s) - \xi(t) \le x\} = (2\pi s)^{-1/2} \int_{-\infty}^{x} e^{-z^2/2s}\, dz].$$

Show that almost all sample functions of $\xi(t)$ are continuous.

1.32. Let $\{\xi(t); t \in [0, 1]\}$ be a real Gaussian process with $E\{\xi(t)\} = 0$ and covariance function

$$C(s, t) = \min\{s, t\} - st.$$

Show that its sample functions are (a.s.) continuous on $[0, 1]$.

1.33. Let $\{\xi(t); t \in [a, b]\}$ be a real process with (a.s.) continuous sample paths. Show that $\xi(t)$ is measurable.

1.34. If all the sample functions of a real stochastic process are Borel measurable, does this imply measurability of the random process?

CHAPTER 2

The Poisson Process and Its Ramifications

2.1. Introduction

We begin by describing in an informal fashion the subject matter of this chapter. The part of the general theory of stochastic processes dealing with countable sets of points randomly distributed on the real line or in an arbitrary space (for instance, Cartesian d-dimensional space) is called the "Theory of Point Processes." Of all point processes, those on the real line have been most widely studied. Notwithstanding their relatively simple structure, they form building blocks in a variety of industrial, biological, geophysical, and engineering applications. The following example describes a general situation which in a natural fashion introduces a point process on a line.

EXAMPLE 2.1.1. Let A be an event which reoccurs repeatedly in time so that the time interval between any two consecutive occurrences of A has a random length. Assume that our observation of the event A begins at time $t = 0$ and that $0 < t_1 < t_2 < \cdots$ are instances of the first, second, etc., occurrence of A. These instances may be regarded as a set of points randomly distributed on the half-line $R_+ = [0, \infty)$. An alternative interpretation of this situation is that the event A occurs at points of a randomly selected countable subsets $\omega \subset R_+$, where

$$\omega: (t_1, t_2, \ldots).$$

The random event A may be the rainfall occurrence at a given site. Due to stochastic nature of the rainfall phenomenon, its arrival times at a given location may be regarded as points randomly distributed on R_+. Other specific cases of A could be the earthquake occurrences, volcanic eruptions, flooding of a city, and so on.

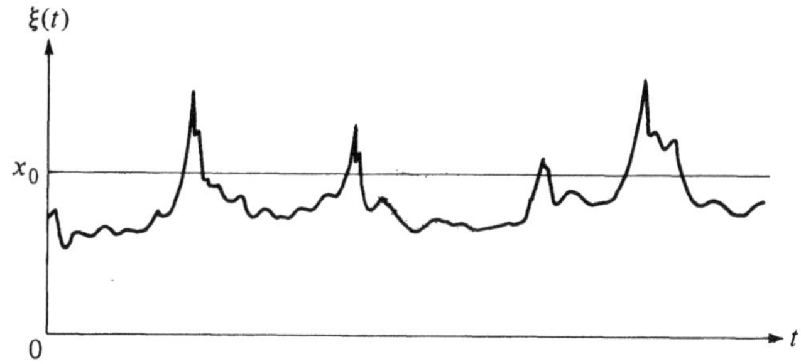

Figure 2.1. Crossings of the level x_0 by the process $\xi(t)$.

Point processes generated by crossings of a fixed level by a stochastic process are of particular interest in engineering. The interest in crossing problems dates back to the original paper of M. Kac (1943) and of S.O. Rice (1945). As an example in hydrology, consider the discharge rate $\xi(t)$ of a streamflow at a given site. Because the surface runoff flows vary randomly in time, $\xi(t)$ is a continuous parameter stochastic process. The crossing points of a fixed level x_0 by $\xi(t)$ are randomly distributed (see Figure 2.1). Of particular interest in various applications is asymptotic behavior of this point process when $x_0 \to +\infty$.

2.2. Simple Point Process on R_+

Set $R_+ = [0, \infty)$ and denote by $\mathscr{R}_+$ the σ-algebra of Borel subsets of R_+. Let Ω be the collection of all infinite countable subsets of R_+ which are locally finite and do not contain zero as an element. In other words, every $\omega \in \Omega$ is of the form

$$\omega: (t_1, t_2, \ldots), \quad 0 < t_1 < t_2 < \cdots, \tag{2.2.1}$$

and such that for any compact $K \subset R_+$

$$\#(\omega \cap K) < \infty,$$

where $\#$ denotes the number of elements in the set. It is clear from this definition that $t_n \to +\infty$ as $n \to \infty$.

Let $\varepsilon_x(\cdot)$ be the Dirac measure on $\mathscr{R}_+$ concentrated at x, i.e.,

$$\forall B \in \mathscr{R}_+, \quad \varepsilon_x(B) = \begin{cases} 1 & \text{if } x \in B \\ 0 & \text{if } x \notin B, \end{cases}$$

and consider the Borel–Radon measure

$$\mu(\cdot) = \sum_{t \in \omega} \varepsilon_t(\cdot). \tag{2.2.2}$$

This measure is called a "point measure" on R_+. It clearly takes only non-negative integer values and

$$\mu(K) < \infty$$

on any compact $K \subset R_+$. Note that every $\omega \in \Omega$ is a closed subset of R_+ because its intersection with any compact subset of R_+ is a finite set.

Let $\{\tau_n\}_1^\infty$ be a sequence of coordinate mappings on Ω, i.e.,

$$\tau_n: \Omega \to R_+,$$

specified as follows: for any ω: $(t_1, t_2, \ldots)$,

$$\tau_n(\omega) = t_n, \quad n = 1, 2, \ldots. \tag{2.2.3}$$

From this it readily follows that on Ω

$$0 < \tau_1(\cdot) < \tau_2(\cdot) < \cdots \tag{2.2.4}$$

and that $\tau_n(\cdot) \to \infty$ as $n \to \infty$. Finally, we denote by $\mathscr{B}$ the least σ-algebra of subsets of Ω with respect to which all τ_n are measurable.

We now give a definition of a point process on R_+.

Definition 2.2.1. A random measure η on $\mathscr{R}_+$,

$$\eta: \mathscr{R}_+ \times \Omega \to N_+,$$

where $N_+ = \{0, 1, \ldots\}$, defined by, $\forall B \in \mathscr{R}_+$

$$\eta(B) = \sum_{k=1}^{\infty} \varepsilon_{\tau_k}(B) \tag{2.2.5}$$

is called a "point process" on R_+.

From (2.2.3) and (2.2.5), it is clear that for every fixed $\omega \in \Omega$, $\eta(B, \omega) = \mu(B)$, where $\mu(\cdot)$ is a point measure defined by (2.2.2). In addition, for any $t \in R_+$ and $\omega \in \Omega$,

$$\eta(\{t\}, \omega) = \begin{cases} 0 & \text{if } t \notin \omega \\ 1 & \text{if } t \in \omega, \end{cases} \tag{2.2.6}$$

which means that the point process does not have multiple points. Such a point process is called "simple."

When $B = (0, t]$, we will write

$$N(t) = \eta((0, t], \cdot), \qquad N(0) \equiv 0. \tag{2.2.7}$$

From (2.2.5) we readily have that

$$N(t) = \sum_{k=1}^{\infty} I_{(0,t]}(\tau_k). \tag{2.2.8}$$

The stochastic process $\{N(t); t \geq 0\}$ is called the "counting random function" of the point process η. It is easy to see from (2.2.8) that every realization of

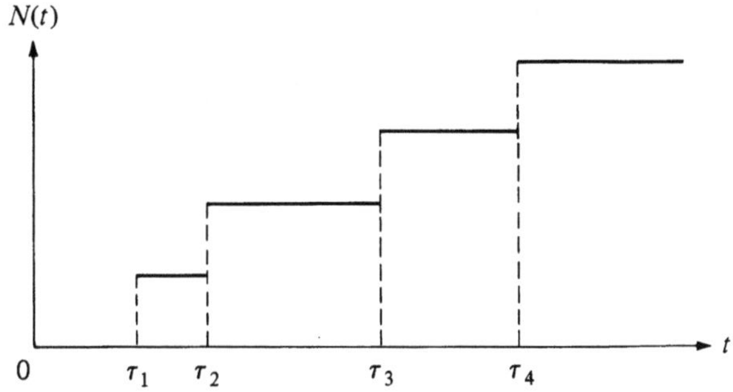

Figure 2.2. A sample function of $N(t)$.

$N(t)$ is a nondecreasing step function, continuous from the right, with unit jumps at each of its discontinuity points. In Figure 2.2, a sample function of $N(t)$ is depicted.

The r.v.'s $\{\tau_n\}_1^\infty$ are usually called "arrival times" of the point process. The following relation between the counting function $N(t)$ and its arrival times τ_n is easy to see:

$$\{\tau_n \le t\} = \{N(t) \ge n\}. \tag{2.2.9}$$

Let P be a probability measure on $\{\Omega, \mathscr{B}\}$ and set

$$F_n(t) = P\{\tau_n \le t\}, \qquad \Lambda(t) = E\{N(t)\}. \tag{2.2.10}$$

Then from (2.2.8), it clearly follows that

$$\Lambda(t) = \sum_{k=1}^{\infty} F_k(t). \tag{2.2.11}$$

2.3. Some Auxiliary Results

In this section, we discuss some inequalities involving sequences of independent Bernoulli r.v.'s. These inequalities, which are of independent interest, will considerably simplify many proofs in forthcoming sections.

Lemma 2.3.1. *Let* $\{I_k\}_1^n$ *be independent Bernoulli* r.v.'s; *then,*

$$P\left\{\sum_{i=1}^{n} I_i > k\right\} \le \left(P\left\{\sum_{i=1}^{n} I_i > 0\right\}\right)^{k+1}. \tag{2.3.1}$$

PROOF. First, we have the following equality:

$$P\left\{\sum_{i=1}^{n} I_i > k\right\} = \sum_{l=1}^{n} P\left\{\sum_{i=1}^{l-1} I_i = 0, I_l = 1, \sum_{i>l}^{n} I_i > k-1\right\}. \tag{2.3.2}$$

Because $\{I_i\}_1^n$ are independent r.v.'s, we readily obtain from (2.3.2) that

$$P\left\{\sum_{i=1}^{n} I_i > k\right\} \le P\left\{\sum_{i=1}^{n} I_i > k-1\right\} \sum_{l=1}^{n} P\left\{\sum_{i=1}^{l-1} I_i = 0, I_l = 1\right\}. \quad (2.3.3)$$

On the other hand,

$$\begin{aligned} P\left\{\sum_{i=1}^{n} I_i > 0\right\} &= 1 - P\{I_1 = 0, \ldots, I_n = 0\} \\ &= 1 - P\{I_1 = 0, \ldots, I_{n-1} = 0\} + P\{I_1 = 0, \ldots, I_{n-1} = 0, I_n = 1\} \\ &= \cdots = 1 - P\{I_1 = 0\} + P\{I_1 = 0, I_2 = 1\} + \cdots \\ &\quad + P\{I_1 = 0, \ldots, I_{n-1} = 0, I_n = 1\} = \sum_{l=1}^{n} P\left\{\sum_{i=1}^{l-1} I_i = 0, I_l = 1\right\}. \end{aligned}$$

From this and (2.3.3) we obtain the recursion

$$P\left\{\sum_{i=1}^{n} I_i > k\right\} \le P\left\{\sum_{i=1}^{n} I_i > k-1\right\} P\left\{\sum_{i=1}^{n} I_i > 0\right\}$$

which proves (2.3.1). □

Let X and Y be r.v.'s on a probability space $\{\Omega, \mathscr{B}, P\}$ with values in some abstract space $\{S, \mathscr{L}\}$.

Definition 2.3.1. The total variation distance $d(X, Y)$ between X and Y is defined by

$$d(X, Y) = \|P_X - P_Y\| = \sup_{D \in \mathscr{L}} |P_X(D) - P_Y(D)|, \quad (2.3.4)$$

where P_X and P_Y are, respectively, the distributions of X and Y on $\{S, \mathscr{L}\}$.

Remark 2.3.1. When X and Y are real-valued r.v.'s, the metric

$$d_0(X, Y) = \|F_X - F_Y\| = \sup_{x} |F_X(x) - F_Y(x)| \quad (2.3.5)$$

is often useful, where F_X and F_Y are the distribution functions (d.f.'s) of X and Y, respectively. Clearly,

$$d_0(X, Y) \le d(X, Y).$$

Finally, if X and Y are integer-valued,

$$d(X, Y) = \frac{1}{2} \sum_{k=0}^{\infty} |P\{X = k\} - P\{Y = k\}|. \quad (2.3.6)$$

The next result is known as the coupling lemma.

Lemma 2.3.2

$$\sup_{D \in \mathscr{L}} |P_X(D) - P_Y(D)| \le P\{X \ne Y\}. \quad (2.3.7)$$

PROOF. We have

$$\begin{aligned}P\{X \in D\} - P\{Y \in D\} &= P\{X \in D\} - P\{Y \in D, X \in D\} - P\{Y \in D, X \notin D\}\\ &\le P\{X \in D, Y \notin D\} \le P\{X \neq Y\}\end{aligned}$$

for all $D \in \mathscr{L}$, which proves the assertion. □

Remark 2.3.2. Concerning the coupling inequality (2.3.7), the following should be pointed out. We can use any joint distribution of (X, Y) in inequality (2.3.7) as long as its marginals are P_X and P_Y. Thus, to get a sharp bound in (2.3.7), we will select, from all joint distributions of (X, Y) with the same marginals P_X and P_Y, one that has the least probability, $P\{X \neq Y\}$.

Application of the coupling lemma hinges on our ability to calculate $P\{X \neq Y\}$, which is not always a simple task.

The next result in this respect is of particular interest to us. Its purpose is to determine simple exact upper bounds for the distance $d_0(X, Y)$ and for $d(X, Y)$, in the case when X is a sum of independent Bernoulli r.v.'s and Y a Poisson r.v. suitably chosen to approximate X in distribution.

Lemma 2.3.3 (Le Cam). *Let $\{I_i\}_1^n$ be independent Bernoulli r.v.'s with*

$$E\{I_i\} = p_i, \quad i = 1, 2, \ldots$$

and Y a Poisson r.v. with $E\{Y\} = \sum_1^n p_i$. Then

$$d\left(\sum_{i=1}^n I_i, Y\right) \le \sum_{i=1}^n p_i^2. \tag{2.3.8}$$

PROOF. We can write

$$Y = Y_1 + \cdots + Y_n,$$

where Y_i are independent Poisson r.v.'s with $E\{Y_i\} = p_i$. Then from the Coupling Lemma 2.3.2, we have

$$\begin{aligned}d\left(\sum_{i=1}^n I_i, Y\right) &= d\left(\sum_{i=1}^n I_i, \sum_{i=1}^n Y_i\right) \le P\left\{\sum_{i=1}^n I_i \neq \sum_{i=1}^n Y_i\right\}\\ &\le P\left(\bigcup_{i=1}^n \{I_i \neq Y_i\}\right) \le \sum_{i=1}^n P\{I_i \neq Y_i\}.\end{aligned} \tag{2.3.9}$$

To evaluate $P\{I_i \neq Y_i\}$, several methods are available. The following one is due to Serfling (1975). Let Z_i also be a Bernoulli r.v. independent of Y_i and such that

$$E\{Z_i\} = 1 - (1 - p_i)e^{p_i}.$$

Set

$$I_i = I_{\{Y_i \ge 1\}} + I_{\{Y_i = 0\}} I_{\{Z_i = 1\}}.$$

Then

$$\begin{aligned} P\{I_i \neq Y_i\} &= P\{Y_i \geq 2\} + P\{Y_i = 0, I_i = 1\} \\ &= 1 - e^{-p_i} - p_i e^{-p_i} + e^{-p_i}[1 - e^{p_i}(1 - p_i)] \\ &\leq 1 - p_i(1 - p_i) - (1 - p_i) = p_i^2. \end{aligned} \tag{2.3.10}$$

This and (2.3.9) prove the assertion. □

2.4. Definition of a Poisson Process

In this section we discuss one of the most important point processes on R_+, the Poisson point process. We first give its definition.

Definition 2.4.1. A simple point process on R_+, with counting random function $\{N(t); t \geq 0\}$, is called a "Poisson process" if

a. $\{N(t); t \geq 0\}$ has independent increments,
b. $\{N(t); t \geq 0\}$ is stochastically continuous.

As before, set $\Lambda(t) = E\{N(t)\}$; we shall show that for all $t < \infty$, $\Lambda(t) < \infty$.

Lemma 2.4.1. *For all $t \geq 0$,*

$$\Lambda(t) \leq \frac{P\{N(t) > 0\}}{1 - P\{N(t) > 0\}}. \tag{2.4.1}$$

PROOF: Consider

$$0 = t_{n0} < t_{n1} < \cdots < t_{nn} = t \tag{2.4.2}$$

so that

$$\max_i (t_{ni} - t_{n,i-1}) \to 0 \tag{2.4.3}$$

as $n \to \infty$. Set

$$\Psi(s) = I_{[1,\infty)}(s); \tag{2.4.4}$$

then we clearly have

$$N(t) = \lim_{n\to\infty} \sum_{i=1}^{n} \Psi(N(t_{ni}) - N(t_{n,i-1})) \quad \text{(a.s.)}. \tag{2.4.5}$$

This follows from the stochastic continuity of $N(t)$ and the fact that the sum on the right-hand side of (2.4.5) is nondecreasing in n. Next, write

$$\Psi_{ni} \equiv \Psi(N(t_{ni}) - N(t_{n,i-1})). \tag{2.4.6}$$

Clearly, $\{\Psi_{ni}\}_1^n$ is a sequence of independent Bernoulli r.v.'s. Thus, according to Lemma 2.3.1, we have

$$P\left\{\sum_{i=1}^{n} \Psi_{ni} > k\right\} \le \left(P\left\{\sum_{i=1}^{n} \Psi_{ni} > 0\right\}\right)^{k+1}.$$

Finally, by letting $n \to \infty$ in this inequality and invoking (2.4.5), we obtain

$$P\{N(t) > k\} \le (P\{N(t) > 0\})^{k+1}$$

or [see (2.2.9)],

$$P\{\tau_{k+1} \le t\} \le (P\{N(t) > 0\})^{k+1}.$$

This and Equation (2.2.10) prove the assertion. □

Corollary 2.4.1. *From* (2.4.1), *it clearly follows that* $\Lambda(t)$ *is finite and continuous at every* $t \ge 0$. *As a matter of fact, for any* $t \ge 0$ *and* $s \ge 0$, *it follows from the Lebesgue dominated convergence theorem and stochastic continuity of* $N(t)$ *that*

$$\lim_{s\to 0} \{\Lambda(t+s) - \Lambda(t)\} = E\left\{\lim_{s\to 0} (N(t+s) - N(t))\right\} = 0,$$

which implies right continuity of $\Lambda(t)$ *at any* $t < \infty$. *In the same fashion, one can prove continuity of* $\Lambda(t)$ *at any* $t \ge 0$ *from the left.*

Set

$$\Lambda(t_0, t_1) = \Lambda(t_1) - \Lambda(t_0), \quad 0 \le t_0 < t_1 < \infty; \tag{2.4.7}$$

then we have the following result:

Proposition 2.4.1. *For any* $0 \le t_0 < t_1 < \infty$ *and* $n = , 1, \ldots,$

$$P\{N(t_1) - N(t_0) = n\} = \exp[-\Lambda(t_0, t_1)] \frac{\{\Lambda(t_0, t_1)\}^n}{n!}. \tag{2.4.8}$$

PROOF. Consider the partition of $[t_0, t_1]$

$$t_0 = t_{n0} < t_{n1} < \cdots < t_{nn} = t_1,$$

where $\max_k (t_{nk} - t_{n,k-1}) \to 0$ as $n \to \infty$, and set

$$\Psi_{ni} = \Psi(N(t_{ni}) - N(t_{n,i-1})),$$

where $\Psi(\cdot)$ is defined by Equation (2.4.4). As in the previous lemma, we have that

$$\sum_{i=1}^{n} \Psi_{ni} \to N(t_1) - N(t_0) \quad \text{(a.s.)}.$$

Set

$$p_{ni} = P\{\Psi_{ni} = 1\}.$$

Because by assumption $N(t)$ is stochastically continuous,

$$\sup_i p_{ni} \leq \sup_i P\{N(t_{ni}) - N(t_{n,i-1}) \geq 1\} \to 0 \tag{2.4.9}$$

as $n \to \infty$. Hence, $p_{ni} \to 0$ as $n \to \infty$ uniformly in i.

Suppose now that as $n \to \infty$,

$$\sum_{i=1}^{n} p_{ni} \to L(t_0, t_1) < \infty, \tag{2.4.10}$$

and consider sequences $\{X_{ni}\}_1^n$ of independent Poisson r.v.'s with $E\{X_{ni}\} = p_{ni}$.

Invoking Le Cam's Inequality (2.3.8), we have

$$d\left(\sum_{i=1}^{n} \Psi_{ni}, \sum_{i=1}^{n} X_{ni}\right) \leq \sum_{i=1}^{n} p_{ni}^2.$$

However, due to (2.4.9) and (2.4.10),

$$\sum_{i=1}^{n} p_{ni}^2 \leq \max_k p_{nk} \sum_{i=1}^{n} p_{ni} \to 0$$

as $n \to \infty$ and the assertion follows due to the fact that

$$\sum_{i=1}^{n} X_{ni} \xrightarrow{d} Y,$$

where Y has a Poisson distribution with $E\{Y\} = L(t_0, t_1)$. □

Remark 2.4.1. The previous proposition was proved assuming that (2.4.10) holds. We shall now prove this hypothesis.

Lemma 2.4.2. *Under the conditions* a *and* b *of Definition* 2.4.1,

$$\lim_{n\to\infty} \sum_{i=1}^{n} p_{ni} < \infty. \tag{2.4.11}$$

Proof. From the definition of Ψ_{nk}, it follows that there exists at least one integer $\nu = 1, 2, \ldots, n$ such that

$$P\left\{\sum_{k=1}^{n} \Psi_{nk} = \nu\right\} > 0. \tag{2.4.12}$$

Because $\{\Psi_{nk}\}_1^n$ are independent Bernoulli r.v.'s, we have

$$\begin{aligned} P\left\{\sum_{k=1}^{n} \Psi_{nk} = \nu\right\} &= \sum_{1\leq i_1<\cdots<i_\nu\leq n}\cdots\sum P\{\Psi_{ni_1}=1,\ldots,\Psi_{ni_\nu}=1,\Psi_{ni_{\nu+1}}=0,\ldots,\Psi_{ni_n}=0\} \\ &= \prod_{i=1}^{n} P\{\Psi_{ni}=0\} \sum_{1\leq i_1<\cdots<i_\nu\leq n}\cdots\sum \prod_{r=1}^{\nu} P\{\Psi_{ni_r}=1\}\Big/\prod_{r=1}^{\nu} P\{\Psi_{ni_r}=0\} \\ &\leq \frac{\prod_{i=1}^{n} P\{\Psi_{ni}=0\}}{(1-\sup_i p_{ni})^\nu} \frac{\left(\sum_1^n p_{ni}\right)^\nu}{\nu!}. \end{aligned}$$

Now, invoking the inequality

$$\prod_{i=1}^{n} P\{\Psi_{ni} = 0\} = \prod_{i=1}^{n} (1 - p_{ni}) \leq \exp\left(-\sum_{1}^{n} p_{ni}\right),$$

we obtain that

$$P\left\{\sum_{k=1}^{n} \Psi_{nk} = v\right\} \leq C \exp\left(-\sum_{1}^{n} p_{ni}\right) \frac{(\sum_{1}^{n} p_{ni})^v}{v!}.$$

Therefore, if $\sum_{i=1}^{n} p_{ni} \to \infty$ as $n \to \infty$, the right-hand side of the last inequality converges to zero, which contradicts (2.4.12). This proves the lemma. □

Remark 2.4.2. When

$$\Lambda(t) = \lambda t, \quad \lambda > 0,$$

the Poisson process is said to be time homogeneous with mean rate λ.

Remark 2.4.3. Any Poisson process can be transformed into a time homogeneous Poisson process. Indeed, let $\Lambda(t) = E\{N(t)\}$ and denote by Λ^{-1} the right continuous inverse of Λ, i.e., for all $u \geq 0$,

$$\Lambda^{-1}(u) = \inf\{t; \Lambda(t) > u\}. \tag{2.4.13}$$

Because $\Lambda(t) \to \infty$ as $t \to \infty$, $\Lambda^{-1}(u)$ is defined for all $u \geq 0$ and, furthermore, satisfies

$$\Lambda(\Lambda^{-1}(u)) = u, \qquad \Lambda(t) > u \quad \text{if } t > \Lambda^{-1}(u).$$

Therefore, the stochastic process $\{N_0(t); t \geq 0\}$ defined by

$$N_0(t) = N(\Lambda^{-1}(t))$$

is a homogeneous Poisson process with

$$E\{N_0(t)\} = \Lambda(\Lambda^{-1}(t)) = t.$$

Remark 2.4.4. Any Poisson process is also a Markov process.

2.5. Arrival Times $\{\tau_k\}$

Let $\{\tau_n\}_1^\infty$ be the sequence of "arrival times" of a Poisson point process $\{N(t); t \geq 0\}$ with $\Lambda(t) = E\{N(t)\}$, and set $F_n(t) = P\{\tau_n \leq t\}$. Then, from (2.2.9) and (2.4.8), we obtain that

$$\begin{aligned} F_n(t) &= 1 - \exp[-\Lambda(t)] \sum_{k=0}^{n-1} \frac{[\Lambda(t)]^k}{k!} \\ &= \frac{1}{\Gamma(n)} \int_0^t \exp[-\Lambda(s)][\Lambda(s)]^{n-1}\, d\Lambda(s), \end{aligned} \tag{2.5.1}$$

where $\Gamma(n) = (n-1)!$

An important property of the arrival times of a Poisson process can be described as follows. Given that exactly n events have occurred in $[t_0, t_1]$, these n points are distributed throughout the interval $[t_0, t_1]$ as n points selected randomly (and independently) from this interval according to the probability distribution

$$\frac{d\Lambda(s)}{\Lambda(t_1) - \Lambda(t_0)}, \quad t_0 \le s \le t_1. \tag{2.5.2}$$

This is established in the following proposition.

Proposition 2.5.1. *Let $t_0 = s_0 < s_1 < \cdots < s_r = t_1$ be a partition of $[t_0, t_1]$, and $k_1, k_2, \ldots, k_r$ non-negative integers. Set $k_1 + \cdots + k_r = k$; then*

$$P\{N(s_i) - N(s_{i-1}) = k_i, i = 1, \ldots, r | N(t_1) - N(t_0) = k\}$$
$$= \frac{k!}{k_1! \ldots k_r!} \prod_1^r \left(\frac{\Lambda(s_{i-1}, s_i)}{\Lambda(t_0, t_1)} \right)^{k_i}. \tag{2.5.3}$$

PROOF. This follows from the fact that

$$\{N(s_i) - N(s_{i-1}) = k_i, i = 1, \ldots, r\} \subset \{N(t_1) - N(t_0) = k\}. \qquad \square$$

To this proposition one may also give the following interpretation: Let Z be a r.v. with support $[t_0, t_1]$ such that

$$P\{Z \le t\} = \frac{\Lambda(t_0, t)}{\Lambda(t_0, t_1)}, \quad t_0 \le t \le t_1, \tag{2.5.4}$$

and consider n independent copies $Z_1, \ldots, Z_n$ of Z. Denote by $Z_{n1}, \ldots, Z_{nn}$ the corresponding sequence of order statistics. Then, the following is a variation of Proposition 2.5.1. Given that $N(t_1) - N(t_0) = n$, the joint distribution of arrival time $\{\tau_i^*\}_1^n$, $t_0 < \tau_1^* < \cdots < \tau_n^* \le t_1$ is the same as of $\{Z_{ni}\}_1^n$. As a matter of fact, after some straightforward calculations, one obtains that, for any $t_0 < u_1 < \cdots < u_n < t_1$,

$$P\{\tau_1^* \in du_1, \ldots, \tau_n^* \in du_n | N(t_1) - N(t_0) = n\} = \frac{n!}{(\Lambda(t_0, t_1))^n} \prod_1^n d\Lambda(u_j). \tag{2.5.5}$$

But then, as is well known, the right-hand side of (2.5.5) is the joint distribution of $(Z_{n1}, \ldots, Z_{nn})$.

The next example shows the usefulness of this result.

EXAMPLE 2.5.1. For every $t > 0$ and $n = 1, 2, \ldots$,

$$E\left\{ \sum_{k=1}^n \tau_k | N(t) = n \right\} = \left(t - \frac{1}{\Lambda(t)} \int_0^t \Lambda(s)\, ds \right) n.$$

Indeed, according to Proposition 2.5.1 and Equation (2.5.4), we have

$$E\left\{\sum_{k=1}^{n}\tau_k|N(t)=n\right\}=E\left\{\sum_{k=1}^{n}Z_{nk}\right\}=E\left\{\sum_{k=1}^{n}Z_k\right\}=nE\{Z_1\}=\frac{n}{\Lambda(t)}\int_0^t s\,d\Lambda(s).$$

As every process with independent increments, a Poisson process is also a Markov process. It is easy to see that, for any $0 < s < t < \infty$ and $0 \le i \le j$, its transition probability is given by

$$P_{ij}(s,t) = P\{N(t) = j|N(s) = i\} = P\{N(t) - N(s) = j - i\}. \quad (2.5.6)$$

It is of some interest to investigate how the Markov stochastic structure of $\{N(t); t \ge 0\}$ reflects on the stochastic structure of the process $\{\tau_n\}_1^\infty$ and vice versa. In the rest of this section, this question will be discussed in some detail. But first, note that, under relatively mild regularity conditions on $\Lambda(t)$,

$$P\{N(t+s) - N(t) \ge 2\} = o(s) \quad (2.5.7)$$

as $s \to 0$. We now have the following result:

Proposition 2.5.2. *For all* $0 < s < t$ *and* $0 \le i \le j$,

$$P_{ij}(s,t) = P\{N(t) = j|\tau_i = s\}. \quad (2.5.8)$$

PROOF. For $i = 1, 2, \ldots,$

$$\begin{aligned} P\{s < \tau_i \le s + \Delta s\} &= P\{N(s) \le i - 1, N(s+\Delta s) \ge i\} \\ &= P\{N(s) = i - 1, N(s + \Delta s) = i\} + o(\Delta s) \end{aligned}$$

as $\Delta s \to 0$. Thus,

$$\begin{aligned} &P_{ij}(s,t)P\{s < \tau_i \le s + \Delta s\} \\ &\quad = P_{ij}(s,t)P\{N(s) = i - 1, N(s + \Delta s) = i\} + o(\Delta s) \\ &\quad = P\{N(s) = i - 1, N(s + \Delta s) = i, N(t) = j\} + o(\Delta s) \\ &\quad = P\{N(t) = j, s < \tau_j \le s + \Delta s\} + o(\Delta s). \end{aligned}$$

From this we conclude that

$$P_{ij}(s,t) = P\{N(t) = j|s < \tau_j \le s + \Delta s\} + o(1).$$

By letting $\Delta s \to 0$ the assertion follows. □

Remark 2.5.1. The proof of (2.5.6) is based on two features of the process $N(t)$: its Markov property and condition (2.5.7). Therefore, the proposition holds for any point process whose counting random function is a Markov process and which satisfies condition (2.5.7).

Corollary 2.5.1. *Set* $T_i = \tau_1$ *and* $T_n = \tau_n - \tau_{n-1}$ *for* $n \ge 2$. *Then, if* $\Lambda(t) = \lambda t$ *[i.e.,* $N(t)$ *is a homogeneous Poisson process],* $\{T_n\}_1^\infty$ *is an i.i.d. sequence of r.v.'s with common* d.f.

$$P\{T_1 \le t\} = 1 - \exp(-\lambda t).$$

PROOF. The proof is quite simple. From (2.5.6), it follows that

$$1 - P_{jj}(s, s + u) = P\{N(s + u) \geq j + 1 | N(s) = j\} = P\{\tau_{j+1} \leq s + u | \tau_j = s\}$$
$$= P\{T_{j+1} \leq u | \tau_j = s\}.$$

On the other hand,

$$P_{jj}(s, s + u) = \exp(-\lambda u),$$

which proves the assertion. □

2.6. Markov Property of $N(t)$ and Its Implications

Let $\{N(t); t \geq 0\}$ be a Markov process. The aim of this section is to show that in such a case the corresponding sequence of arrival times $\{\tau_j\}_1^\infty$ is also a Markov process. The converse, however, does not hold in general. In other words, if $\{\tau_j\}_1^\infty$ is a Markov process, it does not necessarily follow from this that $N(t)$ has the Markov property. An exception is, of course, the Poisson process.

In what follows we shall prove these two statements. Our elementary method of proof will require that the condition

$$P\{N(t + \Delta t) - N(t) \geq 2\} = o(\Delta t) \tag{2.6.1}$$

as $\Delta t \to 0$ holds.

Proposition 2.6.1. *Let $\{N(t); t \geq 0\}$ be a Markov process with transition probability $P_{ij}(s, t)$. If condition (2.6.1) holds, then $\{\tau_j\}_1^\infty$ is also a Markov process with transition probability*

$$P\{\tau_{j+1} \leq t | \tau_j = s\} = 1 - P_{jj}(s, t). \tag{2.6.2}$$

PROOF. From (2.5.8), it readily follows that

$$\sum_{k=j+1}^{\infty} P_{ik}(s, t) = P\{\tau_{j+1} \leq t | \tau_j = s\}$$

which proves (2.6.2). Next,

$$P\{\tau_1 \in dt\} = P\{N(t) = 0, N(t + dt) = 1\}$$
$$= P\{N(t) = 0\} P_{0,1}(t, t + dt). \tag{2.6.3}$$

We also have, for all $j = 1, 2, \ldots,$

$$P\{\tau_{j+1} \in dt | \tau_j = s\} = P\{N(t) = j, N(t + dt) = j + 1 | \tau_j = s\}$$
$$= P_{jj}(s, t) P_{j,j+1}(t, t + dt). \tag{2.6.4}$$

Finally, for $0 < t_1 < \cdots < t_n < \infty$ and $n = 1, 2, \ldots$ arbitrary, we obtain, after some straightforward calculations, that

$$\begin{aligned} P\{\tau_1 \in dt_1, \ldots, \tau_n \in dt_n\} \\ &= P\{N(t_1) = 0, N(t_1 + dt_1) = 1, \ldots, N(t_n) = n - 1, N(t_n + dt_n) = n\} \\ &= P\{N(t_1) = 0\} P_{01}(t_1, t_1 + dt_1) P_{11}(t_1, t_2) P_{12}(t_2, t_2 + dt_2) \ldots \\ &\quad P_{n-1,n-1}(t_{n-1}, t_n) P_{n-1,n}(t_n, t_n + dt_n). \end{aligned}$$

From this, (2.6.3), and (2.6.4), it follows that

$$\begin{aligned} P\{\tau_1 \in dt_1, \ldots, \tau_n \in dt_n\} = P\{\tau_1 \in dt_1\} P\{\tau_2 \in dt_2 | \tau_1 = t_1\} \ldots \\ P\{\tau_n \in dt_n | \tau_{n-1} = t_{n-1}\}. \end{aligned} \tag{2.6.5}$$

This proves the proposition. □

Corollary 2.6.1. *If $N(t)$ is a Poisson process with $\Lambda(t) = E\{N(t)\}$, it clearly follows from (2.5.6) and (2.6.2) that $\{\tau_j\}_1^\infty$ is a Markov process with stationary transition probability*

$$P\{\tau_{j+1} \leq t | \tau_j = s\} = 1 - \exp[-\Lambda(s, t)]. \tag{2.6.6}$$

In fact, as we have indicated before, one can prove much more. To this end we need the following auxiliary result.

Lemma 2.6.1. *Let $\{\tau_j\}_1^\infty$ be a Markov chain with stationary transition probability (2.6.6). Then, for any $0 < t_1 < \cdots < t_n < t$ and $n = 1, 2, \ldots,$*

$$P\{\tau_1 \in dt_1, \ldots, \tau_n \in dt_n | N(t) = n\} = n! \prod_1^n \frac{d\Lambda(t_j)}{\Lambda(t)}. \tag{2.6.7}$$

Proof. After some simple calculations, we deduce that

$$\begin{aligned} P\{\tau_1 \in dt_1, \ldots, \tau_n \in dt_n | N(t) = n\} &= \frac{P\{\tau_1 \in dt_1, \ldots, \tau_n \in dt_n, \tau_{n+1} > t\}}{P\{N(t) = n\}} \\ &= \frac{\exp(-\Lambda(t))}{P\{N(t) = n\}} \prod_1^n d\Lambda(t_j). \end{aligned} \tag{2.6.8}$$

On the other hand,

$$\begin{aligned} P\{\tau_n \in dt\} &= P\{0 < \tau_1 < \cdots < \tau_{n-1} < t, \tau_n \in dt\} \\ &= \int_{0 < t_1 < \cdots < t_{n-1} < t} \cdots \int P\{\tau_1 \in dt_1, \ldots, \tau_{n-1} \in dt_{n-1}, \tau_n \in dt\}. \end{aligned}$$

From this and (2.6.6), we have

$$P\{\tau_n \in dt\} = \int \cdots \int_{0<t_1<\cdots<t_{n-1}<t} \exp[-\Lambda(t)] \left(\prod_1^{n-1} d\Lambda(t_j) \right) d\Lambda(t)$$

$$= \exp[-\Lambda(t)] \frac{\{\Lambda(t)\}^{n-1}}{(n-1)!} d\Lambda(t). \tag{2.6.9}$$

Therefore, for all $t > 0$ and $n = 0, 1, \ldots,$

$$P\{N(t) = n\} = \exp[-\Lambda(t)] \frac{\{\Lambda(t)\}^n}{n!}. \tag{2.6.10}$$

This and (2.5.1) prove the lemma. □

Proposition 2.6.2. *Let $\{N(t); t \geq 0\}$ be a simple point process with $\Lambda(t) = E\{N(t)\}$. A necessary and sufficient condition for $N(t)$ to be a Poisson process is that its sequence of arrival times $\{\tau_j\}_1^\infty$ is a Markov process with stationary transition probability* (2.6.6).

PROOF. Necessity of the condition is obvious. Its sufficiency will be amply illustrated by showing that $N(s)$ and $N(t) - N(s)$ are independent r.v.'s. Thus, let $0 < s < t$ and $k, n = 0, 1, \ldots$ be arbitrary and write

$$P\{N(s) = k, N(t) - N(s) = n\}$$
$$= P\{N(s) = k, N(t) - N(s) = n | N(t) = k + n\} P\{N(t) = k + n\}. \tag{2.6.11}$$

Now, according to Lemma 2.6.1, given $N(t) = n + k$, the k arrival times in $(0, s]$ and n in $(s, t]$ are distributed as $k + n$ independent r.v.'s with common d.f.

$$\frac{\Lambda(u)}{\Lambda(t)}, \quad 0 \leq u \leq 1.$$

Hence,

$$P\{N(s) = k, N(t) - N(s) = n | N(t) = n + k\} = \binom{n+k}{k} \left(\frac{\Lambda(s)}{\Lambda(t)} \right)^k \left(\frac{\Lambda(s,t)}{\Lambda(t)} \right)^{n-k}.$$

This, (2.6.10) and (2.6.11) yield

$$P\{N(s) = k, N(t) - N(s) = n\} = \exp[-\Lambda(s)] \frac{[\Lambda(s)]^k}{k!} \exp[-\Lambda(s,t)] \frac{[\Lambda(s,t)]^n}{n!},$$

which proves the assertion. □

Denote by

$$T_1 = \tau_1, \qquad T_2 = \tau_2 - \tau_1, \ldots \tag{2.6.12}$$

the sequence of interarrival times of a point process. It is often of interest to know the stochastic structure of $\{T_j\}_1^\infty$. As we have seen in Corollary 2.5.1, this is an i.i.d. sequence with common negative exponential distribution if $N(t)$ is a homogeneous Poisson process. If the Poisson process is not homogeneous, the r.v.'s $\{T_k\}_1^\infty$ are neither independent nor identically distributed. In the case when $N(t)$ is a "pure birth process," i.e., a Markov process with stationary transition probability, the stochastic structure of $\{T_j\}_1^\infty$ is easy to determine.

Proposition 2.6.3. *Suppose that $\{N(t); t \geq 0\}$ is a Markov process with stationary transition probability, i.e.,*

$$P_{ij}(s, s + u) = P_{ij}(u), \tag{2.6.13}$$

then $\{T_j\}_1^\infty$ is a sequence of independent r.v.'s with

$$P\{T_{j+1} \leq u\} = 1 - \exp(-\lambda_j u), \quad j = 0, 1, \ldots. \tag{2.6.14}$$

Proof. From (2.6.13), it follows that, for all $t, u > 0$ and $j = 0, 1, \ldots,$

$$\begin{aligned} P_{jj}(u) &= P\{N(t + u) = j | N(t) = j\} = P\{N(t + u) = j | \tau_j = t\} \\ &= 1 - P\{\tau_{j+1} \leq t + u | \tau_j = t\} = 1 - P\{T_{j+1} \leq u | \tau_j = t\}. \end{aligned} \tag{2.6.15}$$

This clearly implies that T_{j+1} is independent of τ_j for all $j = 1, 2, \ldots,$ which proves that $\{T_j\}_1^\infty$ is a sequence of independent r.v.'s with

$$P\{T_{j+1} \leq t\} = 1 - P_{jj}(t). \tag{1.6.16}$$

Next, let $0 < s < t$ be arbitrary and consider

$$\begin{aligned} P_{jj}(t) &= P\{N(s + t) = j | N(s) = j\} \\ &= P\{N(t) = j, N(s + t) = j | N(s) = j\} \\ &= P_{jj}(t - s) P_{jj}(s). \end{aligned}$$

If we set $u = t - s$, we obtain

$$P_{jj}(s + u) = P_{jj}(u) P_{jj}(s), \tag{2.6.17}$$

which is the Cauchy's functional equation. The function $P_{jj}(u)$ is nonincreasing with $P_{jj}(0) = 1$ for all $j = 0, 1, \ldots$. In addition, for all $u, s > 0$ and $j = , 1, \ldots,$

$$P_{jj}(u + s) - P_{jj}(u) = P_{jj}(u)(P_{jj}(s) - 1) \to 0$$

as $s \to 0$, which implies that $P_{jj}(u)$ is continuous for all $u \geq 0$. In such a case, the only solution of (2.6.17) is

$$P_{jj}(u) = \exp(-\lambda_j u), \quad j = 0, 1, \ldots. \tag{2.6.18}$$

This concludes the proof of the proposition. □

2.7. Doubly Stochastic Poisson Process

To motivate the consideration of a doubly stochastic Poisson process, we shall begin by the following example.

EXAMPLE 2.7.1. A deep space probe was sent to investigate the volcanic activities of the Jupiter's moon Io. Stationed high above the moon in a stationary orbit, the probe counts the number of volcanic eruptions on the moon's surface. However, if hit by a meteorite, the functioning of the probe would stop. Therefore, its lifetime is a r.v., say T.

Suppose that the number of volcanic eruptions $N(t)$ is a homogeneous Poisson process with intensity λ. Clearly, T and $N(t)$ are independent r.v.'s. The total number of eruptions detected during the probe's lifetime is $N(T)$. This is a r.v. with the distribution

$$P\{N(T)=n\} = \int_0^\infty P\{N(t)=n\}\,dH(t) = \frac{1}{n!}\int_0^\infty \exp(-\lambda t)(\lambda t)^n\,dH(t),$$

where $H(t) = P\{T \le t\}$.

Suppose now that a meteorite impact at a random time T does not destroy the probe but only reduces its sensitivity, so that after T, the intensity of the Poisson process becomes $\lambda_0 < \lambda$. To simplify the problem, we assume that the second impact by a meteorite is not very likely. In such a case, the intensity of the process $N(t)$ is a r.v. specified as follows:

$$\lambda(s,\cdot) = \lambda I_{\{T>s\}} + \lambda_0 I_{\{T\le s\}}. \tag{2.7.1}$$

Hence,

$$\Lambda(t,\cdot) = \int_0^t \lambda(s,\cdot)\,ds = \lambda\min(T,t) + \lambda_0(t-\min(T,t)). \tag{2.7.2}$$

Denote by $N_0(t)$ the homogeneous Poisson process with $\lambda = 1$. Then, if $N(t)$ is an arbitrary Poisson process with $E\{N(t)\} = \Lambda(t)$, according to Remark 2.4.3,

$$(N_0 \circ \Lambda)(t) = N(t).$$

Therefore, the probability that the probe will detect n volcanic eruptions in $(0, t]$ is

$$\begin{aligned}
P\{N_0(\Lambda(t,\cdot))=n\} &= P\{N_0(\lambda t)=n, T>t\} + P\{N_0(\lambda T+\lambda_0(t-T))=n, T\le t\}\\
&= \exp(-\lambda t)\frac{(\lambda t)^n}{n!}[1-H(t)]\\
&\quad + \int_0^t \exp\{-[\lambda s+\lambda_0(t-s)]\}\frac{[\lambda s+\lambda_0(t-s)]^n}{n!}\,dH(t)\\
&= E\left\{\exp[-\Lambda(t,\cdot)]\frac{[\Lambda(t,\cdot)]^n}{n!}\right\}.
\end{aligned}$$

We now give the following definition of a doubly stochastic Poisson process.

Definition 2.7.1. Let $\{\Lambda(t, \cdot); t \geq 0\}$ be a stochastic process on $\{\Omega, \mathscr{B}, P\}$ whose realizations are strictly increasing functions such that $\Lambda(0, \cdot) = 0$ (a.e.). Let $N_0(t)$ be the Poisson process on the same probability space with $E\{N_0(t)\} = t$, independent of $\Lambda(t, \cdot)$. A point process $\{N(t); t \geq 0\}$ is called a doubly stochastic Poisson (or Cox) process if it has the same distribution as

$$(N_0 \circ \Lambda)(t) = N_0(\Lambda(t, \cdot)). \tag{2.7.3}$$

An exhaustive study on this subject can be found in a monograph by Grandell (1976).

Example 2.7.2. When

$$\Lambda(t, \cdot) = Zt,$$

where Z is a non-negative r.v., the doubly stochastic Poisson process is called a "weighted or mixed" Poisson process. It can be shown that the weighted Poisson process is a Markov process with the transition probability

$$P_{ij}(s, t) = \frac{(t - s)^{j-i}}{(j - i)!} \frac{E\{Z^j e^{-Zt}\}}{E\{Z^i e^{-Zs}\}}. \tag{2.7.4}$$

2.8. Thinning of a Point Process

To provide the motivation for studying this particular procedure, we consider the following example.

Example 2.8.1. Suppose we observe the occurrence of the rainfall at a particular location. Suppose we begin our observation at time $t = 0$ and let $0 < \tau_1 < \tau_2 < \cdots$ be the times of the first, second, etc., rainfall after $t = 0$ at this particular site. The sequence of r.v.'s $\{\tau_n\}_1^\infty$ represents a point process on $R_+ = (0, \infty)$.

Suppose now that for some practical purposes not all rainfall events arriving at the given location are of interest but only those which are sufficiently large to cause flooding. If only their arrival times are recorded, the corresponding point process can be visualized as a point process obtained from the point process $\{\tau_n\}_1^\infty$ by deletion of a certain number of its points. The new point process is called a thinned point process.

In this section we are concerned with the so-called "independent thinning," which can be described as follows. Let $\{N(t); t \geq 0\}$ be an arbitrary simple point process. It undergoes independent thinning if each of its points is deleted with probability p, $0 < p < 1$, and retained with probability $q = 1 - p$, independently for each point. The retained set of points forms a thinned point process.

Let $\eta_q(\cdot)$ be the thinned version after the scale contraction by a factor q. In other words, for $A \in \mathscr{R}_+$,

$$\eta_q(A) \text{ is the thinning of } \eta(q^{-1}A)$$

[see (2.2.5) for the definition of η], where

$$q^{-1}A = \{q^{-1}x; x \in A\}.$$

The point process $\eta_q(\cdot)$ resulting from both thinning and contraction of the time scale will have $\eta_q(A) = n$ say, only if exactly n of the $\eta(q^{-1}A)$ points in the set $q^{-1}A$ from the original process are retained in the thinned process.

It is of some interest to investigate the asymptotic behavior of $\eta_q(\cdot)$ as $q \to 0$. In the rest of this section we shall investigate the weak convergence of $\eta_q(\cdot)$ as $q \to 0$ (weak convergence means convergence in finite-dimensional distributions). The next result has been known for some time (see Belaeyev, 1963).

Proposition 2.8.1. *The point process $\eta_q(\cdot)$, obtained by independent thinning and contraction from a point process $\eta(\cdot)$ on R^+, converges weakly to a homogeneous Poisson process with intensity $\lambda \in (0, \infty)$, if and only if for all $0 < t < \infty$*

$$qN\left(\frac{t}{q}\right) \xrightarrow{p} \lambda t \tag{2.8.1}$$

as $q \to 0$, where $N(\cdot)$ is the counting random function of $\eta(\cdot)$.

Proof. The proof is essentially due to Westcott (1976). Define

$$\Phi_q(s, A) = E(\exp\{-s\eta_q(A)\}),$$
$$\Psi_q(s, A) = E(\exp\{-2s\eta(q^{-1}A)\}),$$

where $s > 0$; then

$$\Phi_q(s, A) = E(E(\exp\{-s\eta_q(A)\}|\eta(q^{-1}A))).$$

The thinning mechanism implies that

$$\begin{aligned} &E(\exp\{-s\eta_q(A)\}|\eta(q^{-1}A)) \\ &= \sum_{k=0}^{\eta(q^{-1}A)} e^{-sk} P\{\eta_q(A) = k|\eta(q^{-1}A)\} \\ &= \sum_{k=0}^{\eta(q^{-1}A)} e^{-sk} \binom{\eta(q^{-1}A)}{k} q^k(1 - q)^{\eta(q^{-1}A)-k} = (1 - q(1 - e^{-s}))^{\eta(q^{-1}A)}. \end{aligned}$$

Thus,

$$\Phi_q(s, A) = E(\exp\{\eta(q^{-1}A)\ln(1 - q(1 - e^{-s}))\}). \tag{2.8.2}$$

But for any $0 < \varepsilon < \frac{1}{2}$

$$q(1 - e^{-s}) < -\log\{1 - q(1 - e^{-s})\} < q(1 - e^{-s})(1 + \varepsilon) \qquad (2.8.3)$$

if $q < \varepsilon$. Because $q \to 0$, we may choose $\varepsilon > 0$ arbitrarily small and deduce from (2.8.2) and (2.8.3) that

$$\Phi_q(s, A) \quad \text{and} \quad \Psi_q(1 - e^{-s}, A)$$

have the same limit, if any, as $q \to 0$.

We shall first prove necessity of (2.8.1). To this end note that $\Psi_q(s', (0, t])$ converges to $e^{-\lambda s' t}$ for all $0 \leq s' < 1$ as $q \to 0$. Now apply the continuity theorem for Laplace transform (Feller, 1971, p. 408), remembering that such functions are analytic, and hence determined, in their domain of definition, by values in an interval.

To prove sufficiency of the condition (2.8.1), note that it implies

$$q\eta(q^{-1}A) \xrightarrow{p} \lambda \|A\|$$

for all $A \in J$, the ring generated by intervals $(0, t]$. Here, $\|\cdot\|$ denotes the Lebesgue measure. So for such A,

$$\Phi_q(s, A) \to \exp\{-\lambda \|A\| (1 - e^{-s})\} \qquad (2.8.4)$$

as $q \to 0$. But from a result by Rényi (1967), the Poisson process is uniquely determined by the fact that

$$P\{\eta(A) = n\} = e^{-\lambda \|A\|} \frac{(\lambda \|A\|)^n}{n!}$$

for all $A \in J$, and (2.8.4) implies all convergent subsequences of finite-dimensional distributions possess this property. Thus, they all have the same limit, which proves the assertion. □

2.9. Marked Point Processes

Consider a simple point process on R_+ with arrival times $\{\tau_j\}_1^\infty$ and the counting random measure η, all defined on a probability space $\{\Omega, \mathscr{B}, P\}$. Let $\{\xi_j\}_1^\infty$ be a sequence of real-valued r.v.'s on the same probability space. The bivariate sequence of r.v.'s

$$\{(\tau_j, \xi_j)\}_1^\infty \qquad (2.9.1)$$

represents a point process on $R_+ \times R$, which is called "a marked point process."

The marked point process (2.9.1) is completely characterized by the nonnegative integer-valued random measure Q on $\mathscr{R}_+ \times \mathscr{R}$ defined by

$$Q(\cdot) = \sum_{j=1}^{\infty} \varepsilon_{(\tau_j, \xi_j)}(\cdot), \qquad (2.9.2)$$

where for any $G \in \mathscr{R}_+ \times \mathscr{R}$

$$\varepsilon_{(x,y)}(G) = \begin{cases} 1 & \text{if } (x,y) \in G \\ 0 & \text{if } (x,y) \notin G. \end{cases} \tag{2.9.3}$$

From (2.9.2) it readily follows that for any $A \in \mathscr{R}_+$

$$Q(A \times R) = \eta(A).$$

The counting random function $N^*(t, D)$ of the marked point process (2.9.1) is defined by

$$N^*(t, D) = Q((0, t] \times D), \quad D \in \mathscr{R}.$$

Clearly,

$$N^*(t, D) = \begin{cases} 0 & \text{on } \{N(t) = 0\} \\ \sum_{j=1}^{\infty} I_{\{\tau_j \le t, \xi_j \in D\}}(\cdot) & \text{on } N\{(t) \ge 1\}, \end{cases} \tag{2.9.4}$$

where $N(t)$ stands for the counting random function of the point process $\{\tau_j\}_1^\infty$.

From (2.9.4) we have that

$$\begin{aligned} N^*(t, D) &= \sum_{j=1}^{\infty} I_{\{N(t) \ge j\}} I_D(\xi_j) \\ &= \sum_{j=1}^{\infty} \left(\sum_{k=j}^{\infty} I_{\{N(t)=k\}} \right) I_D(\xi_j) \\ &= \sum_{k=1}^{\infty} \left(\sum_{j=1}^{k} I_D(\xi_j) \right) I_{\{N(t)=k\}}. \end{aligned}$$

In other words,

$$N^*(t, D) = \sum_{j=1}^{N(t)} I_D(\xi_j), \quad \text{on } \{N(t) \ge 1\}, \tag{2.9.5}$$

and $N^*(t, D) = 0$ on $\{N(t) = 0\}$. In addition,

$$N^*(t, R) = N(t).$$

It is clear from (2.9.5) that for any $D \in \mathscr{R}$ fixed, $N^*(t, D)$ is a stochastic process with nondecreasing step sample functions with unit jumps at those points $\tau_{k_1}, \tau_{k_2}, \ldots$ $(1 \le \tau_{k_1} < \tau_{k_2} < \cdots)$ for which $\xi_{k_j} \in D$. This implies that $N^*(t, D)$ is a thinning of the original point process $\{\tau_j\}_1^\infty$.

Denote by $P_\xi(\cdot)$ the distribution of a r.v. ξ and, as before, set $\Lambda(t) = E\{N(t)\}$.

Proposition 2.9.1. *Assume that*

a. $\{N(t); t \ge 0\}$ *is a Poisson process.*
b. $\{\xi_j\}_1^\infty$ *is a sequence of i.i.d. r.v.'s, with common distribution* $P_\xi(\cdot)$, *independent of* $\{\tau_j\}_1^\infty$.

Then $\{N^*(t, D); t \geq 0\}$ *represents a Poisson process with*

$$E\{N^*(t, D)\} = \Lambda(t) P_\xi(D)$$

if $P_\xi(D) > 0$.

Proof. There are many ways to prove this assertion. The following amply illustrates one of the methods of proof. For any $0 < t_1 < t_2 < \infty$ and $0 \leq k \leq n$,

$$\begin{aligned} &P\{N^*(t_1, D) = k, N^*(t_2, D) = n\} \\ &= P\left\{\sum_{i=1}^{N(t_1)} I_D(\xi_i) = k, \sum_{j=1}^{N(t_2)} I_D(\xi_j) = n\right\} \\ &= \sum_{l=k}^{\infty} \sum_{r=n-k}^{\infty} P\left\{\sum_{i=1}^{l} I_D(\xi_i) = k, \sum_{j=1}^{r} I_D(\xi_{l+j}) = n-k\right\} \\ &\quad \times P\{N(t_1) = l\} P\{N(t_2) - N(t_1) = r\}. \end{aligned}$$

Invoking conditions of the proposition, we obtain

$$P\left\{\sum_{i=1}^{l} I_D(\xi_i) = k\right\} = \binom{l}{k} (P_\xi(D))^k [1 - P_\xi(D)]^{l-k},$$

$$P\left\{\sum_{j=1}^{r} I_D(\xi_j) = n - k\right\} = \binom{r}{n-k} (P_\xi(D))^{n-k} [1 - P_\xi(D)]^{r-n+k},$$

$$P\{N(t_1) = l\} = \exp[-\Lambda(t_1)] \frac{(\Lambda(t_1))^l}{l!},$$

$$P\{N(t_2) - N(t_1) = r\} = \exp[-\Lambda(t_1, t_2)] \frac{(\Lambda(t_1, t_2))^r}{r!}.$$

From this, after some straightforward calculations, we have

$$\begin{aligned} &P\{N^*(t_1, D) = k, N^*(t_2, D) = n\} \\ &= \exp[-\Lambda(t_1) P_\xi(D)] \frac{(\Lambda(t_1) P_\xi(D))^k}{k!} \\ &\quad \times \exp[-\Lambda(t_1, t_2) P_\xi(D)] \frac{(\Lambda(t_1, t_2) P_\xi(D))^{n-k}}{(n-k)!}. \end{aligned}$$

Therefore,

$$\begin{aligned} &P\{N^*(t_1, D) = k, N^*(t_2, D) - N^*(t_1, D) = r\} \\ &= P\{N^*(t_1, D) = k, N^*(t_2, D) = r + k\} \\ &= \exp[-\Lambda(t_1) P_\xi(D)] \frac{(\Lambda(t_1) P_\xi(D))^k}{k!} \\ &\quad \times \exp[-\Lambda(t_1, t_2) P_\xi(D)] \frac{(\Lambda(t_1, t_2) P_\xi(D))^r}{r!}, \end{aligned}$$

from which we conclude that $\{N^*(t, D); t \geq 0\}$ is a Poisson process with $E\{N^*(t, D)\} = \Lambda(t)P_\xi(D)$. This proves the assertion. □

2.10. Modeling of Floods

In spite of the experience accumulated over many centuries in dealing with floods, losses in property and lives not only increased considerably in recent times but all indications are that they will increase even more in the future. How and why? Is it because the rainfall–runoff relationships have changed, or that hydrological factors responsible for floods have multiplied all of sudden? There is plenty of evidence that this increase is not due to a dramatic shift in the natural balance. Instead, the escalation of flood damage around the world is a result of new factors emerging gradually in our societies. We shall indicate one of them.

In many of the highly industrialized and densely populated areas of the world, a reduction of the natural retention area of the flood plain has taken place. Due to this fact, the flood waves have increased in amplitude and accelerated, resulting in more flood damage downstream than had ever been anticipated before. In fact, there are reaches of some European rivers in which the last few years have repeatedly brought floods which exceeded the 100-year design flood, on which the designs of various hydraulic structures and flood protection works were based.

Denote by $\xi(t)$ the discharge rate of a streamflow at a given site. Clearly, $\{\xi(t); t \geq 0\}$ is a non-negative stochastic process. We can assume, without violating our physical intuition, that almost all sample functions of $\xi(t)$ are continuous. Then, according to Proposition 1.6.1, there exists a version which is separable and measurable.

Set

$$\chi(t) = \sup_{0 \leq s \leq t} \xi(s). \tag{2.10.1}$$

It is apparent that $\chi(t_1) \leq \chi(t_2)$ for all $0 \leq t_1 < t_2$. For the purpose of flood modeling, it is essential to determine

$$\Phi(x, t) = P\{\chi(t) \leq x\}. \tag{2.10.2}$$

Due to the separability property, the function $\Phi(x, t)$ is well defined. Unfortunately, we know very little about the stochastic process $\xi(t)$ and its stochastic structure which makes an evaluation of $\Phi(x, t)$ extremely difficult. For this reason, the common approach in practice to the problem of estimation of the distribution $\Phi(x, t)$ is based on an empirical procedure [the method of best curve fit to the observed values of the maximum $\chi(t)$]. This clearly is not satisfactory and we will attempt a different approach.

Our approach is based on the following rationale. Because the floods are our concern, then only those flows that can cause flooding are of interest to

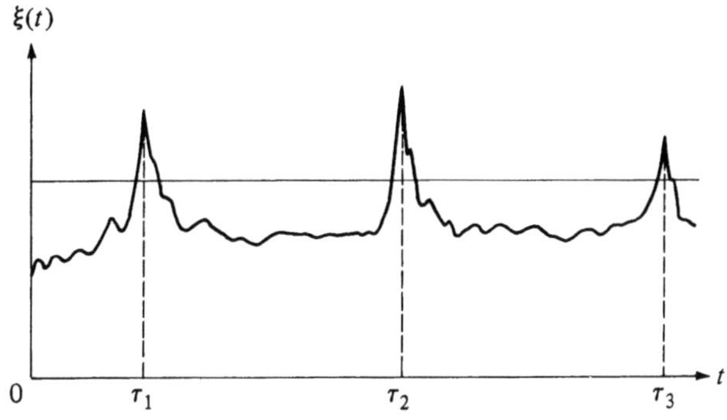

Figure 2.3. A realization of the flow process $\xi(s)$.

us. For this reason we will confine our attention to those flows which exceed a certain threshold level x_0 (see Figure 2.3). The truncated part of the process $\xi(s)$ above x_0 that lies between an upcrossing and the corresponding down-crossing of this threshold is called an "exceedance" or an "excursion." The time point at which the nth exceedance achieves its maximum is denoted by τ_n, $n = 1, 2, \ldots$. These exceedances are, of course, caused by the rainfalls.

It is quite clear that $\{\tau_j\}_1^\infty$ is a sequence of r.v.'s such that

$$0 < \tau_1 < \tau_2 < \cdots \quad \text{and} \quad \tau_n \to \infty \text{ as } n \to \infty. \tag{2.10.3}$$

Thus, $\{\tau_j\}_1^\infty$ represents a simple point process obtained by the thinning of the point process associated with the rainfall occurrences at the given location. It seems "intuitively plausible" to assume that we are dealing with independent thinning. If x_0 is high enough, the point process $\{\tau_j\}_1^\infty$ should have a Poisson distribution, which has been confirmed by observation records.

Let $\{N(t); t \geq 0\}$ be the counting function of the point process $\{\tau_j\}_1^\infty$ which is assumed to be Poisson with $E\{N(t)\} = \Lambda(t)$ and set

$$X_n = \xi(\tau_n) - x_0. \tag{2.10.4}$$

For the sake of simplicity, we will assume that $\{X_k\}_1^\infty$ is an i.i.d. sequence of r.v.'s, independent of $\{\tau_k\}_1^\infty$. In such a case, it follows from Proposition 2.9.3 that $\{(X_k, \tau_k)\}_1^\infty$ is a marked Poisson process.

Denote by $N^*(t, D)$ its counting function, and set

$$E\{N^*(t, D)\} = \Lambda(t) P_X(D),$$

where

$$\Lambda(t) = E\{N(t)\} \quad \text{and} \quad P\{X_1 \leq x\} = H(x). \tag{2.10.5}$$

Write

$$\chi^*(0) = 0, \qquad \chi^*(t) = \sup\{X_k; \tau_k \leq t\}, \tag{2.10.6}$$

$$F_t(x) = P\{\chi^*(t) \leq x\}, \quad x \geq 0. \tag{2.10.7}$$

It is easy to see that, for every $x \geq 0$ and $t \geq 0$,

$$\{\chi^*(t) \leq x\} = \{N^*(t, (x, \infty)) = 0\}. \tag{2.10.8}$$

From this, we obtain

$$F_t(x) = \exp\{-\Lambda(t)[1 - H(x)\}. \tag{2.10.9}$$

Next, we will investigate some properties of the stochastic process $\{\chi^*(t); t \geq 0\}$. For any $0 \leq s < t$, set

$$\chi^*(s, t) = \sup\{X_j; \tau_j \in (s, t]\}. \tag{2.10.10}$$

Because the increments of $N^*(t, D)$ are independent, we have

$$\begin{aligned} P\{\chi(t) \leq x\} &= P\{\chi(s) \leq x, \chi(s, t) \leq x\} \\ &= P\{N^*(s, (x, \infty)) = 0, N^*(t, (x, \infty)) - N^*(s, (x, \infty)) = 0\} \\ &= P\{\chi^*(s) \leq x\} P\{\chi^*(s, t) \leq x\}. \end{aligned} \tag{2.10.11}$$

Hence,

$$P\{\chi^*(s, t) \leq x\} = \frac{P\{\chi^*(t) \leq x\}}{P\{\chi^*(s) \leq x\}}. \tag{2.10.12}$$

Using (2.10.11), we can easily show that $\chi^*(t)$ is a Markov process. Indeed, for any

$$0 < t_1 < \cdots < t_n < t \quad \text{and} \quad 0 < x_1 \leq \cdots \leq x_n < x,$$

$n = 1, 2, \ldots,$ we have

$$\begin{aligned} &P\{\chi^*(t) \leq x | \chi^*(t_1) = x_1, \ldots, \chi^*(t_n) = x_n\} \\ &\qquad = P\{\chi^*(t_n) \leq x, \chi^*(t_n, t) \leq x | \chi^*(t_1) = x_1, \ldots, \chi^*(t_n) = x_n\} \\ &\qquad = P\{\chi^*(t_n, t) \leq x\}. \end{aligned}$$

This proves the assertion.

An extensive discussion on this topic can be found in the work of Todorovic (1982).

Problems and Complements

2.1. A switchboard receives on average two calls per minute. If it is known that the number of calls is represented by a homogeneous Poisson process, find
 (i) the probability that 10 calls will arrive in 5 minutes;
 (ii) the probability that the third call comes during the second minute.

2.2. An orbiting satellite is subject to bombardment by small particles which arrive according to a homogeneous Poisson process. Let $p > 0$ be the conditional probability that a particle which has hit the satellite will also hit a given instrument on it. Determine the probability that in $(0, t]$:

(i) the particular instrument will be hit exactly k times;
(ii) the instrument will be hit at least once.

2.3. Let $\{N(t); t \geq 0\}$ be a homogeneous Poisson process. Show that for any $0 < s < t$ and $0 < k < n$

$$P\{N(s) = k | N(t) = n\} = \binom{n}{k}\left(\frac{s}{t}\right)^k\left(1 - \frac{s}{t}\right)^{n-k}.$$

2.4. A life insurance company has established that the average rate of the policy holders whose insurance is less than \$100,000 is a constant p. Assume that arrival times of insurance claims represent a homogeneous Poisson process. If death has nothing to do with the value of a particular policy, find:
(i) the probability that there will be exactly k claims, each less than \$100,000, in an interval $(0, t]$;
(ii) the probability that all the claims in $(0, t]$ will be higher than or equal to \$100,000.

2.5. Suppose that all car drivers of a city are categorized in several groups, say A_1, $A_2, \ldots, A_n$, according to their driving abilities. It is established that every car accident for drivers in group A_i carries a small probability p of fatality. Suppose that accidents (for members from A_i) occur according to a homogeneous Poisson process. If the random event "fatal accident" is independent of the process, determine:
(i) the probability that a driver from A_i will stay alive during $(0, t]$;
(ii) the expected time of the fatal accident.

2.6. Buses arrive in accordance with a homogeneous Poisson process. Somebody arrives at the bus stop at time t. Find the distribution of the waiting time for the next bus.

2.7. Let $\{N(t); t \geq 0\}$ be a homogeneous Poisson process and $0 < s < t$. Show that, for any $k = 1, 2, \ldots, n$,

$$P\{\tau_k \leq s | N(t) = n\} = \sum_{j=k}^{n}\binom{n}{j}\left(\frac{s}{t}\right)^j\left(1 - \frac{s}{t}\right)^{n-j}.$$

2.8. Suppose that the number of eggs laid by an insect in an interval of time $(0, t]$ is a homogeneous Poisson process with mean rate λ. Assume that the probability of one egg hatching is $p > 0$. If $X_1(t)$ is the number of hatched eggs and $X_2(t)$ the number of unhatched eggs in $(0, t]$, show that (assuming independence of eggs) $X_1(t)$ and $X_2(t)$ are independent Poisson processes with mean rates λp and $\lambda(1 - p)$, respectively.

2.9. Let $\{N(t); t \geq 0\}$ be a Poisson process with $E\{N(t)\} = \lambda t$. Show that

$$\frac{1}{t}N(t) \to \lambda \quad \text{(a.s.)}$$

as $t \to \infty$.

2.10. Let $\{N(t); t \geq 0\}$ be a Poisson process with $E\{N(t)\} = \Lambda(t)$. Show that [see (2.6.12)]

$$P\{T_{n+1} > t | \tau_1 = s, \ldots, \tau_n = s_n\} = \exp\{-\Lambda(t + s_n) + \Lambda(s_n)\},$$

where $s_1 < \cdots < s_n$ and $t > 0$.

2.11. Let $\{N(t); t \geq 0\}$ be a Poisson process with $E\{N(t)\} = \lambda t$. It undergoes independent thinning with a probability p (see Section 2.8). Let $N_1(t)$ be the number of recorded points and $N_2(t)$ the number of deleted points. Show that $N_1(t)$ and $N_2(t)$ are independent Poisson processes with $E\{N_1(t)\} = \lambda pt$.

2.12. Show that

$$E\left(\exp\left\{-\sum_{i=1}^{\infty} f(\tau_i)\right\}\right) = \exp\left\{-\int_0^{\infty} \{1 - \exp[-f(s)]\}\, d\Lambda(s)\right\},$$

where $\{\tau_n\}$ is the sequence of arrival times of a Poisson process $N(t)$ with $E\{N(t)\} = \Lambda(t)$ and $f(\cdot) \geq 0$.

2.13. Let $\{N_i(t)\}_1^n$ be n independent Poisson processes with $E\{N_i(t)\} = \lambda t$. Show that

$$\lim_{n\to\infty} \sum_{k=0}^{\infty} f\left(x + \frac{k}{n}\right) \frac{(\lambda n t)^k}{k!} \exp(-\lambda n t) = f(x + \lambda t),$$

where $f(\cdot)$ is a continuous function.

2.14. Let $X_{in}, \ldots, X_{nn}$ be a sequence of independent Bernoulli r.v.'s such that

$$P\{X_{in} = 1\} = p_{in}, \qquad P\{X_{in} = 0\} = q_{in}, \qquad p_{in} + q_{in} = 1,$$

for all $i = 1, 2, \ldots, n$, and

$$\sum_{i=1}^{n} p_{in} = \lambda \quad \text{(independent of } n\text{)}.$$

(As an example consider a town in which an arsonist secretly lives, and if there are n houses, excluding his, p_{kn} is the probability of a particular house being burned down.) Show that

$$S_n = \sum_{i=1}^{n} X_{in} \xrightarrow{\mathscr{L}} N,$$

where N is a Poisson r.v. with $E\{N\} = \lambda$.

2.15. Let $\{\xi_n\}_1^\infty$ be a sequence of independent r.v.'s such that

$$P\{\xi_n = k\} = e^{-\lambda_n} \frac{(\lambda_n)^k}{k!}$$

and $\lambda_n \to +\infty$ as $n \to \infty$. Show that

$$P\left\{\frac{\xi_n - \lambda_n}{\sqrt{\lambda_n}} \leq x\right\} \to (2\pi)^{-1/2} \int_{-\infty}^{x} e^{-1/2 z^2}\, dz.$$

2.16. Let $\{N(t); t \geq 0\}$ be a Poisson process with $E\{N(t)\} = \Lambda(t)$. Determine the probability density of $(\tau_1, \ldots, \tau_n)$ and of $\tau_n - \tau_{n-1}$, $n = 1, 2, \ldots$.

2.17. Let $\{N(t); t \geq 0\}$ be the counting random function of a simple point process. If $N(t)$ is a homogeneous Markov process (see Proposition 2.6.3), determine $P\{N(t) = k\}$ and $P_{ij}(u)$ [$N(t)$ is called the pure birth process].

2.18. For a pure birth process (see the previous problem), find necessary and sufficient conditions for

$$\sum_{n=0}^{\infty} P\{N(t) = n\} = 1$$

to hold.

2.19. Let $\{N(t); t \geq 0\}$ be a nonhomogeneous Markov process such that the limit

$$\lambda_n(t) = \lim_{h \to 0} \frac{1}{h} P\{N(t+h) = n+1 | N(t) = n\}$$

exists; in the case when

$$\lambda_n(t) = \varphi_0(t) + n\varphi_1(t),$$

$N(t)$ is called a Pòlya process. Determine

$$P\{N(t) = n\}$$

in the case when

$$\varphi_0(t) = \frac{\alpha}{1 + \alpha t \lambda}, \qquad \varphi_1(t) = \frac{\alpha\lambda}{1 + \alpha t \lambda}, \qquad \alpha \geq 0, \quad \lambda > 0,$$

and show that

$$\sum_{k=0}^{\infty} P\{N(t) = k\} = 1.$$

2.20. Let $\mathscr{L}$ be the smallest ring of subsets of $R_+ = [0, \infty)$ (i.e., if $A, B \in \mathscr{L}$, $A \cup B \in \mathscr{L}$, and $A - B \in \mathscr{L}$) containing all intervals of the form $(a, b] \subset R_+$. Let $\mu(\cdot)$ be a nonatomic Borel–Radon measure on R_+ (i.e., μ is non-negative and finite on compact subsets of R_+). If, for each $D \in \mathscr{L}$,

$$P\{\eta(D) = n\} = \exp[-\mu(D)] \frac{[\mu(D)]^n}{n!},$$

the point process η [see (2.2.5)] is a Poisson process [This is due to Rényi (1967).]

2.21. (Continuation) If, for each $D \in \mathscr{L}$,

$$P\{\eta(D) = 0\} = \exp[-\mu(D)]$$

and $P\{\eta(\Gamma) \geq 2\} \leq \mu(\Gamma)\varphi(\mu(\Gamma))$, where $\varphi(\cdot) \geq 0$ is increasing and such that $\varphi(x) \downarrow 0$ as $x \downarrow 0$, then $\eta(\cdot)$ is a Poisson process.

2.22. Let $\{N(t); t \geq 0\}$ be a Poisson process with $E\{N(t)\} = \Lambda(t)$ and arrival times $\{\tau_n\}_1^\infty$. Consider $\{(\tau_n, \xi_n)\}_1^\infty$, where $\{\xi_n\}_1^\infty$ is an i.i.d. sequence with common d.f. $H(x) = P\{\xi_n \leq x\}$, independent of $N(t)$.
(i) Show that $\{\chi(t); t \geq 0\}$ is a Markov process, where

$$\chi(t) = \max\{\xi_k; \tau_k \leq t\}.$$

(ii) Let $T(t)$ be the time instant in $(0, t]$ where the maximum $\chi(t)$ is achieved. Find the d.f. of $T(t)$.

2.23. (Continuation) Determine

$$P\{\chi(t) \leq x, T(t) \leq s\}.$$

CHAPTER 3

Elements of Brownian Motion

3.1. Definitions and Preliminaries

In Chapter 1 (Section 1.6, Example 1.3), we discussed in some detail the nature and causes of the random motion of a small colloidal-size particle submerged in water. According to kinetic theory, this movement is due to the thermal diffusion of the water molecules, which are incessantly bombarding the particle, forcing it to move constantly in a zigzag path. The phenomenon was named "Brownian motion" after R. Brown, an English botanist who was first to observe it. In 1904, H. Poincaré explained that large particles submerged in water do not move, notwithstanding a huge number of impacts from all directions by the molecules of the surrounding medium, simply because, according to the theory of large numbers, they neutralize each other.

The qualitative analysis of the motion was given independently by Einstein (1905) and Smoluchowski in 1916. However, a rigorous mathematical formulation of the problem was formulated a decade later by N. Wiener. To acknowledge his contribution, the stochastic process, which represents a mathematical model of the motion, is called the Wiener process. In the modern literature, the term Brownian motion process is used more often than the Wiener process. In this book, we shall use both names equally.

Throughout this chapter we shall confine ourselves to the one-dimensional aspect of the motion. Suppose we begin our observation of the wandering particle at time $t = 0$ and assume it is at point $x = 0$. Denote by $\xi(t)$ its position on the line at time $t > 0$. Due to the chaotic nature of the motion $\xi(t)$ is a r.v. for every $t > 0$ with $\xi(0) = 0$. Examining further the nature of this phenomenon, it seems reasonable to suppose that $E\{\xi(t)\} = 0$ for all $t \geq 0$. Finally, if the temperature of the water remains constant, the distribution of any increment $\xi(t+s)-\xi(t)$ should not depend on t. This gives rise to the following definition:

Definition 3.1.1. A standard Brownian motion or Wiener process is a stochastic process $\{\xi(t); t \geq 0\}$ on a probability space $\{\Omega, \mathscr{B}, P\}$ having the following properties:

(i) $\xi(0) = 0$ (a.s.);
(ii) $\{\xi(t); t \geq 0\}$ has independent increments;
(iii) for any $0 \leq s < t$,

$$P\{\xi(t) - \xi(s) \leq x\} = \frac{1}{\sqrt{2\pi(t-s)}} \int_{-\infty}^{x} \exp\left(-\frac{u^2}{2(t-s)}\right) du. \quad (3.1.1)$$

Remark 3.1.1. As any process with independent increments, $\xi(t)$ is a Markov process; the stationary transition probability function of Brownian motion is

$$P\{\xi(t+\tau) \leq x \mid \xi(t) = y\} = \frac{1}{\sqrt{2\pi\tau}} \int_{-\infty}^{x-y} \exp\left(-\frac{u^2}{2\tau}\right) du. \quad (3.1.2)$$

From (3.1.1), it follows that for any $0 \leq s \leq t$

$$\operatorname{Var}\{\xi(t) - \xi(s)\} = t - s. \quad (3.1.3)$$

On the other hand,

$$\begin{aligned} E\{\xi(s)\xi(t)\} &= E\{\xi(s)[\xi(t) - \xi(s)] + \xi^2(s)\} \\ &= E\{\xi^2(s)\} = s. \end{aligned}$$

In other words for two arbitrary non-negative numbers u and v,

$$E\{\xi(u)\xi(v)\} = \min\{u, v\}. \quad (3.1.4)$$

The increments of a standard Brownian motion are stationary in the sense that the distribution of $\xi(t+\tau) - \xi(t)$ depends only on τ.

From the fact that $\xi(t)$ has independent increments and (3.1.1), we can easily determine the joint probability density of $(\xi(t_1), \ldots, (t_n))$ for any $0 < t_1 < \cdots < t_n$. It can be readily seen that this density is given by the expression

$$f_{t_1,\ldots,t_n}(x_1,\ldots,x_n) = f_{t_1}(x_1) f_{t_2-t_1}(x_2 - x_1) \ldots f_{t_n-t_{n-1}}(x_n - x_{n-1}), \quad (3.1.5)$$

where

$$f_t(x) = \frac{1}{\sqrt{2\pi t}} \exp\left(-\frac{x^2}{2t}\right). \quad (3.1.6)$$

The system of distributions (3.1.5) clearly satisfies the necessary consistency conditions of the Kolmogorov theorem 1.4.1 This assures the existence of a standard Brownian motion.

Roughly speaking, there are two types of statements to be made about any stochastic process. The first one deals with its distributional properties and the second one with sample path properties. Concerning the distributional properties of Brownian motion, Equations (3.1.5) and (3.1.6) provide enough information in this respect. As far as the sample path properties are concerned, we will show that almost all of them are continuous functions. To this end, we need the following lemma.

Lemma 3.1.1. Let X be a r.v. with $N(0, \sigma^2)$ distribution, then

$$E\{X^4\} = 3\sigma^4.$$

PROOF.

$$\begin{aligned} E\{X^4\} &= \frac{1}{\sigma\sqrt{2\pi}} \int_{-\infty}^{\infty} x^4 \exp\left(-\frac{x^2}{2\sigma^2}\right) dx = \sigma^4 \sqrt{\frac{2}{\pi}} \int_0^{\infty} x^4 \exp\left(-\frac{x^2}{2}\right) dx \\ &= \sigma^4 \frac{4}{\sqrt{\pi}} \int_0^{\infty} u^{3/2} e^{-u}\, du = \sigma^4 \frac{4}{\sqrt{\pi}} \Gamma\left(\frac{5}{2}\right) = 3\sigma^4. \end{aligned}$$ □

We now can prove the following statement.

Proposition 3.1.1. *With probability* 1, *every sample function of a separable Brownian process is uniformly continuous on every finite interval.*

PROOF. Let $\{\xi(t); t \geq 0\}$ be a separable version of a standard Brownian motion (which, according to Doob's proposition 1.7.2, always exists). Then according to (3.1.1) and Lemma 3.1.1 for any $t \geq 0$

$$E(\xi(t+h) - \xi(t))^4 = 3h^2.$$

From this and Theorem 1.9.1 (with $C = 3$, $\beta = 1$, and $\alpha = 4$), the assertion follows. □

Before concluding this section, we will mention that the process $\{X(t); t \geq 0\}$, defined as

$$X(t) = \alpha t + \xi(t), \tag{3.1.7}$$

is called a Brownian motion with drift α. On the other hand, the stochastic process $\{Z(t); 0 \leq t \leq 1\}$, specified as

$$Z(t) = \xi(t) - t\xi(1), \tag{3.1.8}$$

is called a Brownian bridge (or tied down Brownian motion). In both cases, $\xi(t)$ is, of course, a standard Brownian motion.

From (3.1.8), we see the following: $Z(0) = Z(1) = 0$ (a.s.); in addition, the Brownian bridge is a Gaussian stochastic process. In other words, for any $0 < t_1 < \cdots < t_n \leq 1$, $(Z(t_1), \ldots, Z(t_n))$ has a multivariate normal distribution.

3.2. Hitting Times

Let $\{\xi(t); t \geq 0\}$ be a separable version of a standard Brownian motion process. Denote by τ_x the first instant $t > 0$ at which $\xi(t) = x$ (the hitting time of x). More formally,

$$\tau_x(\omega) = \inf\{t > 0; \xi(t, \omega) = x\}. \tag{3.2.1}$$

Clearly,

$$\{\tau_x \geq t\} = \left\{\sup_{0 \leq s \leq t} \xi(s) \leq x\right\} \tag{3.2.2}$$

(see Figure 3.1). From this, we conclude, taking into account that the $\xi(t)$ is separable, that τ_x is a r.v. for all $x \in R$ such that $0 < \tau_{x_1} < \tau_{x_2}$ (a.s.) for all $x_1 < x_2$. It is also clear that τ_x and τ_{-x} have the same distribution.

Denote by τ_x^0 the time of the first crossing of the level x by the process $\xi(t)$, i.e.,

$$\tau_x^0 = \inf\{t > 0; \xi(t) > x\}. \tag{3.2.3}$$

Our physical intuition compels us to deduce that $\tau_x \leq \tau_x^0$. We will actually show that $\tau_x = \tau_x^0$ (a.s.).

We now embark on the problem of determining the distribution of τ_x.

Proposition 3.2.1. *For all* $x \in R$,

$$P\{\tau_x \leq t\} = 2P\{\xi(t) \geq x\} \tag{3.2.4}$$

and $E\{\tau_x\} = \infty$ *if* $x \neq 0$.

PROOF. Suppose first that $x > 0$. From (3.2.1), we readily deduce that for an instant $t \geq 0$

$$\xi(t) \geq x \quad \Rightarrow \quad \tau_x \leq t.$$

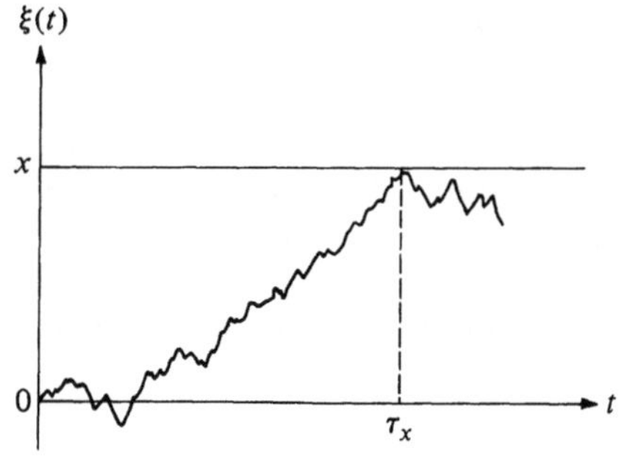

Figure 3.1. Random hitting time τ_x.

Consequently,

$$\{\xi(t) \geq x\} \subset \{\tau_x \leq t\}. \tag{3.2.5}$$

Next, because for all $0 \leq s < t$ and $x \in R$,

$$P\{\xi(t) \leq x | \xi(s) = x\} = P\{\xi(t) \geq x | \xi(s) = x\},$$

which is easy to prove using (3.1.2), it seems quite clear that

$$P\{\xi(t) \leq x | \tau_x \leq t\} = P\{\xi(t) \geq x | \tau_x \leq t\} \tag{3.2.6}$$

because at the instant τ_x the process was at x.

Therefore, taking into account (3.2.5), we obtain

$$P\{\xi(t) \leq x | \tau_x \leq t\} = \frac{P\{\xi(t) \geq x\}}{P\{\tau_x \leq t\}}. \tag{3.2.7}$$

This and (3.2.6) then yield

$$\begin{aligned} P\{\tau_x \leq t\} &= 2P\{\xi(t) \geq x\} \\ &= (2/\pi)^{1/2} \int_{x/\sqrt{t}}^{\infty} e^{-1/2z^2}\, dz, \end{aligned} \tag{3.2.8}$$

which clearly shows that $\tau_x < \infty$ (a.s.) for all $x > 0$. On the other hand,

$$\begin{aligned} E\{\tau_x\} &= \int_0^\infty P\{\tau_x > s\}\, ds = \int_0^\infty \left\{1 - (2/\pi)^{1/2} \int_{x/\sqrt{s}}^{\infty} e^{-1/2z^2}\, dz\right\} ds \\ &= (2/\pi)^{1/2} \int_0^\infty ds \int_0^{x/\sqrt{s}} e^{-1/2z^2}\, dz = (2/\pi)^{1/2} \int_0^\infty e^{-1/2z^2}\, dz \int_0^{x^2/z^2} ds \\ &= x^2 (2/\pi)^{1/2} \int_0^\infty \frac{1}{z^2} e^{-1/2z^2}\, dz \geq x^2 (2/\pi)^{1/2} e^{-1/2} \int_0^1 \frac{dz}{z^2} = \infty. \end{aligned}$$

This completes the proof of the proposition in the case when $x > 0$. If $x < 0$, the assertion follows from the fact that τ_x and τ_{-x} have the same distribution. Thus, from (3.2.8), we have

$$P\{\tau_x \leq t\} = (2/\pi)^{1/2} \int_{|x|/\sqrt{t}}^{\infty} e^{-1/2z^2}\, dz. \qquad \square$$

Corollary 3.2.1. *From Proposition* 3.2.1, *we deduce that given* $x \neq 0$, *the standard Brownian motion process will hit* x *with probability* 1. *However, the average length of time to achieve this is infinite.*

Concerning the r.v.'s τ_x and τ_x^0, we have the following result.

Lemma 3.2.1. *For all* $x \in R$, $\tau_x = \tau_x^0$ (a.s.)

Proof. Because $\xi(t)$ and $-\xi(t)$ have the same distribution, it suffices to prove the statement for $x > 0$. Clearly, $\tau_x \leq \tau_x^0$; hence, what we have to show is that

$P\{\tau_x < \tau_x^0\} = 0$. Obviously,

$$\{\tau_x < \tau_x^0\} \subset \bigcup_{i=1}^{\infty} \bigcup_{n=1}^{\infty} \left\{ \sup_{0 \le s \le i/n} \xi(s) = x \right\}. \tag{3.2.9}$$

Therefore, to prove the lemma it suffices to show that, for all $x > 0$, $P\{\sup_{0<s\le t} \xi(s) = x\} = 0$. To this end, note that for $t_1 < t$,

$$P\left\{ \sup_{0 \le s \le t} \xi(s) = x \right\} = P\left\{ \sup_{0 \le s \le t_1} \xi(s) = x,\ \sup_{t_1 < s \le t} \xi(s) \le x \right\}$$

$$+ P\left\{ \sup_{0 \le s \le t_1} \xi(s) \le x,\ \sup_{t_1 \le s \le t} \xi(s) = x \right\}$$

$$\le P\left\{ \sup_{0 \le s \le t_1} \xi(s) = x \right\} + \int_{-\infty}^{x} P\left\{ \xi(t_1) \in dy,\ \sup_{t_1 \le s \le t} \xi(s) - \xi(t_1) = x - y \right\}$$

$$= P\left\{ \sup_{0 \le s \le t_1} \xi(s) = x \right\} + \int_{-\infty}^{x} P\{\xi(t_1) \in dy\} P\left\{ \sup_{t_1 \le s \le t} \xi(s) - \xi(t_1) = x - y \right\}.$$

However, $P\{\sup_{t_1 \le s \le t} \xi(s) - \xi(t_1) = x - y\} = 0$ for almost all y (with respect to Lebesgue measure). Thus,

$$P\left\{ \sup_{0 \le s \le t} \xi(s) = x \right\} \le P\left\{ \sup_{0 \le s \le t_1} \xi(s) = x \right\}.$$

In other words, $P\{\sup_{0 \le s \le t} \xi(s) = x\}$ does not decrease as $t \downarrow 0$. On the other hand, due to the continuity of $\xi(t)$, we have for all $\varepsilon > 0$

$$P\left\{ \max_{0 \le s \le t} \xi(s) = \varepsilon \right\} \to 0 \quad \text{as } t \downarrow 0.$$

Thus,

$$P\left\{ \sup_{0 \le s \le t} \xi(s) = x \right\} \le \lim_{t \to 0} P\left\{ \sup_{0 \le s \le t} \xi(s) > \frac{1}{2} x \right\} = 0.$$

From this and (3.2.9), the assertion follows. □

Remark 3.2.1. The last inequality follows from

$$\left\{ \sup_{0 \le s \le t} \xi(s) = x \right\} \subset \left\{ \sup_{0 \le s \le t} \xi(s) > \frac{1}{2} x \right\}.$$

3.3. Extremes of $\xi(t)$

The result (3.2.4) is the key to many properties of the Brownian sample functions. For instance, from (3.2.2) and (3.2.4), we obtain that

$$P\{\chi(t) > x\} = (2/\pi)^{1/2} \int_{x/\sqrt{t}}^{\infty} e^{-1/2z^2}\, dz, \tag{3.3.1}$$

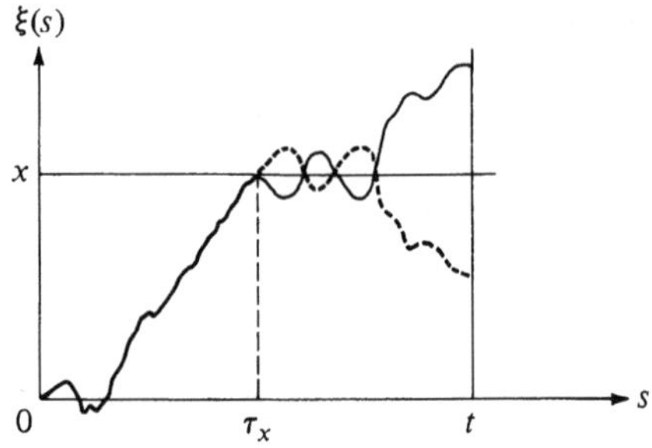

Figure 3.2. The reflection principle.

where

$$\chi(t) = \sup_{0 \le s \le t} \xi(s). \tag{3.3.2}$$

We will give one more proof of (3.3.1) using the so-called reflection principle.

The following argument is not easy to formalize, although it seems intuitively quite clear. We have

$$P\{\chi(t) > x, \xi(t) > x\} = P\{\chi(t) > x, \xi(t) < x\}. \tag{3.3.3}$$

This can be explained heuristically as follows: For every sample path of the process $\xi(t)$ which hits the level x before time t, but finishes above this level at time t, there is another "equally probable trajectory" (shown by the dotted line in Figure 3.2) that hits x before t and such that $\xi(t) < x$. This is the meaning of (3.3.3). On the other hand,

$$\begin{aligned} P\{\chi(t) > x\} &= P\{\chi(t) > x, \xi(t) > x\} + P\{\chi(t) > x, \xi(t) < x\} \\ &= 2P\{\chi(t) > x, \xi(t) > x\} \end{aligned}$$

due to (3.3.3). The proof of (3.3.1) now follows from the fact that

$$P\{\chi(t) > x, \xi(t) > x\} = P\{\xi(t) > x\}$$

and from (3.1.1).

Remark 3.3.1. Equation (3.3.1) leads to a very interesting conclusion: For all $t > 0$,

$$P\{\chi(t) > 0\} = 1.$$

On the other hand,

$$\begin{aligned} P\left\{\inf_{0 \le s \le t} \xi(s) < 0\right\} &= P\left\{-\sup_{0 \le s \le t} (-\xi(s)) < 0\right\} \\ &= P\left\{\sup_{0 \le s \le t} (-\xi(s)) > 0\right\}. \end{aligned}$$

But due to the symmetry of $\xi(t)$,

$$P\left\{\sup_{0\le s\le t} \xi(s) > 0\right\} = P\left\{\sup_{0\le s\le t} (-\xi(s)) > 0\right\}.$$

Consequently, for all $t > 0$,

$$P\left\{\inf_{0\le s\le t} \xi(s) < 0\right\} = 1.$$

Therefore, starting from point $x = 0$ at time $t = 0$, a trajectory of a standard Brownian motion process will intersect with the t axis infinitely many times in $(0, t]$ for any $t > 0$. This shows how irregular sample paths of a Brownian motion process are, and yet, almost surely they are continuous functions.

The following proposition gives the joint distribution of $(\chi(t), \tau_x)$.

Proposition 3.3.1. *For any $t > 0$ and $x \le y$,*

$$P\{\chi(t) \le y, \tau_x \le u\} = \left(\frac{x}{\pi}\right)\int_0^u \int_0^{y-x} \exp\left\{-\frac{1}{2}\left(\frac{v^2}{t-s} + \frac{x^2}{s}\right)\right\}\frac{dv\,ds}{s\sqrt{s(t-s)}}. \tag{3.3.4}$$

Proof. The following argument is not rigorous but it can be made so using the strong Markov property concept which will be discussed later.

Because the Brownian motion process has independent increments and the distribution of $\xi(s+t) - \xi(s)$ does not depend on s, it seems intuitively clear that once it reaches a point x at time τ_x, it behaves after τ_x as if x is its starting point. Therefore, given that $\tau_x = s$, where $0 < s \le t$, we have, for all $0 \le x \le y$,

$$\begin{aligned}P\{\chi(t) \le y \mid \tau_x = s\} &= P\left\{\sup_{s\le u\le t} \xi(u) \le y \mid \tau_x = s\right\}\\ &= P\left\{x + \sup_{0\le u\le t-s} \xi(u) \le y\right\} = P\{\chi(t-s) \le y - x\}.\end{aligned}$$

From this, (3.3.1), and (3.2.8), we obtain

$$\begin{aligned}P\{\chi(t) \le y, \tau_x \le u\} &= \int_0^u P\{\chi(t) \le y \mid \tau_x = s\}\, dP\{\tau_x \le s\}\\ &= \left(\frac{x}{2}\right)\int_0^u \left(\frac{2}{\pi(t-s)}\right)^{1/2} \int_0^{y-x} \exp\left\{\frac{-v^2}{2(t-s)}\right\}\\ &\quad \times \left(\frac{2}{\pi s^3}\right)^{1/2} \exp\left(-\frac{x^2}{2s}\right) dv\, ds,\end{aligned}$$

which proves the assertion. □

Denote by $T(t)$ the epoch of the largest value of $\xi(s)$ in $[0, t]$. In the sequel we will attempt to determine the joint distribution of $(\chi(t), T(t))$.

Proposition 3.3.2. *For all* $0 < u < t$ *and* $x > 0$,

$$P\{\chi(t) \in dx, T(t) \in du\} = \frac{x}{\pi u\sqrt{u(t-u)}} \exp\left(-\frac{x^2}{2u}\right) dx\, du. \tag{3.3.5}$$

Proof. Our proof of the proposition is based on the observation that

$$T(t) = \tau_x \quad \text{on the set } \{\chi(t) = x\}.$$

Consequently, for any $0 < s < t$,

$$\begin{aligned} P\{\chi(t) \in dx, \tau_x \in ds\} &= P\{\tau_x \in ds | \chi(t) = x\} P\{\chi(t) \in dx\} \\ &= P\{T(t) \in ds | \chi(t) = x\} P\{\chi(t) \in dx\} \\ &= P\{T(t) \in ds, \chi(t) \in dx\}. \end{aligned} \tag{3.3.6}$$

But from (3.3.4), we have

$$P\{\chi(t) \in dy, \tau_x \in du\} = \frac{x}{\pi u\sqrt{u(t-u)}} \exp\left[-\frac{1}{2}\left(\frac{(y-x)^2}{t-u} + \frac{x^2}{u}\right)\right] du\, dy. \tag{3.3.7}$$

Replacing y with x in this equation, it follows from (3.3.6) that, for $0 < u < t$ and $x > 0$,

$$P\{\chi(t) \in dx, T(t) \in du\} = \frac{x}{\pi u\sqrt{u(t-u)}} \exp\left(-\frac{x^2}{2u}\right) du\, dx,$$

which proves the assertion. □

Corollary 3.3.1. *From the last equation, we have*

$$P\{T(t) \in du\} = \int_0^\infty P\{\chi(t) \in dx, T \in du\} = \frac{du}{\pi\sqrt{u(t-u)}}. \tag{3.3.8}$$

Therefore,

$$\begin{aligned} P\{T(t) \le s\} &= \frac{1}{\pi} \int_0^s \frac{du}{\sqrt{u(t-u)}} \\ &= \frac{2}{\pi} \arcsin \sqrt{\frac{s}{t}}. \end{aligned}$$

This is the famous arc sin law. In Figure 3.3, a graphical depiction of its probability density is given.

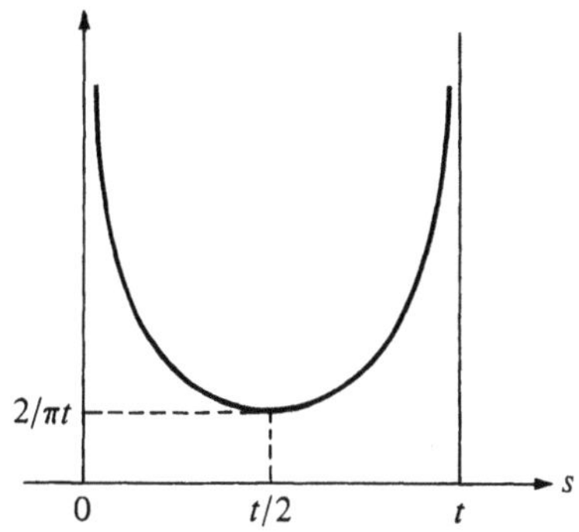

Figure 3.3. Graphical presentation of arc sin law.

3.4. Some Properties of the Brownian Paths

The Brownian motion process is governed by subtle and remarkable principles. Some feeling for the nature of the process may be obtained from considerations of the local properties of its paths.

Proposition 3.4.1. *Let $\{\xi(t); t \geq 0\}$ be separable standard Brownian motion process, then almost all its sample functions are not differentiable at any $t \geq 0$.*

PROOF. We want to show that for all $t \geq 0$ and $h > 0$,

$$P\left\{\overline{\lim_{h\to 0+}} \frac{\xi(t+h)-\xi(t)}{h} = \infty\right\} = 1. \tag{3.4.1}$$

To prove this, note that for any $0 < h < \delta$,

$$\sup_{0<h<\delta} \frac{\xi(t+h)-\xi(t)}{h} \geq \frac{1}{\delta} \sup_{0<h<\delta} (\xi(t+h)-\xi(t)).$$

From this and (3.3.1), we have for any $x > 0$,

$$\begin{aligned} P\left\{\sup_{0<h<\delta} \frac{\xi(t+h)-\xi(t)}{h} > x\right\} &\geq P\left\{\sup_{0<h<\delta} (\xi(t+h)-\xi(t)) > \delta x\right\} \\ &= P\left\{\sup_{0<h<\delta} \xi(h) > \delta x\right\} \\ &= \left(\frac{2}{\pi}\right)^{1/2} \int_{x\sqrt{\delta}}^{\infty} e^{-1/2z^2}\, dz \to 1 \end{aligned}$$

as $\delta \to 0$. From this, (3.4.1) follows. Because $\xi(t)$ and $-\xi(t)$ have the same distribution, it follows also that

$$P\left\{\varlimsup_{h\to 0+}\frac{\xi(t+h)-\xi(t)}{h}=-\infty\right\}=1.$$

This proves the assertion. □

Corollary 3.4.1. *For almost all ω, $\xi(\cdot,\omega)$ does not have bounded variation on any interval of $[0,\infty)$. Hence, the graph of $\xi(t,\omega)$ is not rectifiable. Furthermore, for almost every ω, there is no interval $(a,b)\subset[0,\infty)$ on which $\xi(\cdot,\omega)$ is monotone. Indeed, if $\xi(\cdot,\omega)$ were of bounded variation on a finite interval, it would be differentiable almost everywhere (Lebesgue measure) there. Also, if $\xi(\cdot,\omega)$ were monotone on an interval, it would be differentiable at almost every point of this interval.*

In the previous section (see Remark 3.3.1) we have shown that

$$P\left\{\sup_{0\le s\le h}\xi(s)>0\right\}=P\left\{\inf_{0\le s\le h}\xi(s)<0\right\}=1$$

for any arbitrary small $h>0$. The next result yields properties of Brownian sample functions for very large t.

Proposition 3.4.2

$$P\left\{\sup_{0\le t<\infty}\xi(t)=+\infty\right\}=P\left\{\inf_{0\le t<\infty}\xi(t)=-\infty\right\}=1. \tag{3.4.2}$$

Proof. Let $k>0$ be an integer. Then taking into account (3.3.1),

$$P\left\{\sup_{0\le t<\infty}\xi(t)>k\right\}\ge P\left\{\sup_{0\le s\le t}\xi(s)>k\right\}$$

$$=(2/\pi)^{1/2}\int_{k/\sqrt{t}}^{\infty}e^{-1/2z^2}\,dz\to(2/\pi)^{1/2}\int_{0}^{\infty}e^{-1/2z^2}\,dz=1$$

as $t\to\infty$. Thus,

$$P\left\{\sup_{0\le t<\infty}\xi(t)=\infty\right\}=P\left\{\bigcap_{k=1}^{\infty}\left\{\sup_{0\le t<\infty}\xi(t)>k\right\}\right\}=1.$$

On the other hand,

$$P\left\{\inf_{0\le t<\infty}\xi(t)=-\infty\right\}=P\left\{\sup_{0\le t<\infty}(-\xi(t))=\infty\right\}=1,$$

which proves the proposition. □

According to the Corollary 3.4.1, almost all sample paths of a Brownian motion process have infinite variation in any arbitrary small interval of time. In contrast to this unpleasant feature of the process, a very sharp positive statement can be made about the so-called "quadratic variation" of Brownian sample paths, as the following result clearly shows:

Proposition 3.4.3. *Let* $0 = t_{n0} < t_{n1} < \cdots < t_{nn} = t$ *be a partition of the interval* $[0, t]$, *such that* $\max_i (t_{n,i+1} - t_{ni}) \to 0$ *as* $n \to \infty$. *Then*

$$\underset{n\to\infty}{\text{l.i.m.}} \sum_{i=0}^{n-1} (\xi(t_{n,i+1}) - \xi(t_{ni}))^2 = t \tag{3.4.3}$$

and

$$P\left\{\lim_{n\to\infty} \sum_{i=0}^{n-1} (\xi(t_{n,i+1}) - \xi(t_{ni}))^2 = t\right\} = 1, \tag{3.4.4}$$

where l.i.m. indicates the convergence in quadratic mean.

Proof. From (3.1.3), we have

$$E(\xi(t_{n,i+1}) - \xi(t_{ni}))^2 = t_{n,i+1} - t_{ni}, \tag{4.3.5}$$

so that

$$E \sum_{i=0}^{n-1} (\xi(t_{n,i+1}) - \xi(t_{ni}))^2 = t. \tag{3.4.6}$$

Therefore, to prove (3.4.3) it is sufficient to show that

$$\text{Var}\left\{\sum_{i=0}^{n-1} (\xi(t_{n,i+1}) - \xi(t_{ni}))^2\right\} \to 0$$

as $n \to \infty$. But from Lemma 3.1.1 and (3.4.5), we obtain

$$\text{Var}\left\{\sum_{i=0}^{n-1} (\xi(t_{n,i+1}) - \xi(t_{ni}))^2\right\} = \sum_{i=0}^{n-1} \text{Var}\left\{(\xi(t_{n,i+1}) - \xi(t_{ni}))^2\right\}$$

$$= \sum_{i=0}^{n-1} \{E(\xi(t_{n,i+1}) - \xi(t_{ni}))^4 - (t_{n,i+1} - t_{ni})^2\}$$

$$= 2\sum_{i=0}^{n-1} (t_{n,i+1} - t_{ni})^2 \le 2 \sup_{0\le i\le n} (t_{n,i+1} - t_{ni}) \sum_{i=0}^{n-1} (t_{n,i+1} - t_{ni})$$

$$= 2t \sup_{0\le i\le n} (t_{n,i+1} - t_{ni}) \to 0$$

as $n \to \infty$, which proves (3.4.3).

To prove the second part of the proposition, set

$$t_{ni} = (i/n)t$$

and consider

$$Y_n = \sum_{i=0}^{n-1} \left\{[\xi(t_{n,i+1}) - \xi(t_{ni})]^2 - \frac{t}{n}\right\}.$$

By the Markov inequality, we have

$$P\{|Y_n| > \varepsilon\} \leq \frac{1}{\varepsilon^4} E\{Y_n\}^4. \tag{3.4.7}$$

After some elementary calculations, using the fact that for any $0 \leq s - t$

$$E(\xi(t) - \xi(s))^{2k} = 1 \cdot 3 \cdots (2k-1)(t-s)^k, \qquad k = 1, 2, \ldots, \tag{3.4.8}$$

we deduce that

$$E\{Y_n^4\} \leq C\left(\frac{t}{n}\right)^3 t,$$

where $C > 0$ is a constant. From this and (3.4.7), it follows that, for any $\varepsilon > 0$,

$$P\{|Y_n| > \varepsilon \text{ i.o.}\} \leq Ct^4 \sum_{n=1}^{\infty} \frac{1}{n^3} < \infty.$$

Consequently, $Y_n \to 0$ (a.s.) as $n \to \infty$. This completes the proof. □

Corollary 3.4.2. *By means of this result, we can show once more that almost all sample functions of a Brownian process are not rectifiable. First, we have the inequality*

$$\sum_{i=0}^{n-1} (\xi(t_{n,i+1}) - \xi(t_{ni}))^2 \leq \sup_{0 \leq i \leq n-1} |\xi(t_{n,i+1}) - \xi(t_{ni})| \sum_{i=0}^{n-1} |\xi(t_{n,i+1}) - \xi(t_{ni})|.$$

But, from (3.3.1), *we have that*

$$\sup_{0 \leq i \leq n-1} |\xi(t_{n,i+1}) - \xi(t_{ni})| \to 0 \quad \text{(a.s.)}$$

From this and Proposition 3.4.3, *it clearly follows that*

$$\sum_{i=0}^{n-1} |\xi(t_{n,i+1}) - \xi(t_{ni})| \to \infty \quad \text{(a.s.)},$$

which is the conclusion of the corollary.

3.5. Law of the Iterated Logarithm

One of the most basic propositions in probability theory is the strong law of large numbers. It states that if $\{\xi_i\}_1^\infty$ is an i.i.d. sequence of r.v.'s with $E|\xi_1| < \infty$ and $E\{\xi_1\} = 0$, then

$$S_n/n \to 0 \quad \text{(a.s.) as } n \to \infty,$$

where $S_n = \xi_1 + \cdots + \xi_n$. Roughly speaking, it means that for any $\varepsilon > 0$, $|S_n|$ will be less than $n\varepsilon$ if n is sufficiently large, so that S_n oscillates with an amplitude less then $n\varepsilon$.

In many situations, it is of some interest to have more precise information on the rate of growth of the sum S_n as $n \to \infty$. This is given in the form of the celebrated law of the iterated logarithm, which is considered as a crowning achievement of the classical probability theory. The following proposition is a precise formulation of the law of the iterated logarithm.

Proposition 3.5.1. *Let $\{\xi_i\}_1^\infty$ be an i.i.d. sequence of r.v.'s with $E(\xi_1) = 0$ and $E(\xi_1^2) = 1$. Then*

$$P\left\{\limsup_{n\to\infty} \frac{S_n}{\sqrt{2n \ln \ln n}} = 1\right\} = 1,$$

$$P\left\{\liminf_{n\to\infty} \frac{S_n}{\sqrt{2n \ln \ln n}} = -1\right\} = 1.$$

This result for bounded r.v.'s was first obtained by Khintchine in 1924. Later, it was generalized by Kolmogorov and Feller. Under the conditions of the proposition, it was proved by Hartman and Wintner (1941). In this section, we will establish an analogous result for a standard Brownian motion process. But first we will prove the following auxiliary result.

Lemma 3.5.1. *Let a r.v. X have the $N(0, 1)$ distribution. Then, for any $x > 0$,*

$$\frac{1}{\sqrt{2\pi}}\left(\frac{1}{x} - \frac{1}{x^3}\right)e^{-x^2/2} < P\{X > x\} < \frac{1}{x\sqrt{2\pi}}e^{-x^2/2}. \tag{3.5.1}$$

PROOF. For all $y > 0$,

$$\left(1 - \frac{3}{y^4}\right)e^{-y^2/2} < e^{-y^2/2} < \left(1 + \frac{1}{y^2}\right)e^{-y^2/2}.$$

Multiplying by $1/\sqrt{2\pi}$ and integrating over (x, ∞), where $x > 0$, the assertion follows. □

Corollary 3.5.1. *For $x > 0$ sufficiently large,*

$$P\{X > x\} \sim \frac{1}{x\sqrt{2\pi}}e^{-x^2/2} \tag{3.5.2}$$

and, for all $x > 1$,

$$P\{X > x\} < e^{-x^2/2}. \tag{3.5.3}$$

This result follows readily from Lemma 3.5.1.

We are now ready to embark on the problem of proving the law of the iterated logarithm for a Brownian motion process. We will first prove the so-called local law of the iterated logarithm.

Proposition 3.5.1. *Let $\{\xi(t); t \geq 0\}$ be a standard Brownian motion process. Then*

$$P\left\{\limsup_{t\to 0} \frac{\xi(t)}{\sqrt{2t \ln\ln t^{-1}}} = 1\right\} = 1, \tag{3.5.4}$$

$$P\left\{\liminf_{t\to 0} \frac{\xi(t)}{\sqrt{2t \ln\ln t^{-1}}} = -1\right\} = 1. \tag{3.5.5}$$

PROOF. We clearly have

$$P\{\xi(t) > x\} = (1/2\pi)^{1/2} \int_{x/\sqrt{t}}^{\infty} e^{-u^2/2}\, du.$$

From this and (3.5.2), it follows that

$$P\{\xi(t) > x\} \sim \frac{t}{x\sqrt{2\pi}} e^{-x^2/2} \tag{3.5.6}$$

as $x/\sqrt{t} \to +\infty$. Let $0 < b < 1$ and consider the random event B_k defined by

$$B_k = \left\{\sup_{0 \leq s \leq b^k} \xi(s) > (1 + \varepsilon)x_{k+1}\right\},$$

where $\varepsilon > 0$ is arbitrarily small and

$$x_k = \sqrt{b^k \ln\ln b^{-k}}.$$

Then as $k \to \infty$ it is clear that $x_k/\sqrt{b^k} \to \infty$. Therefore, from (3.2.2), (3.2.4), and (3.5.6), we have, as $k \to \infty$,

$$\begin{aligned}
P(B_k) &= 2P\{\xi(b^k) > (1 + \varepsilon)x_{k+1}\} \\
&\sim \left(\frac{2}{\pi}\right)^{1/2} \frac{\sqrt{b^k}}{(1 + \varepsilon)x_{k+1}} \exp\left(-\frac{(1 + \varepsilon)^2 x_{k+1}^2}{2b^k}\right) \\
&= \frac{1}{1 + \varepsilon}(\pi b \ln b^{-k-1})^{-1/2}(\ln b^{-k-1})^{-b(1+\varepsilon)^2} \\
&= \frac{(\ln b^{-1})^{b(1+\varepsilon)^2}}{(1 + \varepsilon)\sqrt{\pi b}} \frac{1}{(k + 1)^{b(1+\varepsilon)^2}\sqrt{\ln(k + 1) + \ln\ln b^{-1}}}.
\end{aligned}$$

Next, set $b = (1 + \varepsilon)^{-1}$; then clearly

$$P(B_k) \sim \frac{C(\varepsilon)}{k^{1+\varepsilon}\sqrt{\ln k}}, \tag{3.5.7}$$

where

$$C(\varepsilon) = \frac{[\ln(1 + \varepsilon)]^{(1+\varepsilon)}}{\sqrt{\pi(1 + \varepsilon)}}.$$

Because the series $\sum_1^\infty 1/k^{1+\varepsilon}$ converges, it follows readily from (3.5.7) that

$$\sum_{k=1}^{\infty} P(B_k) < \infty$$

and, consequently, by the Borel–Cantelli lemma,

$$P\{B_k \text{ i.o.}\} = 0.$$

Therefore, for all sufficiently small $t > 0$ (for $t < b^k$ for some k)

$$\xi(t) < (1 + \varepsilon)\sqrt{2t \ln \ln t^{-1}}$$

or, equivalently,

$$P\left\{\limsup_{t \to 0} \frac{\xi(t)}{\sqrt{2t \ln \ln t^{-1}}} < 1 + \varepsilon\right\} = 1$$

for all $\varepsilon > 0$.

Let us now show that

$$P\left\{\limsup_{t \to 0} \frac{\xi(t)}{\sqrt{2t \ln \ln t^{-1}}} > 1 - \varepsilon\right\} = 1. \tag{3.5.8}$$

To this end, define

$$D_k = \left\{\xi(b^k) - \xi(b^{k+1}) > \left(1 - \frac{\varepsilon}{2}\right)x_k\right\}.$$

Clearly, $\{D_k\}_1^{\infty}$ is a sequence of independent events. Because the distribution of $\xi(b^k) - \xi(b^{k+1})$ is equal to the distribution of $\xi(b^k(1 - b))$, we have for large k that

$$\begin{aligned} P(D_k) &= \frac{1}{\sqrt{2\pi}} \int_{(1-\varepsilon/2)x_k/b^k(1-b)}^{\infty} e^{-u^2/2}\, du \\ &\sim \frac{\sqrt{b^k(1 - b)}}{\sqrt{2\pi}(1 - \varepsilon/2)x_k} \exp\left(-\frac{(1 - \varepsilon/2)^2 2b^k \ln \ln b^{-k}}{2b^k(1 - b)}\right). \end{aligned}$$

From this, after some straightforward calculations, we obtain, as $k \to \infty$, that

$$P(D_k) \sim \frac{\sqrt{1 - b}}{2(1 - \varepsilon/2)\sqrt{\pi}} \frac{k^{-((1-(\varepsilon/2))/(1-b))}}{\sqrt{\ln k}}.$$

From this, it follows that, for $b \leq \varepsilon/2$, the sum

$$\sum_{1}^{\infty} P(D_k) = \infty.$$

Thus, invoking the second Borel–Cantelli lemma, we have

$$P\{D_k \text{ i.o.}\} = 1. \tag{3.5.9}$$

Further, due to symmetry of $\xi(t)$, it follows from (3.5.3) that for sufficiently large k and arbitrary $\delta > 0$

$$\xi(b^{k+1}) > -(1 + \delta)x_k.$$

This and (3.5.8) then imply that the events

$$\xi(b^{k+1}) + [\xi(b^k) - \xi(b^{k+1})] = \xi(b^k) > -(1+\delta)x_k + \left(1 - \frac{1}{2}\varepsilon\right)x_k$$
$$= -(1+\delta)x_k + x_k\varepsilon/2 + (1-\varepsilon)x_k$$

occur infinitely often. But, the right-hand side of this inequality can be made larger than $(1-\varepsilon)x_k$ (by a proper choice of b). For instance, this is the case if b is such that

$$-(1+\delta)x_k + \varepsilon x_k/2 > 0,$$

which holds if

$$(1+\delta)\sqrt{b\frac{\ln\ln a^{-k-1}}{\ln\ln a^{-k}}} < \frac{\varepsilon}{2},$$

which is true if b is sufficiently small. This proves (3.5.8) and because $\varepsilon > 0$ is arbitrarily small, relation (3.5.4) follows.

To prove (3.5.5), note that due to the symmetry of $\xi(t)$,

$$\limsup_{t\to 0} \frac{-\xi(t)}{\sqrt{2t\ln\ln t^{-1}}} = 1 \quad \text{(a.s.)}$$

$$\Rightarrow \liminf_{t\to 0} \frac{\xi(t)}{\sqrt{2t\ln\ln t^{-1}}} = -1 \quad \text{(a.s.).}$$

This is the desired result. □

Next, we shall prove the following result.

Lemma 2.5.2. *If* $\{\xi(t); t \geq 0\}$ *is a standard Brownian motion process, so is*

$$t\xi\left(\frac{1}{t}\right).$$

Proof. Let $0 \leq u < t$ and consider

$$E\left(\exp\left\{\theta i\left[t\xi\left(\frac{1}{t}\right) - u\xi\left(\frac{1}{u}\right)\right]\right\}\right)$$
$$= E\exp\left\{\theta i\left(\xi\left(\frac{1}{t}\right)(t-u) - u\left[\xi\left(\frac{1}{u}\right) - \xi\left(\frac{1}{t}\right)\right]\right)\right\}$$
$$= \exp\left\{-\frac{\theta^2}{2}(t-u)^2\frac{1}{t} - \frac{(\theta u)^2}{2}\left(\frac{1}{u} - \frac{1}{t}\right)\right\}$$
$$= \exp\left\{-\frac{\theta^2}{2}(t-u)\right\},$$

Independence of increments in this case can be established by showing that they are uncorrelated. Specifically,

$$E\left(u\xi\left(\frac{1}{u}\right)\left[t\xi\left(\frac{1}{t}\right) - u\xi\left(\frac{1}{u}\right)\right]\right)$$
$$= E\left(ut\xi\left(\frac{1}{u}\right)\xi\left(\frac{1}{t}\right) - u^2\xi^2\left(\frac{1}{u}\right)\right) = ut\left(\frac{1}{t}\right) - u^2\left(\frac{1}{u}\right) = 0.$$

This proves the lemma. □

Proposition 3.5.2. *Let* $\{\xi(t); t \geq 0\}$ *be a standard Brownian motion process; then*

$$P\left\{\limsup_{t\to\infty} \frac{\xi(t)}{\sqrt{2t \ln \ln t}} = 1\right\} = 1,$$
$$P\left\{\liminf_{t\to\infty} \frac{\xi(t)}{\sqrt{2t \ln \ln t}} = -1\right\} = 1.$$

PROOF. This follows from the fact that

$$\limsup_{t\to\infty} \frac{\xi(t)}{\sqrt{2t \ln \ln t}} = \limsup_{u\to 0} \frac{u\xi(1/u)}{\sqrt{2u \ln \ln(1/u)}} = 1 \qquad \text{(a.s.)},$$

and so on.

3.6. Some Extensions

Every Brownian process has independent and normally distributed increments. These are the defining features of the process. Also, every separable version has, with probability 1, continuous sample paths. It is remarkable that, in a certain sense, the converse also holds. It other words, every stochastic process with independent increments is Gaussian if its sample functions are continuous with probability 1. This result is due to Doob. We now give a precise formulation of this statement.

Proposition 3.6.1. *Let* $\{w(t); t \geq 0\}$ *be a stochastic process with independent increments. If its sample functions are continuous with probability* 1, *then its increments are normally distributed.*

PROOF. We will show that, under the conditions of the proposition, $w(t)$ is normally distributed for every $t \geq 0$. Because, for any $0 \leq s < t$,

$$w(t) = w(s) + [w(t) - w(s)],$$

it will then follow from the Cramér theorem that $w(t) - w(s)$ has a normal distribution.

The stochastic process $w(t)$ is clearly separable (see Proposition 1.10.1). Because every sample path of $w(t)$ is continuous with probability 1, it is also uniformly continuous in every finite subinterval of $[0, \infty)$. Hence, for every $\varepsilon > 0$, there exists a $\delta = \delta(\varepsilon)$ such that

$$P\left\{\sup_{|u-v|<\delta} |w(u) - w(v)| \geq \varepsilon\right\} < \varepsilon, \quad u, v \in [0, t].$$

Let $\varepsilon_1 > \varepsilon_2 > \cdots$ and $\varepsilon_n \to 0$ as $n \to \infty$. Consider

$$0 = t_{n0} < t_{n1} < \cdots < t_{nk_n} = t,$$

where

$$t_{ni} - t_{n,i-1} = \frac{t}{k_n} \quad \text{and} \quad \frac{t}{k_n} < \delta(\varepsilon_n).$$

Set

$$Y_{ni} = \begin{cases} 0 & \text{if } |w(t_{ni}) - w(t_{n,i-1})| \geq \varepsilon_n \\ w(t_{ni}) - w(t_{n,i-1}) & \text{if } |w(t_{ni}) - w(t_{n,i-1})| < \varepsilon_n. \end{cases}$$

Clearly, then,

$$P\left\{w(t) \neq \sum_1^{k_n} Y_{ni}\right\} = P\left(\bigcup_{i=1}^{k_n} \{|w(t_{ni}) - w(t_{n,i-1})| \geq \varepsilon_n\}\right)$$

$$= P\left\{\sup_i |w(t_{ni}) - w(t_{n,i-1})| \geq \varepsilon_n\right\} < \varepsilon_n.$$

Now, using the independence of $\{Y_{ni}\}$, we have

$$E(e^{i\theta w(t)}) = \lim_{n\to\infty} E\left(\exp\left(i\theta \sum_1^{k_n} Y_{ni}\right)\right).$$

Set

$$\alpha_{ni} = E(Y_{ni}), \qquad V_{ni} = \operatorname{Var}\{Y_{ni}\},$$

$$\alpha_n = \sum_1^{k_n} \alpha_{ni}, \qquad V_n = \sum_1^{k_n} V_{ni}.$$

If $\alpha_n \to \alpha$ and $V_n \to \sigma^2$, where α and σ are finite, we obtain

$$E(e^{i\theta w(t)}) = \lim_{n\to\infty} \exp(i\theta\alpha_n) \prod_1^{k_n} E(\exp[i\theta(Y_{ni} - \alpha_{ni})])$$

$$= e^{i\theta\alpha} \lim_{n\to\infty} \prod_1^{k_n} \left(1 - \frac{\theta^2}{2} V_{ni} + V_{ni} O(\varepsilon_n)\right)$$

$$= e^{i\theta\alpha} \lim_{n\to\infty} \prod_1^{k_n} \exp\left(-\frac{\theta^2}{2} V_{ni} + V_{ni} O(\varepsilon_n)\right)$$

$$= e^{i\theta\alpha} \lim_{n\to\infty} \left[\exp\left(-\frac{\theta^2}{2} V_n\right)\right].$$

Therefore,

$$E(e^{i\theta w(t)}) = \exp\left(i\theta\alpha - \frac{\theta^2}{2}\sigma^2\right),$$

which is the desired result. □

3.7. The Ornstein–Uhlenbeck Process

The strange irregular motion of a small particle submerged in liquid, caused by molecular bombardment, was first described mathematically by L. Bachelier in 1900. He went so far as to note the Markov property of the process. In 1905, A. Einstein and, independently, M. Smolukhowski proposed theories of the motion which could be used, for instance, to evaluate molecular diameters.

A rigorous mathematical theory of Brownian motion was developed by Wiener in 1923 (the Wiener process). This theory, however, makes no pretence of having any real connections with physical Brownian motion—no particle can follow a typical sample path of the Wiener process. In 1930, Ornstein and Uhlenbeck proposed yet another process, somewhat similar to the Wiener process but more closely related to physical reality. The foundation for their work was laid down 22 years earlier by P. Langevin, whose theory will be discussed briefly here.

The theory of Brownian motion developed by Einstein and Smoluchowski were not based on Newtonian mechanics. Langevin's approach, on the other hand, relies heavily on Newton's Second Law of Motion. In what follows we will give a brief account of Langevin's model.

Denote by m the mass of a Brownian particle suspended in liquid and let $v(t)$ be its velocity at time t. There are two forces acting on this particle. One is the frictional force exerted by liquid, which according to Stoke's law, is given by $-\beta v(t)$, where $\beta > 0$ is a constant which depends of the viscosity of the liquid and on the particle's mass and diameter.

The second force acting on the particle is due to the effect of molecular bombardment. It produces instantaneous random changes in the acceleration of the particle. Denote this force by $w(t)$. Then, according to Newton's Second Law of Motion, we have

$$m\Delta v(t) = -\beta v(t)\Delta t + \Delta w(t). \tag{3.7.1}$$

We assume that $w(0) = 0$ and that the following conditions hold:

(i) the stochastic process $\{w(t); t \geq 0\}$ has independent increments;
(ii) the distribution of $w(t+s) - w(t)$ depends only on s; and
(iii) the sample paths of $w(t)$ are continuous with probability 1.

But then, according to Proposition 3.6.1, $w(t)$ is a Brownian motion process, possibly with drift. Assuming then that $E\{w(t)\} \equiv 0$ (no drift) and putting $E\{w(t)\}^2 = \sigma^2 t$, we can write

$$w(t) = \sigma\xi(t), \tag{3.7.2}$$

where $\xi(t)$ is standard Brownian motion. With this equation (3.7.1) becomes

$$m\Delta v(t) = -\beta v(t)\Delta t + \sigma\Delta\xi(t).$$

Dividing by Δt and letting $\Delta t \to 0$, we obtain

$$m\frac{dv(t)}{dt} = -\beta v(t) + \sigma\frac{d\xi(t)}{dt}, \tag{3.7.3}$$

which is called "the Langevin equation."

The unpleasant thing here is that this equation contains the derivative of the Brownian motion process which, as we know very well, does not exist. Therefore, the equation does not formally make sense. The problem of finding a proper stochastic interpretation of the Langevin equation was resolved by Doob in 1942, in the following fashion. Write Equation (3.7.3) as

$$m\,dv(t) = -\beta v(t)\,dt + \sigma\,d\xi(t) \tag{3.7.4}$$

and try to give these differentials a suitable interpretation. We will interpret (3.7.4) to mean that, with probability 1,

$$m\int_a^b f(t)\,dv(t) = -\beta\int_a^b f(t)v(t)\,dt + \sigma\int_a^b f(t)\,d\xi(t) \tag{3.7.5}$$

for all $0 \le a < b < \infty$ and $f(\cdot)$ a nonrandom continuous function on $[a, b]$. As we shall see in the next section, all these integrals exist when the stochastic processes are continuous with probability 1.

Finally, if in (3.7.5) we put $a = 0$, $b = t$, and $f(t) = e^{\alpha t}$, where $\alpha = \beta/m$, we obtain

$$\int_0^t d(e^{\alpha s}v(s)) = \frac{\sigma}{m}\int_0^t e^{\alpha s}\,d\xi(s).$$

From this, assuming that $v(0) = v_0\sigma/m$ (constant) we readily deduce that, with probability 1,

$$v(t) = \frac{\sigma}{m}\left(v_0 e^{-\alpha t} + \int_0^t e^{-\alpha(t-s)}\,d\xi(s)\right). \tag{3.7.6}$$

Therefore, the velocity $v(t)$ of a Brownian particle is the stochastic process defined by (3.7.6).

Definition 3.7.1. The stochastic process $\{v(t); t \ge 0\}$, where $v(t)$ is given by (3.7.6), is called the Ornstein–Uhlenbeck process. Integrating by parts, (3.7.6) can also be written as

$$v(t) = \frac{\sigma}{m}\left(v_0 e^{-\alpha t} + \xi(t) - \alpha e^{-\alpha t}\int_0^t e^{\alpha s}\xi(s)\,ds\right). \tag{3.7.7}$$

From this definition, we obtain immediately that the average velocity of a Brownian particle (according to the Ornstein–Uhlenbeck model) is

$$E\{v(t)\} = \frac{\sigma}{m} v_0 e^{-\alpha t}. \tag{3.7.8}$$

To determine its covariance, let $t \geq s$. Then

$$\begin{aligned}\operatorname{Cov}(v(t), v(s)) &= \left(\frac{\sigma}{m}\right)^2 E\left[\left(e^{-\alpha t}\int_0^t e^{\alpha\tau}\, d\xi(\tau)\right)\left(e^{-\alpha s}\int_0^s e^{\alpha\tau}\, d\xi(\tau)\right)\right]\\ &= \left(\frac{\sigma}{m}\right)^2 e^{-\alpha(t+s)}\int_0^s e^{2\alpha\tau}\, d\tau\\ &= \left(\frac{\sigma}{m}\right)^2 e^{-\alpha(t+s)}\frac{e^{2\alpha s}-1}{2\alpha}\\ &= \left(\frac{\sigma}{m}\right)^2 \frac{e^{-\alpha(t-s)} - e^{-\alpha(t+s)}}{2\alpha}.\end{aligned} \tag{3.7.9}$$

For $t = s$, we obtain

$$\operatorname{Var}\{v(t)\} = \left(\frac{\sigma}{m}\right)^2 \frac{1 - e^{-2\alpha t}}{2\alpha}. \tag{3.7.10}$$

From (3.7.6), we readily deduce that the Ornstein–Uhlenbeck process is Gaussian. This follows from the fact that the integral

$$\int_0^t e^{-\alpha s}\, d\xi(s)$$

is the limit of sums

$$\sum_i e^{\alpha t_i}[\xi(t_{i+1}) - \xi(t_i)]$$

of independent normally distributed r.v.'s. All these results can be summarized as follows.

Proposition 3.7.1. *The Ornstein–Uhlenbeck process* $\{v(t); t \geq 0\}$ *is a Gaussian process with*

$$E\{v(t)\} = \frac{\sigma}{m} v_0 e^{-\alpha t},$$

$$\operatorname{Cov}(v(t), v(s)) = \left(\frac{\sigma}{m}\right)^2 \frac{e^{-\alpha|t-s|} - e^{-\alpha(t+s)}}{2\alpha}.$$

To summarize, the solution $v(t)$ of the Langevin equation (3.7.4) is called the Ornstein–Uhlenbeck process. It is a model for the velocity of a Brownian particle. The "derivative" of $\xi(t)$, which formally does not exist, is called a (Gaussian) white noise (the reason for this will be explained later). However, because

$$x(t) = x_0 + \int_0^t v(s)\,ds$$

is the displacement of the particle, $x(t)$ is the physical Brownian motion. Therefore, $v(t)$ is the physical noise.

By letting $t \to \infty$, we obtain from (3.7.8) and (3.7.10) that

$$E\{v(t)\} \to 0 \quad \text{and} \quad \operatorname{Var}\{v(t)\} \to \left(\frac{\sigma}{m}\right)^2 \frac{1}{2\alpha} = \rho. \tag{3.7.11}$$

Thus,

$$v(t) \xrightarrow{d} v(\infty) \quad \text{and} \quad v(\infty) \sim N(0, \rho).$$

Now, let U be an $N(0, \rho)$ r.v. independent of $\{v(t); t \geq 0\}$, and consider

$$Z(t) = e^{-\alpha t}\left(\frac{\sigma}{m}\int_0^t e^{\alpha s}\,d\xi(s) + U\right). \tag{3.7.12}$$

Then the following result holds.

Proposition 3.7.2. *$\{Z(t); t \geq 0\}$ given by (3.7.12) is a stationary Gaussian process with*

$$\operatorname{Cov}(s, t) = \rho e^{-\alpha|t-s|}. \tag{3.7.13}$$

Proof. The proof is straightforward. Normality follows from the fact that the r.v.'s in (3.7.12) are all normal. The covariance is obtained by direct computation. □

Proposition 3.7.3. *$\{Z(t); t \geq 0\}$ given by (3.7.12) is a Markov process with stationary transition probability*

$$P\{Z(s+t) \in B \mid Z(s) = x\}$$
$$= \frac{1}{\sqrt{2\pi\rho(1 - e^{-2\alpha t})}}\int_B \exp\left\{-\frac{(u - xe^{-\alpha t})^2}{2\rho(1 - e^{-2\alpha t})}\right\} du \tag{3.7.14}$$

and (a.s.) continuous sample functions.

Proof. According to (3.7.13), for all $0 \leq s < t$ and $u > 0$,

$$E\{(Z(t+u) - e^{-\alpha u}Z(t))Z(s)\} = \rho e^{-\alpha(t+u-s)} - \rho e^{-\alpha u}e^{-\alpha(t-s)} = 0.$$

Therefore, $Z(t+u) - e^{-\alpha u}Z(t)$ is independent of all $Z(s)$ if $s \leq t$. On the other hand, for any $0 < s_1 < \cdots < s_n < s$,

$$P\{Z(s+t) \in B \mid Z(s_1) = x_1, \ldots, Z(s_n) = x_n, Z(s) = x\}$$
$$= P\{Z(s+t) - e^{-\alpha t}Z(s) \in B - e^{-\alpha t}x \mid Z(s_1) = x_1, \ldots, Z(s_n) = x_n, Z(s) = x\}$$
$$= P\{Z(s+t) - e^{-\alpha t}Z(s) \in B - xe^{-\alpha t}\}.$$

But, the r.v. $Z(s+t) - e^{-\alpha t}Z(s)$ is normal with mean zero and

$$E(Z(s+t) - e^{-\alpha t}Z(s))^2 = E(Z(s+t) - e^{-\alpha t}Z(s))(Z(s+t) - e^{-\alpha t}Z(s))$$
$$= E\{(Z(s+t) - e^{-\alpha t}Z(s))Z(s+t)\} = \rho(1 - e^{-2\alpha t}).$$

The continuity of sample paths follows from (3.7.7). This completes the proof of the assertion. □

3.8. Stochastic Integration

Let $\{x(t); t \in [a,b]\}$ and $\{y(t); t \in [a,b]\}$ be real stochastic processes on a probability space $\{\Omega, \mathscr{B}, P\}$. The fundamental problem of stochastic integration is, roughly speaking, to give a sensible interpretation to the expression

$$\int_a^b x(s)\,dy(s). \tag{3.8.1}$$

If the process $\{y(t); t \in [a,b]\}$ is not of bounded variation, the integral (3.8.1) cannot be defined pathwise (i.e., for each $\omega \in \Omega$ separately) as an ordinary Stieltjes integral. Thus, a pathwise Stieltjes approach breaks down.

It was Itô (1944), extending the work of Wiener, who discovered a way of defining stochastic integrals of the form

$$\int_a^b h(s,\omega)\,d\xi(s,\omega), \tag{3.8.2}$$

where ξ is a Brownian process and h a suitable random function, defined on the same probability space.

EXAMPLE 3.8.1. To illustrate Itô's approach, let us determine the integral

$$\int_a^b \xi(s)\,d\xi(s),$$

where $\{\xi(s); s \in [a,b]\}$ is a standard Brownian motion. To this end, consider

$$a = t_{n0} < t_{n1} < \cdots < t_{nn} = b,$$

where

$$\max_i (t_{n,i+1} - t_{ni}) \to 0 \quad \text{as } n \to \infty.$$

We now define

$$\int_a^b \xi(s)\,d\xi(s) = \underset{n\to\infty}{\text{l.i.m.}} \sum_{i=0}^{n-1} \xi(t_{ni})[\xi(t_{n,i+1}) - \xi(t_{ni})].$$

Because

$$\sum_{i=0}^{n-1} \xi(t_{ni})[\xi(t_{n,i+1}) - \xi(t_{ni})]$$
$$= \frac{1}{2}\sum_{i=0}^{n-1} [\xi^2(t_{n,i+1}) - \xi^2(t_{ni})] - \frac{1}{2}\sum_{i=1}^{n-1} [\xi(t_{n,i+1}) - \xi(t_{ni})]^2$$
$$= \frac{1}{2}[\xi^2(b) - \xi^2(a)] - \frac{1}{2}\sum_{i=1}^{n-1} [\xi(t_{n,i+1}) - \xi(t_{ni})]^2,$$

we obtain, invoking Proposition 3.4.3, that

$$\int_a^b \xi(s)\,d\xi(s) = \tfrac{1}{2}[\xi^2(b) - \xi^2(a)] - \tfrac{1}{2}(b - a).$$

Many attempts have been made to generalize Itô's integral. The first generalization consisted of replacing the Brownian motion process with a square integrable martingale. Kunita and Watanabe (1967) introduced the concept of a local continuous martingale and a stochastic integral with respect to it. Now the latest result in the theory is that one cannot integrate with respect to anything more general than a semimartingale.

In this section, we discuss somewhat informally a version of the concept of a stochastic integral. Our aim is to present some basic facts about stochastic integration to justify the mathematical operation of the previous section in connection with the solution of the Langevin equation.

Let $\{\zeta(t); t \in [a,b]\}$ be a second-order random process (see Definition 1.5.8) such that $E\{\zeta(t)\} = 0$ and $h(\cdot)$ be a real nonrandom function on $[a,b]$. In the sequel, we will define the concept of Riemann stochastic integral of $h(\cdot)$ with respect to ζ,

$$\int_a^b h(s)\zeta(s)\,ds, \tag{3.8.3}$$

for a suitable class of functions $h(\cdot)$.

Let

$$a = t_{n0} < t_{n1} < \cdots < t_{nn} = b$$

be a partition of $[a,b]$ such that $\sup_i (t_{n,i+1} - t_{ni}) \to 0$ as $n \to \infty$. Consider the sequence of r.v.'s $\{U_n\}_1^\infty$ defined by

$$U_n = \sum_{i=0}^{n-1} h(t_{ni})\zeta(t_{ni})(t_{n,i+1} - t_{ni}).$$

If $\{U_n\}_1^\infty$ converges in the mean square to a r.v. U, i.e., if

$$E(U_n - U)^2 \to 0 \quad \text{as } n \to \infty, \tag{3.8.4}$$

then we call U the Riemann integral of $h(\cdot)$ with respect to ζ and write

$$U = \int_a^b h(s)\zeta(s)\,ds.$$

Note that (3.8.4) is equivalent to

$$E(U_n - U_k)^2 \to 0 \quad \text{as } \min\{k, n\} \to \infty.$$

Next we discuss conditions under which the limit (3.8.4) exists.

Proposition 3.8.1. *For the integral* (3.8.3) *to exist, it is sufficient that the integral*

$$\int_a^b \int_a^b h(s)h(t)C(s,t)\,ds\,dt \tag{3.8.5}$$

exists, where

$$C(s,t) = E\{\zeta(s)\zeta(t)\}.$$

Proof. Assume that the integral (3.8.5) exists and consider

$$E(U_n - U_k)^2 = EU_n^2 - 2EU_nU_k + EU_k^2.$$

After some straightforward calculations, we find that

$$\lim_{n\to\infty} EU_n^2 = \int_a^b \int_a^b h(s)h(t)C(s,t)\,ds\,dt, \tag{3.8.6}$$

$$\lim_{\min(n,k)\to\infty} EU_nU_k = \int_a^b \int_a^b h(s)h(t)C(s,t)\,ds\,dt. \tag{3.8.7}$$

Therefore,

$$\lim_{\min(n,k)\to\infty} E(U_n - U_k)^2 = 0,$$

which proves the sufficiency of (3.8.5). □

Corollary 3.8.1. *If we assume that the function $h(\cdot)$ is of bounded variation and $C(s,t)$ is continuous, then the integral*

$$\int_a^b \int_a^b C(s,t)\,dh(s)\,dh(t) \tag{3.8.8}$$

exists.

Now, consider

$$\begin{aligned}
&\sum_{i=0}^{n-1} h(t_{ni})[\zeta(t_{n,i+1}) - \zeta(t_{ni})] \\
&\quad = \sum_{i=0}^{n-1} h(t_{ni})\zeta(t_{n,i+1}) - \sum_{i=0}^{n-1} h(t_{ni})\zeta(t_{ni}) \\
&\quad = \sum_{j=1}^{n} h(t_{n,j-1})\zeta(t_{nj}) + h(b)\zeta(b) - h(a)\zeta(a) - \sum_{i=1}^{n} h(t_{ni})\zeta(t_{ni}) \\
&\quad = h(b)\zeta(b) - h(a)\zeta(a) - \sum_{n=1}^{n} \zeta(t_{ni})[h(t_{ni}) - h(t_{n,i-1})].
\end{aligned} \tag{3.8.9}$$

But, because by assumption the integral (3.8.8) exists, it follows from Proposition 3.8.1 that

$$\underset{n\to\infty}{\text{l.i.m.}} \sum_{i=1}^{n} \zeta(t_{ni})[h(t_{ni}) - h(t_{n,i+1})] = \int_a^b \zeta(s)\, dh(s)$$

exists. Therefore, because the limit of the right-hand side of (3.8.9) exists, so does the left-hand side and we have

$$\int_a^b h(s)\, d\zeta(s) = h(b)\zeta(b) - h(a)\zeta(a) - \int_a^b \zeta(s)\, dh(s). \tag{3.8.10}$$

Example 3.8.2. Let $\{\xi(t); t \geq 0\}$ be a standard Brownian motion process. For which functions $h(\cdot)$ does the integral

$$\int_0^t h(s)\xi(s)\, ds \tag{3.8.11}$$

exist in the mean square? For this integral to exist, the integral (3.8.5) must exist. Because, in this case,

$$C(s, t) = \min(s, t),$$

we have

$$\begin{aligned}\int_0^t \int_0^t \min(u, v) h(u) h(v)\, du\, dv &= \int_0^t h(v)\, dv \left\{ \int_0^v u h(u)\, du + v \int_v^t h(u)\, du \right\} \\ &= \int_0^t h(v)\, dv \int_0^v u h(u)\, du + \int_0^t v h(v)\, dv \int_v^t h(u)\, du \\ &= \int_0^t h(v)\, dv \int_0^v u h(u)\, du + \int_0^t h(u)\, du \int_0^u v h(v)\, dv \\ &= 2 \int_0^t \int_0^v h(v) u h(u)\, du\, dv.\end{aligned}$$

From this, we deduce that the integral (3.8.11) exists if $h(\cdot) \in L_1$. This, then, implies that the integral

$$\int_0^t h(s)\, d\xi(s)$$

also exists if $h(\cdot)$ is a function of bounded variation.

Problems and Complements

3.1. Determine the correlation function of a standard Brownian motion.

3.2. Let $\{\xi(t); t \geq 0\}$ be a standard Brownian motion. Determine the joint probability density of $\xi(t_1)$ and $\xi(t_2)$, $0 < t_1 < t_2 < 1$, given $\xi(1) = 0$.

3.3. If $\{\xi(t); t \geq 0\}$ is a standard Brownian motion, show that $X(t) = t\xi(t^{-1})$, if $t > 0$ and $X(0) = 0$, is also a standard Brownian motion.

3.4. The reflected Brownian motion is $\{|\xi(t)|; t \geq 0\}$, where $\xi(t)$ is a standard Brownian motion. Show that $|\xi(t)|$ is a Markov process. Find $E\{\xi|(t)|\}$ and $\text{Var}\{|\xi(t)|\}$.

3.5. Let $\{\xi(t); t \geq 0\}$ be a standard Brownian motion. Find the conditional probability density of $\xi(t)$ given that $\xi(t_1) = x_1$, $\xi(t_2) = x_2$, and $t_1 < t < t_2$.

3.6. Let $\{\xi_i(t); t \geq 0\}$, $i = 1, 2$, be two independent standard Brownian motions. Define

$$X(t) = \begin{cases} \xi_1(t), & t \geq 0 \\ \xi_2(-t), & t < 0. \end{cases}$$

Find the covariance function of $\{X(t); -\infty < t < \infty\}$.

3.7. Let $\{\xi(t); t \geq 0\}$ be a standard Brownian motion. Show that

$$X(t) = \frac{1}{c}\xi(ct)$$

is a separable Brownian motion.

3.8. Let $\{Z(t); 0 \leq t \leq 1\}$ be a Brownian bridge [see (3.1.8)]. Find $\text{Cov}(Z(t_1)Z(t_2))$. Define

$$Y(t) = (1 + t)Z\left(\frac{t}{1 + t}\right), \quad t \geq 0.$$

Show that $\{Y(t); t \geq 0\}$ is a standard Brownian motion.

3.9. Let $\{\xi(t); t \geq 0\}$ be a standard Brownian motion and consider

$$Z(t) = \int_0^t \xi(s)\, ds, \quad t \geq 0.$$

(i) What kind of process is $Z(t)$?
(ii) Determine its covariance function.

3.10. Let $\{\xi(t); t \geq 0\}$ be a standard Brownian motion and let τ_x be the first hitting time of x (see Section 3.2.1). Show that

$$\tau_x \stackrel{d}{=} x^2\tau_i.$$

3.11. Let $\chi(t) = \sup_{0 \leq s \leq t} \xi(s)$. Show that, for any $x < y$,

$$P\{\xi(t) \leq x, \chi(t) > y\} = P\{\xi(t) > 2y - x\}.$$

3.12. Show that $\{\xi(\tau_x + t) - \xi(\tau_x); t \geq 0\}$ is a Brownian motion independent of τ_x.

3.13. Let $\{\xi(t); t \geq 0\}$ be a standard Brownian motion and $\chi(t) = \sup\{\xi(s); 0 \leq s \leq t\}$. Show that, for $x < y$ and $y > 0$,

$$P\{\chi(t) \leq y, \xi(t) \in dx\} = \frac{1}{\sqrt{2\pi t}}\left\{\exp\left(-\frac{x^2}{2t}\right) - \exp\left(-\frac{(2y - x)^2}{2t}\right)\right\}dx.$$

3.14. Let $\{\xi(t); t \geq 0\}$ be a standard Brownian motion and τ_x be the first hitting time of a state $x > 0$. Define

$$Z(t) = \begin{cases} \xi(t) & \text{if } t < \tau_x \\ x & \text{if } t \geq \tau_x. \end{cases}$$

Determine $P\{Z(t) \leq y\}$, where $y < x$.

3.15. Show that the processes $|\xi(t)|$ and $\chi(t) - \xi(t)$, where $\xi(t)$ is a standard Brownian motion and $\chi(t) = \sup\{\xi(s); 0 \leq s \leq t\}$, are stochastically equivalent in the wide sense (see Definition 1.3.1).

3.16. Determine the probability

$$P\left\{\min_{t_0 \leq s \leq t_0+t} \xi(s) \leq 0 | \xi(t_0) = x_0\right\}.$$

[This is the probability that $\xi(s) = 0$ at least once in $(t_0, t_0 + t)$ given that $\xi(t_0) = x_0 > 0$.]

3.17. Let $\{\xi(t); t \geq 0\}$ be a standard Brownian motion. Find the probability that $\xi(t)$ has at least one zero in $(t_0, t_0 + t)$.

3.18. Let T^* be the largest zero in $(0, t)$; then

$$P\{T^* \leq u\} = \frac{2}{\pi} \text{arc sin} \sqrt{(u/t)}.$$

3.19. Let $\{\xi(t); t \geq 0\}$ be a standard Brownian motion; then

$$V(t) = e^{-t}\xi(e^{2t})$$

is called an Ornstein–Uhlenbeck process. Show that $V(t)$ is a Gaussian process. Determine its covariance function.

3.20. Let $\{\xi(t); t \geq 0\}$ be a standard Brownian motion and $h(\cdot)$ a real continuous function on $[0, \infty)$. Consider

$$X(t) = \int_0^t h(s)\xi(s)\, ds.$$

What kind of a process is $X(t)$? Determine its mean and variance.

3.21. Let $f(\cdot)$ and $h(\cdot)$ defined on $[0, \infty)$ be differentiable and $\{\xi(t); t \geq 0\}$ be a standard Brownian motion. Show that

$$E\left\{\int_a^b \int_a^b f(s)h(t)\, d\xi(s)\, d\xi(t)\right\} = \int_a^b h(t)f(t)\, dt.$$

3.22. (Continuation) Show that

$$E\left\{\int_a^b \int_c^d h(s)f(t)\, d\xi(s)\, d\xi(t)\right\} = 0 \quad \text{if } a < b \leq c < d.$$

3.23. Verify the identity

$$E\left\{\int_a^b \int_a^c f(s)h(t)\, d\xi(s)\, d\xi(t)\right\} = \int_a^{\min\{b,c\}} h(t)f(t)\, dt.$$

3.24. Let $\{\xi(t); t \geq 0\}$ be a standard Brownian motion. Find the mean and covariance of

$$e^{\alpha t} \int_0^t e^{-\alpha s}\, d\xi(s).$$

3.25. Let $\{\xi(t); t \geq 0\}$ be a standard Brownian motion. Show that, for all $t \geq 0$,

$$|\xi(t)| \stackrel{d}{=} \xi(t) - \inf_{0 \leq s \leq t} \xi(s).$$

3.26. Suppose that the stochastic process $\{X(t); t \geq 0\}$ is a solution of the stochastic differential equation

$$mX'(t) + \alpha X(t) = \xi(t), \qquad X(0) = v_0,$$

where m, α are positive constants and $\xi(t)$ is a standard Brownian motion. What kind of process is $X(t)$? Determine its mean and covariance function.

3.27. Let the process $\{X(t); t \geq 0\}$ be a solution of the following stochastic differential equation:

$$\alpha X'(t) + \beta X(t) = \xi'(t), \qquad X(0) = v_0,$$

where $\alpha, \beta > 0$ are constants, v_0 is a real number, and $\xi(t)$ is a standard Brownian motion. Define precisely what is meant by a solution of this equation. Find the solution satisfying the initial condition $X(0) = v_0$. Find the mean and covariance of $\xi(t)$.

CHAPTER 4

Gaussian Processes

4.1. Review of Elements of Matrix Analysis

In this section we present a review of some basic properties of the square matrices that will be needed throughout this chapter. It is expected that those who read this section have some background in matrix analysis.

Let $M = (\alpha_{ij})$ be a square matrix whose elements (or entries) α_{ij} are, unless otherwise stated, real numbers. If $\alpha_{ii} = 1$ for all i and $\alpha_{ij} = 0$ when $i \neq j$ the square matrix is called the "unit matrix" and denoted by I. As usual we will denote by M' the "transpose" of the matrix M, which is a square matrix $M' = (\beta_{ij})$, such that $\beta_{ij} = \alpha_{ji}$ for all i and j. From this definition, it follows that $I' = I$. When $M = M'$, the square matrix M is said to be "symmetric." The following properties of the "unary" operation $'$ are easy to verify:

$$(M')' = M, \qquad (M_1 + M_2)' = M_1' + M_2', \qquad (M_1 \cdot M_2)' = M_2' \cdot M_1'. \quad (4.1.1)$$

Let $M = (\alpha_{ij})$ be an arbitrary square matrix. In this book, we shall use the symbol $|M|$ to denote the determinant of M. One can verify that

$$|M'| = |M|. \quad (4.1.2)$$

If $|M| = 0$, the square matrix M is said to be "singular"; otherwise, we call it nonsingular.

Let M be a nonsingular square matrix; then there exists the unique square matrix, denoted by M^{-1}, such that

$$M^{-1}M = MM^{-1} = I. \quad (4.1.3)$$

The square matrix M^{-1} is called the "inverse" of M. It is well known that

$$(M^{-1})^{-1} = M, \qquad (M^{-1})' = (M')^{-1}. \quad (4.1.4)$$

The matrix M^{-1} is nonsingular; if M_1 and M_2 are nonsingular so is $M_1 \cdot M_2$

and

$$(M_1 \cdot M_2)^{-1} = M_2^{-1} \cdot M_1^{-1}. \tag{4.1.5}$$

In addition,

$$|M^{-1}| = |M|^{-1}. \tag{4.1.6}$$

We will use the notation $\mathbf{x}$ for a column vector. Then $\mathbf{x}'$ is a row vector. Now let $A = (a_{ij})$ be an $n \times n$ symmetric matrix. The matrix A is said to be "non-negative definite" if the quadratic form

$$\mathbf{x}'A\mathbf{x} = \sum_{i=1}^{n} \sum_{j=1}^{n} a_{ij}x_i x_j \geq 0. \tag{4.1.7}$$

If $\mathbf{x}'A\mathbf{x} = 0$ if and only if $\mathbf{x} = 0$, the square matrix A is said to be "positive definite." A symmetric matrix A is positive definite if and only if there exists a square matrix C such that $|C| > 0$ and

$$C'C = A. \tag{4.1.8}$$

If M is an $n \times n$ matrix, the equation

$$|M - \lambda \mathbf{I}| = 0 \tag{4.1.9}$$

is of degree n with respect to λ. The roots $\lambda_1, \ldots, \lambda_n$ of this equation are called the "eigenvalues" (or "spectral values") of the matrix M. It is a well-known result that every symmetric $n \times n$ matrix A has n real, not necessarily distinct, eigenvalues $\lambda_1, \ldots, \lambda_n$, and that

$$|A| = \prod_{i=1}^{n} \lambda_i. \tag{4.1.10}$$

If all $\lambda_i > 0$, the symmetric matrix A is positive definite.

Finally, a square matrix T is said to be "orthogonal" if

$$T' \cdot T = I. \tag{4.1.11}$$

The next result concerning orthogonal matrices is due to Scheffé. Let A be a symmetric matrix; then there exists an orthogonal matrix T such that

$$T'AT = D, \tag{4.1.12}$$

where $D = (d_{ij})$ is a diagonal matrix; that is, $d_{ij} = 0$ when $i \neq j$. In addition, $d_{ii} = \lambda_j$. Finally, let M be an arbitrary square matrix; then the product

$$M'M = A$$

is a symmetric and non-negative definite square matrix.

4.2. Gaussian Systems

Let

$$X_1, \ldots, X_n \tag{4.2.1}$$

be a sequence of real r.v.'s on a probability space $\{\Omega, \mathscr{B}, P\}$, such that $\{X_i\}_1^n \subset L_2\{\Omega, \mathscr{B}, P\}$. Denote by

$$\mu_j = E(X_j), \qquad \sigma_{ij} = E(X_i - \mu_i)(X_j - \mu_j). \tag{4.2.2}$$

The symmetric $n \times n$ matrix

$$A = (\sigma_{ij}) \tag{4.2.3}$$

is called the covariance matrix. Because

$$\sum_{i=1}^{n}\sum_{j=1}^{n} \sigma_{ij}x_i x_j = E\left(\sum_{i=1}^{n} (X_i - \mu_i)x_i\right)^2 > 0, \tag{4.2.4}$$

the matrix A is positive definite. Thus, $|A| > 0$ and, consequently $|A^{-1}| > 0$.

Denote by $\boldsymbol{\mu}' = (\mu_1, \ldots, \mu_n)$. The system of r.v.'s (4.2.1) is normally distributed if its joint probability density is

$$f(x_1, \ldots, x_n) = K \exp\{-\tfrac{1}{2}(\mathbf{x} - \boldsymbol{\mu})' A^{-1}(\mathbf{x} - \boldsymbol{\mu})\} \tag{4.2.5}$$

and A^{-1} is also a symmetric positive definite matrix.

Now let us determine K. Because A^{-1} is symmetric and positive definite, there exists a square $n \times n$ nonsingular matrix C such that

$$C'C = A^{-1}. \tag{4.2.6}$$

Hence,

$$\begin{aligned}(\mathbf{x} - \boldsymbol{\mu})' A^{-1}(\mathbf{x} - \boldsymbol{\mu}) &= (\mathbf{x} - \boldsymbol{\mu})' C'C(\mathbf{x} - \boldsymbol{\mu}) \\ &= (C(\mathbf{x} - \boldsymbol{\mu}))' \cdot C(\mathbf{x} - \boldsymbol{\mu}).\end{aligned}$$

Set

$$\mathbf{y} = C(\mathbf{x} - \boldsymbol{\mu}), \qquad \mathbf{x} = C^{-1}\mathbf{y} + \boldsymbol{\mu}. \tag{4.2.7}$$

The Jacobian J of this transformation has the form

$$J = \|C^{-1}\|, \tag{4.2.8}$$

where $\|C^{-1}\|$ indicates the absolute value of the determinant $|C^{-1}|$. From (4.2.6), we have

$$|C'||C| = |A^{-1}|, \qquad |C|^2 = |A|^{-1}. \tag{4.2.9}$$

Hence,

$$J = |A|^{1/2}. \tag{4.2.10}$$

Therefore,

$$\int_{-\infty}^{\infty} \cdots \int_{-\infty}^{\infty} f(x_1, \ldots, x_n)\,dx_1 \ldots dx_n = K \int_{-\infty}^{\infty} \cdots \int_{-\infty}^{\infty} e^{-1/2\mathbf{y}'\mathbf{y}} |A|^{1/2}\,dy_1 \ldots dy_n = 1,$$

or

$$K \cdot |A|^{1/2} \left\{\int_{-\infty}^{\infty} e^{-1/2x^2}\,dx\right\}^n = 1.$$

From this, we readily obtain that

$$K = \left(\frac{|A^{-1}|}{(2\pi)^n}\right)^{1/2}, \tag{4.2.11}$$

and, thus, $f(x_1,\ldots,x_n)$ has the form

$$f(x_1,\ldots,x_n) = \left(\frac{|A^{-1}|}{(2\pi)^n}\right)^{1/2} e^{-1/2(\mathbf{x}-\boldsymbol{\mu})'A^{-1}(\mathbf{x}-\boldsymbol{\mu})}. \tag{4.2.12}$$

Denote by

$$\varphi(t_1,\ldots,t_n) = E\left\{\exp\left(i\sum_1^n t_jX_j\right)\right\} = E\{e^{i\mathbf{t}'\mathbf{X}}\} \tag{4.2.13}$$

the characteristic function of the system (4.2.1), where

$$\mathbf{t}' = (t_1,\ldots,t_n), \qquad \mathbf{X}' = (X_1,\ldots,X_n). \tag{4.2.14}$$

The following proposition gives the form of the characteristic function (4.2.13) assuming that the joint probability density of the system (4.2.1) is (4.2.12).

Proposition 4.2.1. *The characteristic function* $\varphi(t_1,\ldots,t_n)$ under (4.2.12) is

$$\varphi(t_1,\ldots,t_n) = \exp(i\mathbf{t}'\boldsymbol{\mu} - \tfrac{1}{2}\mathbf{t}'A\mathbf{t}), \tag{4.2.15}$$

where $\boldsymbol{\mu}' = (\mu_1,\ldots,\mu_n)$.

PROOF. Because the symmetric matrix A is positive definite, there exists a square $n \times n$ nonsingular matrix L such that

$$LL' = A.$$

Then,

$$(\mathbf{x}-\boldsymbol{\mu})'A^{-1}(\mathbf{x}-\boldsymbol{\mu}) = [L^{-1}(\mathbf{x}-\boldsymbol{\mu})]'L^{-1}(\mathbf{x}-\boldsymbol{\mu}),$$

so that

$$\varphi(t_1,\ldots,t_n) = \left(\frac{|A^{-1}|}{(2\pi)^n}\right)^{1/2}$$
$$\times \int_{-\infty}^{\infty}\cdots\int_{-\infty}^{\infty} \exp\{i\mathbf{t}'\mathbf{x} - \tfrac{1}{2}[L^{-1}(\mathbf{x}-\boldsymbol{\mu})]'L^{-1}(\mathbf{x}-\boldsymbol{\mu})\}\,dx_1\ldots dx_n.$$

Now set

$$z = L^{-1}(\mathbf{x}-\boldsymbol{\mu}), \qquad \mathbf{x} = L\mathbf{z} + \boldsymbol{\mu}.$$

The Jacobian of this transformation is

$$J = \|L\| = (|A|)^{1/2}.$$

Therefore,

$$\varphi(t_1,\ldots,t_n) = \left(\frac{|A^{-1}|}{(2\pi)^n}\right)^{1/2} \int_{-\infty}^{\infty} \cdots \int_{-\infty}^{\infty} \exp[i\mathbf{t}'(L\mathbf{z}+\boldsymbol{\mu}) - \tfrac{1}{2}\mathbf{z}'\mathbf{z}]\,|A|^{1/2}\,dz_1\ldots dz_n$$

$$= \frac{\exp(i\mathbf{t}'\boldsymbol{\mu})}{(2\pi)^{n/2}} \int_{-\infty}^{\infty} \cdots \int_{-\infty}^{\infty} \exp(-\tfrac{1}{2}\mathbf{z}'\mathbf{z} + i\mathbf{t}'L\mathbf{z})\,dz_1\ldots dz_n,$$

or if we set $\mathbf{u}' = \mathbf{t}'L$, where $\mathbf{u}' = (u_i,\ldots,u_n)$,

$$= \frac{\exp(i\mathbf{t}'\boldsymbol{\mu})}{(2\pi)^{n/2}} \left\{\prod_{j=1}^{n} \int_{-\infty}^{\infty} \exp(-\tfrac{1}{2}z_j^2 + iu_jz_j)\,dz_j\right\}$$

$$= \frac{\exp(i\mathbf{t}'\boldsymbol{\mu})}{(2\pi)^{n/2}} \left(\prod_{j=1}^{n} \exp(-\tfrac{1}{2}u_j^2)\right) \prod_{j=1}^{n} \left(\int_{-\infty}^{\infty} \exp[-\tfrac{1}{2}(z_j - iu_j)^2]\,dz_j\right)$$

$$= \frac{\exp(i\mathbf{t}'\boldsymbol{\mu})}{(2\pi)^{n/2}} [\exp(-\tfrac{1}{2}\mathbf{u}'\mathbf{u})](2\pi)^{n/2} = \exp(i\mathbf{t}'\boldsymbol{\mu} - \tfrac{1}{2}\mathbf{u}'\mathbf{u})$$

$$= \exp(i\mathbf{t}'\boldsymbol{\mu} - \tfrac{1}{2}\mathbf{t}'LL'\mathbf{t}) = \exp(i\mathbf{t}'\boldsymbol{\mu} - \tfrac{1}{2}\mathbf{t}'A\mathbf{t}).$$

This proves the proposition. □

4.3. Some Characterizations of the Normal Distribution

A system of r.v.'s

$$X_1,\ldots,X_n \tag{4.3.1}$$

defined on a probability space $\{\Omega,\mathscr{B},P\}$ is said to be normal (or Gaussian) if $\{X_i\} \subset L_2\{\Omega,\mathscr{B},P\}$ and its joint probability density is given by Equation (4.2.12). We will show several properties of a Gaussian system.

Let

$$Z = \sum_{i=1}^{n} \alpha_i X_i = \boldsymbol{\alpha}'\mathbf{X}, \tag{4.3.2}$$

where $\alpha_1,\ldots,\alpha_n$ are constants and

$$\mathbf{X}' = (X_1,\ldots,X_n), \qquad \boldsymbol{\alpha}' = (\alpha_1,\ldots,\alpha_n),$$

be a linear combination. We want to show that the r.v. Z is also normally distributed with

$$E\{Z\} = \sum_{i=1}^{n} \alpha_i\mu_i = \boldsymbol{\alpha}'\boldsymbol{\mu},$$

$$\operatorname{Var}\{Z\} = \sum_{i=1}^{n}\sum_{j=1}^{n} \alpha_i\alpha_j\sigma_{ij}.$$

See (4.2.2) for definition of the notation. This will be demonstrated using characteristic functions.

According to (4.2.13), the characteristic function of $\mathbf{X}$ is

$$\varphi(t_1,\ldots,t_n) = E\{\exp(i\mathbf{t}'\mathbf{X})\} = E\left\{\exp\left(i\sum_{k=1}^{n} t_k X_k\right)\right\}$$

$$= \exp\left(i\sum_{k=1}^{n} t_i\mu_i - \frac{1}{2}\sum_{k=1}^{n}\sum_{j=1}^{n} t_k t_j \sigma_{kj}\right). \tag{4.3.3}$$

On the other hand,

$$E\{e^{iuZ}\} = E\left\{\exp\left(iu\sum_{j=1}^{n}\alpha_j X_j\right)\right\}. \tag{4.3.4}$$

From this and (4.3.3), we see that (4.3.4) is a particular case of (4.3.3) obtained by setting

$$t_i = u\alpha_i.$$

Hence,

$$E\{e^{iuZ}\} = \exp\left(iu\sum_{i=1}^{n}\alpha_i\mu_i - \frac{1}{2}u^2\sum_{k=1}^{n}\sum_{i=1}^{n}\alpha_k\alpha_i\sigma_{ki}\right) \tag{4.3.5}$$

and this is the characteristic function of the normal distribution with mean $\sum_1^n \alpha_i\mu_i$ and variance $\sum_1^n\sum_1^n \alpha_k\alpha_j\sigma_{kj}$. This proves our assertion.

Does the converse hold? In other words, if every linear combination of (4.3.1) is normally distributed, does this imply that the system $(X_1,\ldots,X_n)$ has a normal joint distribution? The answer is affirmative as we will show next. To this end, assume that the Z [see (4.3.2)] is normally distributed; then its characteristic function is (4.3.5). Now, consider

$$\varphi(t_1,\ldots,t_n) = E\left\{\exp\left(i\sum_1^n t_j X_j\right)\right\}.$$

This clearly can be obtained from (4.3.5) if in (4.3.5) we set $u\alpha_j = t_j$. Hence,

$$\varphi(t_1,\ldots,t_n) = \exp\left(i\sum_1^n t_j u_j - \frac{1}{2}\sum_1^n\sum_1^n t_j t_k \sigma_{jk}\right)$$

$$= \exp(i\mathbf{t}'\boldsymbol{\mu} - \tfrac{1}{2}\mathbf{t}'A\mathbf{t}). \tag{4.3.6}$$

According to Proposition 4.2.1, this is the characteristic function of the normal distribution. This proves our assertion.

The previous two results can be formulated as follows. A system of n r.v.'s from $L_2\{\Omega,\mathscr{B},P\}$ is Gaussian if and only if every linear combination of its members is normally distributed.

4.4. The Gaussian Process

Let $\{\xi(t); t\in T\}$ be a real-valued stochastic process on $\{\Omega,\mathscr{B},P\}$ such that

$$\{\xi(t); t\in T\} \subset L_2\{\Omega,\mathscr{B},P\}.$$

Set

$$\mu(t) = E\{\xi(t)\} \tag{4.4.1}$$

and

$$C(s,t) = E(\xi(s) - \mu(s))(\xi(t) - \mu(t)), \tag{4.4.2}$$

the mean and the covariance function, respectively, of the process $\xi(t)$.

Definition 4.4.1. The stochastic process $\{\xi(t); t \in T\}$ is said to be a "Gaussian process" if and only if for each $\{t_1, \ldots, t_n\} \subset T$, $n = 1, 2, \ldots$, $(\xi(t_1), \ldots, \xi(t_n))$ is jointly normal.

The covariance function (4.4.2) possesses the following properties:

(i) $C(s,t) = C(t,s)$;
(ii) for any $\{t_1, \ldots, t_n\} \subset T$, $n = 1, 2, \ldots$, and any real numbers $u_1, \ldots, u_n$,

$$\sum_{i=1}^{n} \sum_{j=1}^{n} u_i u_j C(t_i, t_j) \geq 0. \tag{4.4.3}$$

In other words, the covariance function is non-negative definite [see (4.2.4)].

We now show that the symmetry and non-negative definiteness are sufficient conditions for there to exist a Gaussian random process with a specified covariance function. Let $C(s,t)$ be a symmetric non-negative definite function. Consider $\{t_i\}_1^n \subset T$ such that $t_1 < \cdots < t_n$ and specify that $\xi(t_1), \ldots, \xi(t_n)$ is jointly Gaussian with covariance matrix $[C(t_i, t_j)]$, $i, j = 1, \ldots, n$. Then any subfamily is also jointly Gaussian. Thus, this family satisfies the Kolmogorov consistency requirement. It now remains to apply the Kolmogorov theorem.

Example 4.4.1. Let $\{\xi(t); t \geq 0\}$ be a Gaussian process with

$$E\{\xi(t)\} = 0 \quad \text{and} \quad C(s,t) = \min(s,t). \tag{4.4.4}$$

From the definition, it follows that, for any $0 < t_1 < \cdots < t_n$, the system of r.v.'s $(\xi(t_1), \ldots, \xi(t_n))$ has a normal distribution with the covariance matrix

$$A = \begin{bmatrix} t_1, t_1, \ldots, t_1 \\ t_1, t_2, \ldots, t_2 \\ \vdots \\ t_1, t_2, \ldots, t_n \end{bmatrix}. \tag{4.4.5}$$

One can show that this matrix is positive definite. Let us now prove that the process $\xi(t)$ has independent increments. First, from (4.4.4), it readily follows that, for any $0 \leq s < t$,

$$\mathrm{Var}\{\xi(t) - \xi(s)\} = t - s. \tag{4.4.6}$$

On the other hand, for any $0 \leq s_1 < s_2 < s_3 < s_4$.

$$E(\xi(s_2) - \xi(s_1))(\xi(s_4) - \xi(s_3)) = s_2 - s_2 - s_1 + s_1 = 0,$$

which clearly shows that the r.v.'s

$$\xi(t_1), \xi(t_2) - \xi(t_1), \ldots, \xi(t_n) - \xi(t_1) \tag{4.4.7}$$

are uncorrelated. On the other hand, because every linear combination of a Gaussian system is also normally distributed, each $\xi(t_i) - \xi(t_{i-1})$ is a normal r.v. with mean zero and variance given by (4.4.6). Hence, (4.4.7) are jointly Gaussian and, because they are uncorrelated, they must be independent. Therefore, the process $\{\xi(t); t \geq 0\}$ has independent increments, which are normally distributed, and the distribution of $\xi(t+s) - \xi(t)$ does not depend on t. Thus, $\{\xi(t); t \geq 0\}$ is clearly a standard Brownian motion. From this, we may conclude that a standard Brownian motion is a Gaussian stochastic process satisfying condition (4.4.4).

Remark 4.4.1. If $(X_1, \ldots, X_n)$ is a jointly Gaussian collection, then these r.v.'s are mutually independent if and only if the covariance matrix of the system is diagonal. Moreover, because the covariance matrix is diagonal if and only if each couple of X_i and X_j is independent, a Gaussian collection is mutually independent if and only if it is pairwise independent.

4.5. Markov Gaussian Process

In this section, the following result is required: Let $(X_1, \ldots, X_n)$ be a Gaussian system of r.v.'s with joint probability density

$$f(x_1, \ldots, x_n) = \left(\frac{|A^{-1}|}{(2\pi)^n}\right)^{1/2} \exp\left(-\frac{1}{2}\sum_{i=1}^{n}\sum_{j=1}^{n} a_{ij}x_ix_j\right), \tag{4.5.1}$$

where $A^{-1} = [a_{ij}]$ is a symmetric positive definite matrix. The conditional probability density of X_n given that $X_1 = x_1, \ldots, X_{n-1} = x_{n-1}$, denoted by $f(x_n|x_1, \ldots, x_{n-1})$, is defined by

$$\frac{f(x_1, \ldots, x_n)}{\int_{-\infty}^{\infty} f(x_1, \ldots, x_n)\,dx_n} = \frac{\exp(-\frac{1}{2}\sum_{i=1}^{n}\sum_{j=1}^{n} a_{ij}x_ix_j)}{\int_{-\infty}^{\infty}\exp(-\frac{1}{2}\sum_{i=1}^{n}\sum_{j=1}^{n} a_{ij}x_ix_j)\,dx_n}. \tag{4.5.2}$$

After some simple algebraic calculations, one readily obtains that

$$\sum_{i=1}^{n}\sum_{j=1}^{n} a_{ij}x_ix_j = \sum_{i=1}^{n-1}\sum_{j=1}^{n-1} a_{ij}x_ix_j + x_n^2 a_{nn} + 2x_n\sum_{i=1}^{n-1} a_{in}x_i. \tag{4.5.3}$$

From this and (4.5.2), we have

$$\begin{aligned} f(x_n|x_1, \ldots, x_{n-1}) &= \frac{\exp(-\frac{1}{2}a_{nn}x_n^2 - x_n\sum_{i=1}^{n-1} a_{in}x_i)}{\int_{-\infty}^{\infty}\exp(-\frac{1}{2}a_{nn}x_n^2 - x_n\sum_{i=1}^{n-1} a_{in}x_i)\,dx_n} \\ &= \frac{\exp\{-\frac{1}{2}a_{nn}[x_n + \sum_{i=1}^{n-1}(a_{in}/a_{nn})x_i]^2\}}{\int_{-\infty}^{\infty}\exp\{-\frac{1}{2}a_{nn}[x_n + \sum_{i=1}^{n-1}(a_{in}/a_{nn})x_i]^2\}\,dx_n} \\ &= C\exp\left\{-\frac{1}{2}a_{nn}\left[x_n + \sum_{i=1}^{n-1}\left(\frac{a_{in}}{a_{nn}}\right)x_i\right]^2\right\}, \end{aligned} \tag{4.5.4}$$

where C is a norming constant independent of $x_1, \dots, x_{n-1}$. This is clearly a normal probability density with mean

$$-\sum_{i=1}^{n-1}\left(\frac{a_{in}}{a_{nn}}\right)x_i.$$

Thus,

$$E\{X_n|X_1 = x_1, \dots, X_{n-1} = x_{n-1}\} = -\sum_{i=1}^{n-1}\left(\frac{a_{in}}{a_{nn}}\right)x_i. \tag{4.5.5}$$

Let $\{\xi(t); t \in T\}$ be a Gaussian process with

$$E(\xi(t)) = 0 \quad \text{and} \quad C(s,t) = E(\xi(s)\xi(t)),$$

$s, t \in T$, which is also a Markov process. We want to show that the following result holds.

Proposition 4.5.1. *The covariance matrix $C(s,t)$ of a Markov Gaussian process with mean value 0 satisfies the condition that, for any t_1, t_2, $t_3 \in T$ with $t_1 < t_2 < t_3$,*

$$C(t_1, t_3) = \frac{C(t_1,t_2)C(t_2,t_3)}{C(t_2,t_2)}. \tag{4.5.6}$$

Proof. Because $\xi(t)$ is a Markov process, it follows from (4.5.5) that, for any $s < t$ from T,

$$E\{\xi(t)|\xi(s) = x\} = -\left(\frac{a_{n-1,n}}{a_{nn}}\right)x. \tag{4.5.7}$$

To determine $(a_{n-1,n}/a_{nn})$, consider

$$f_{s,t}(x_1, x_2) = \frac{(|A^{-1}|)^{1/2}}{2\pi}\exp(-\tfrac{1}{2}\mathbf{x}'A^{-1}\mathbf{x}),$$

where

$$A^{-1} = \begin{bmatrix} C(s,s) & C(s,t) \\ C(t,s) & C(t,t) \end{bmatrix}^{-1}$$

$$= \frac{1}{C(s,s)C(t,t) - C^2(s,t)}\begin{bmatrix} C(t,t) & -C(s,t) \\ -C(s,t) & C(s,s) \end{bmatrix}.$$

From this and (4.5.7), we conclude that

$$E\{\xi(t)|\xi(s) = x\} = \left(\frac{C(s,t)}{C(s,s)}\right)x. \tag{4.5.8}$$

Finally, according to (1.5.6), we can write

$$C(t_1,t_3) = E(\xi(t_1)\xi(t_3)) = E\{E(\xi(t_1)\xi(t_3)|\xi(t_2))\}$$
$$= E\{E(\xi(t_1)|\xi(t_2))E(\xi(t_3)|\xi(t_2))\}.$$

Now, invoking (4.5.8), we have

$$C(t_1,t_3) = E\left\{\left(\frac{C(t_1,t_2)}{C(t_2,t_2)}\right)\xi(t_2)\cdot\left(\frac{C(t_2,t_3)}{C(t_2,t_2)}\right)\xi(t_2)\right\}$$

$$= \left(\frac{C(t_1,t_2)C(t_2,t_3)}{(C(t_2,t_2))^2}\right)E(\xi(t_2))^2,$$

which proves the assertion. □

According to the last proposition, the covariance function of every Gaussian Markov process with mean zero must satisfy Equation (4.5.6). One can show that the converse is also true. In other words, any mean zero Gaussian process whose covariance function satisfies Equation (4.5.6) must have the Markov property.

To prove the last statement, we need to observe that a zero mean Gaussian process is Markov if and only if, for any $t_1 < t_2 < \cdots < t_n$, $n = 2, 3, \ldots$,

$$E\{\xi(t_n)|\xi(t_1) = x_1, \ldots, \xi(t_{n-1}) = x_{n-1}\} = E\{\xi(t_n)|\xi(t_{n-1}) = x_{n-1}\}. \quad (4.5.9)$$

Necessity of this condition is obvious. To prove its sufficiency, note that from (4.5.5) and (4.5.7), it follows that

$$a_{1n} = a_{2n} = \cdots = a_{n-2,n} = 0,$$

so that, according to (4.5.4),

$$f(x_n|x_1,\ldots,x_{n-1}) = C\exp\left\{-\frac{1}{2}a_{nn}\left[x_n + \left(\frac{a_{n-1,n}}{a_{nn}}\right)x_{n-1}\right]^2\right\}.$$

On the other hand, it is not difficult to verify that

$$f(x_n|x_{n-1}) = C\exp\left\{-\frac{1}{2}a_{nn}\left[x_n + \left(\frac{a_{n-1,n}}{a_{nn}}\right)x_{n-1}\right]^2\right\},$$

which proves the assertion.

Let us now show that (4.5.6) implies (4.5.9). Note that from (4.5.6), it follows that, for any $1 \le k \le n-1$,

$$C(t_k,t_n) = \frac{C(t_k,t_{n-1})C(t_{n-1},t_n)}{C(t_{n-1},t_{n-1})},$$

which implies that

$$E\left\{\left[\xi(t_n) - \left(\frac{C(t_{n-1},t_n)}{C(t_{n-1},t_{n-1})}\right)\xi(t_{n-1})\right]\xi(t_k)\right\} = 0,$$

for all $k = 1, \ldots, n-1$. Thus, because these are normally distributed r.v.'s,

$$\xi(t_n) - \left(\frac{C(t_{n-1},t_n)}{C(t_{n-1},t_{n-1})}\right)\xi(t_{n-1})$$

is independent of $\xi(t_1), \ldots, \xi(t_{n-1})$. Therefore,

$$E\left\{\xi(t_n) - \left(\frac{C(t_{n-1}, t_n)}{C(t_{n-1}, t_{n-1})}\right)\xi(t_{n-1}) | \xi(t_1) = x_1, \ldots, \xi(t_{n-1}) = x_{n-1}\right\} = 0,$$

or

$$E\{\xi(t_n) | \xi(t_1) = x_1, \ldots, \xi(t_{n-1}) = x_{n-1}\} = \left(\frac{C(t_{n-1}, t_n)}{C(t_{n-1}, t_{n-1})}\right) x_{n-1}$$
$$= E\{\xi(t_n) | \xi(t_{n-1}) = x_{n-1}\}.$$

4.6. Stationary Gaussian Process

Let $\{\xi(t); t \in T\}$ be a real-valued Gaussian process such that

$$E\{\xi(t)\} = 0, \quad \forall t \in T. \tag{4.6.1}$$

The marginal distributions of such a stochastic process are completely determined by its covariance function

$$C(s, t) = E\{\xi(s)\xi(t)\}. \tag{4.6.2}$$

According to Definition 1.5.6, marginal distributions of a strictly stationary stochastic process are invariant under time-shifts. When is a Gaussian zero-mean process strictly stationary? From the previous remarks, the condition should be on $C(s, t)$. To simplify our notation, assume that $T = [0, \infty)$; then we have the following result.

Proposition 4.6.1. *A real-valued Gaussian process $\{\xi(t); t \geq 0\}$ with zero-mean and covariance function $C(s, t)$ is strictly stationary if and only if, for any $0 \leq s \leq t$,*

$$C(s, t) = C(0, t - s) = R(t - s). \tag{4.6.3}$$

PROOF. If $\xi(t)$ is strictly stationary, then

$$E\{\xi(s)\xi(t)\} = E\{\xi(s + \tau)\xi(t + \tau)\}$$

or

$$C(s, t) = C(s + \tau, t + \tau).$$

By taking $\tau = -s$ we have

$$C(s, t) = C(0, t - s) = R(t - s).$$

On the other hand, if (4.6.3) holds, the characteristic function of $(\xi(t_1), \ldots, \xi(t_n))$ $(0 \leq t_1 < \cdots < t_n)$ is

$$\varphi(u_1, \ldots, u_n) = \exp\left[-\frac{1}{2}\sum_{i=1}^{n}\sum_{j=1}^{n} u_i u_j R(t_i - t_j)\right]$$

[see (4.3.6)]. But this is also the characteristic function of $(\xi(t_1 + \tau), \ldots, \xi(t_n + \tau))$. This proves the assertion. □

Corollary 4.6.1. *A zero-mean wide sense stationary Gaussian process is also strictly stationary.*

When is a stationary Gaussian process also a Markov process? The following proposition due to Doob provides an answer to this question.

Proposition 4.6.2. *A stochastically continuous stationary Gaussian process $\{\xi(t); t \geq 0\}$ with*

$$E\{\xi(t)\} = 0 \quad \text{and} \quad E\{\xi(t)\}^2 = 1$$

is Markov if and only if

$$R(s) = e^{-y|s|}, \quad y > 0. \tag{4.6.4}$$

PROOF. If $\xi(t)$ is stationary and Gaussian, then according to the condition of the proposition, (4.6.3) and (4.5.6), we have that

$$R(s + t) = R(s)R(t). \tag{4.6.5}$$

Now, because $\xi(t)$ is stochastically continuous, $R(u)$ is continuous at $u = 0$ (prove this); then, because $R(0) = 1$, we have

$$R(t + h) - R(t) = R(t)[R(h) - 1] \to 0 \quad \text{as } h \to 0.$$

This shows that $R(t)$ is continuous at every $t \in R$. Then the unique solution of the Cauchy equation (4.6.5) for $t > 0$ is

$$R(t) = e^{-yt}, \quad y > 0.$$

On the other hand, $R(-t) = R(t)$, which proves necessity of condition (4.6.4).

Conversely, if the covariance of a stationary zero-mean and variance 1 Gaussian process is of the form (4.6.4), then condition (4.5.6) holds. This implies that condition (4.5.9) must also hold, which proves the proposition. □

Problems and Complements

4.1. Let $\mathbf{X} = (X_1, \ldots, X_4)$ be a Gaussian system of r.v.'s with covariance matrix

$$A = \begin{bmatrix} 15 & 3 & 1 & 0 \\ 3 & 16 & 6 & -2 \\ 1 & 6 & 4 & 1 \\ 0 & -2 & 1 & 3 \end{bmatrix}.$$

Determine its probability density for $f(x_1, \ldots, x_4)$ if

$$\mu_1 = 1, 0, \qquad \mu_2 = 0, \qquad \mu_3 = -1, 0, \qquad \mu_4 = 1.$$

4.2. The probability density of the system $\mathbf{X} = (X_1, X_2, X_3)$ is given by

$$f(x_1, x_2, x_3) = \frac{1}{16}\sqrt{\frac{3}{\pi^3}}\exp\left(-\frac{1}{8}[2x_1^2 + 4x_2^2 - 2x_2(x_3 + 5) + (x_3 + 5)^2]\right).$$

Find μ_1, μ_2, μ_3, and the covariance matrix.

4.3. Let (X_1, X_2, X_3, X_4) be a Gaussian system of r.v.'s with $E\{X_i\} = 0$, $i = 1, \ldots, 4$. Show that

$$E\{X_1 X_2 X_3 X_4\} = \sigma_{12}\sigma_{34} + \sigma_{13}\sigma_{24} + \sigma_{14}\sigma_{23},$$

where $\sigma_{ij} = E\{X_i X_j\}$.

4.4. Let $\{X_i\}_1^n$ be an i.i.d. sequence of r.v.'s with each $X_i \sim N(0, 1)$. Let $\{a_i\}_1^n$ and $\{b_i\}_1^n$ be real numbers. Show that

$$Y = \sum_{i=1}^{n} a_i X_i \quad \text{and} \quad Z = \sum_{i=1}^{n} b_i X_i$$

are independent r.v.'s if and only if $\sum_{i=1}^{n} a_i b_i = 0$

4.5. Let A be an $k \times k$ square symmetric nonsingular positive definite matrix. Show that there exists a nonsingular square matrix Γ such that $A = \Gamma\Gamma^{-1}$.

4.6. Let X and Y be i.i.d. r.v.'s with finite variance σ^2. If there exist constants α and β such that $\alpha^2 + \beta^2 = 1$, $\alpha \cdot \beta \neq 0$, and

$$\alpha X + \beta Y \stackrel{d}{=} X,$$

then $X \sim N(0, \sigma)$.

4.7. Let $\{\xi(t); t \geq 0\}$ be a standard Brownian motion. Show that

a. $E\{\xi(t_2) | \xi(t_1) = x_1\} = x_1, \quad t_1 < t_2;$

b. $E\{\xi(t_2) | \xi(t_1) = x_1, \xi(t_3) = x_3\} = x_1 + \left(\dfrac{x_3 - x_1}{t_3 - t_1}\right)(t_2 - t_1),$

for all $0 < t_1 < t_2 < t_3$.

4.8. Let X_1 and X_2 be i.i.d. r.v.'s $X_i \sim N(0, \sigma)$. Consider

$$\xi(t) = X_1 \cos \lambda t + X_2 \sin \lambda t.$$

Find the mean and the covariance function of the stochastic process $\{\xi(t); t \geq 0\}$. Show that the process is Gaussian.

4.9. Let $\{X(t); t \geq 0\}$ be a Gaussian process with

$$E\{X(t)\} = 0 \quad \text{and} \quad E\{X(t)X(t + \tau)\} = C(\tau).$$

Find the covariance function of $\{\eta(t); t \geq 0\}$, where

$$\eta(t) = X(t)X(s + t).$$

4.10. Show that a Brownian motion process $\{\xi(t); t \geq 0\}$ is a Gaussian process.

4.11. Construct (X, Y) which is not normal, but X and Y are normal.

4.12. Let $\{X_n\}_1^\infty$ be a sequence of r.v.'s $X_n \sim N(\mu_n, \sigma_n)$ such that $X_n \xrightarrow{\text{q.m.}} X$. If $\operatorname{Var}\{X\} = \sigma^2 > 0$, show that $X \sim N(\mu, \sigma)$.

4.13. Let X and θ be independent r.v.'s, θ uniform in $[\theta, 2\pi]$, and

$$f_X(x) = \begin{cases} 2x^3 e^{-1/2x^4}, & x \geq 0 \\ 0, & x < 0. \end{cases}$$

Let $\{\xi(t); t \geq 0\}$ be stochastic process defined by

$$\xi(t) = X^2 \cos(2\pi t + \theta).$$

Show that $\xi(t)$ is Gaussian.

4.14. Let $\{X(t); t \geq 0\}$ be a stochastic process specified by

$$X(t) = e^{-t}\xi(e^{2t}),$$

where $\xi(t)$ is a standard Brownian motion process. Show that $X(t)$ is a Gaussian Markov process.

4.15. Let $\{\xi(s); s \geq 0\}$ be a standard Brownian motion. Consider

$$X(t) = \int_0^t \xi(s)\, ds.$$

What kind of process is $\{X(t); t \geq 0\}$? Determine its covariance function.

4.16. Let $\{\xi(t); t \geq 0\}$ be a Gaussian process specified by

$$\xi(t) = X \cos(2\pi t + U),$$

where U is uniform, in $[0, 2\pi]$ independent of x. Determine the distribution of X.

4.17. Let $\{\xi(t); t \geq 0\}$ be a Gaussian process with $E\{\xi(t) = 0\}$ and suppose that $C(s, t) = E\{\xi(s)\xi(t)\}$ is continuous in s and t. Show that

$$X(t) = \int_0^t \alpha(s)\xi(s)\, ds$$

is also a Gaussian process where $\alpha(\cdot)$ is a continuous function.

4.18. Assume that $\{\xi(t); t \geq 0\}$ is a stationary Gaussian process. Show that

$$\int_t^{t+\tau} \xi(s)\, ds = Z(t)$$

is also a stationary Gaussian.

4.19. Let $\{\xi(s); s \geq 0\}$ be a Markov Gaussian process. Show that

$$X(t) = \int_t^{t+\tau} \xi(s)\, ds$$

is not Markovian.

4.20. Let $\{\xi(t); t \geq 0\}$ be a stationary Markov Gaussian process with continuous covariance function $C(s)$. Determine $C(s)$.

4.21. Complete the proof of Proposition 4.6.2.

CHAPTER 5

L_2 Space

5.1. Definitions and Preliminaries

In many applications of the theory of stochastic processes, an important role is played by families of square integrable (second-order) r.v.'s. In this section, we give some basic definitions and prove some fundamental inequalities involving second-order complex-valued r.v.'s.

Let $\{\Omega, \mathscr{B}, P\}$ be a probability space and Z a complex-valued r.v., i.e.,

$$Z = X + iY,$$

where X and Y are real-valued r.v.'s on $\{\Omega, \mathscr{B}, P\}$. As usual, we write $X = \mathrm{Re}\{Z\}$ for the "real" part and $Y = \mathrm{Im}\{Z\}$ for the "imaginary" part of the complex variable Z. The conjugate $\bar{Z}$ of Z is defined by

$$\bar{Z} = X - iY$$

and the modulus $|Z|$ is given by

$$|Z|^2 = (X^2 + Y^2).$$

The obvious connection between the conjugate and the modulus of Z is

$$Z \cdot \bar{Z} = |Z|^2.$$

Definition 5.1.1. A complex-valued r.v. Z on $\{\Omega, \mathscr{B}, P\}$ is called second order if

$$E|Z|^2 < \infty.$$

The family of all such r.v.'s is denoted by

$$L_2 = L_2\{\Omega, \mathscr{B}, P\}.$$

Proposition 5.1.1. *For any* $Z_1, Z_2 \in L_2$,

$$|E\overline{Z}_1 \cdot Z_2|^2 \leq E|Z_1|^2 \cdot E|Z_2|^2. \tag{5.1.1}$$

PROOF. This is Schwarz's inequality. It follows readily from

$$\begin{aligned} 0 &\leq E\left|Z_1 - \left(\frac{EZ_1 \cdot \overline{Z}_2}{E|Z_2|^2}\right)Z_2\right|^2 \\ &= E\left(Z_1 - \left(\frac{EZ_1 \cdot \overline{Z}_2}{E|Z_2|^2}\right)Z_2\right)\left(\overline{Z}_1 - \left(\frac{E\overline{Z}_1 \cdot Z_2}{E|Z_2|^2}\right)\overline{Z}_2\right) \\ &= E|Z_1|^2 - \left(\frac{|EZ_1 \cdot \overline{Z}_2|^2}{E|Z_2|^2}\right). \end{aligned}$$

With the usual pointwise definitions of addition and multiplication by complex constants, L_2 becomes a linear (or vector) space. As such, L_2 makes no provision for such concepts as the norm of Z. We now append this notion to L_2. To this end, we need the following convention in dealing with elements of L_2.

Definition 5.1.2. Any two $Z_1, Z_2 \in L_2$ are to be regarded as equal if

$$Z_1 = Z_2 \quad \text{(a.s.)}. \tag{5.1.2}$$

Definition 5.1.3. For $Z_1, Z_2 \in L_2$, the complex number (Z_1, Z_2) defined by

$$(Z_1, Z_2) = EZ_1 \cdot \overline{Z}_2 \tag{5.1.3}$$

is called the "inner" product.

That (5.1.3) always exists follows trivially from Schwarz's inequality. We obviously have

$$\begin{aligned} &\text{(i) } (cZ_1, Z_2) = c(Z_1, Z_2), \\ &\text{(ii) } (Z_1, Z_2) = \overline{(Z_2, Z_1)}, \\ &\text{(iii) } (Z_1 + Z_2, Z) = (Z_1, Z) + (Z_2, Z). \end{aligned} \tag{5.1.4}$$

Finally, from (5.1.2), it follows that

$$(Z, Z) = 0 \text{ implies } Z = 0 \quad \text{(a.s.)}.$$

Thus, (Z_1, Z_2) satisfies all the requirements for an inner product. Therefore, L_2 is an inner product space.

Remark 5.1.1. The convention (5.1.2) is an informal device for dealing with the problem posed by failure of $(Z, Z) = 0$ to imply $Z = 0$. A formal means of treating this difficulty is to replace the space L_2 with the space of equivalence classes defined by the assertion that two r.v.'s are equivalent if and only if they are equal (a.e.). All subsequent statements would be applied to equivalence classes rather than to the r.v.'s themselves.

The inner product can be used to define another function on L_2.

Definition 5.1.4. For each $Z \in L_2$, the "norm," denoted by $\|Z\|$, is defined by

$$\|Z\| = +(Z, Z)^{1/2}. \tag{5.1.5}$$

Proposition 5.1.2. *For any* $Z_1, Z_2 \in L_2$,

$$\|Z_1 + Z_2\|^2 + \|Z_1 - Z_2\|^2 = 2\|Z_1\|^2 + 2\|Z_2\|^2. \tag{5.1.6}$$

PROOF. This is the Parallelogram Law. To prove it, write

$$\|Z_1 + Z_2\|^2 = (Z_1 + Z_2, Z_1 + Z_2) = \|Z_1\|^2 + (Z_1, Z_2) + (Z_2, Z_1) + \|Z_2\|^2,$$

$$\|Z_1 - Z_2\|^2 = (Z_1 - Z_2, Z_1 - Z_2) = \|Z_1\|^2 + (Z_1, -Z_2) + (-Z_2, Z_1) + \|Z_2\|^2$$

By adding these and employing (5.1.4), the assertion follows. □

Definition 5.1.5. If $Z_1, Z_2 \in L_2$ are such that $(Z_1, Z_2) = 0$, we say that they are "orthogonal" and indicate this by writing $Z_1 \perp Z_2$.

Proposition 5.1.3. *If* $Z_1 \perp Z_2$, *we have*

$$\|Z_1 + Z_2\|^2 = \|Z_1\|^2 + \|Z_2\|^2. \tag{5.1.7}$$

PROOF. Write

$$\begin{aligned}\|Z_1 + Z_2\|^2 &= (Z_1 + Z_2, Z_1 + Z_2)\\ &= \|Z_1\|^2 + (Z_1, Z_2) + (Z_2, Z_1) + \|Z_2\|^2\\ &= \|Z_1\|^2 + \|Z_2\|^2.\end{aligned}$$

Definition 5.1.6. A subset $\mathcal{O} \subset L_2$ is called an "orthogonal collection" if, for any Z_1, $Z_2 \in \mathcal{O}$ such that $Z_1 \neq Z_2$ (a.s.), we have $Z_1 \perp Z_2$. In addition, if, for every $Z \in \mathcal{O}$, we have $\|Z\| = 1$, $\mathcal{O}$ is called an "orthonormal collection."

The following result is known as the Bessel inequality.

Proposition 5.1.4. *If* $\{Z_i\}_1^n \subset L_2$ *is an orthonormal sequence, then, for any* $Z \in L_2$,

$$\sum_{i=1}^{n} |(Z, Z_i)|^2 \leq \|Z\|^2. \tag{5.1.8}$$

PROOF.

$$0 \leq \left\| Z - \sum_{i=1}^{n} (Z, Z_i) Z_i \right\|^2$$

$$= \left(Z - \sum_{i=1}^{n} (Z, Z_i) Z_i, Z - \sum_{i=1}^{n} (Z, Z_i) Z_i \right)$$

$$= \|Z\|^2 - \sum_{i=1}^{n} \overline{(Z,Z_i)}(Z,Z_i) - \sum_{i=1}^{n} (Z,Z_i)(Z_i,Z) + \sum_{i=1}^{n}\sum_{j=1}^{n} (Z,Z_i)\overline{(Z,Z_j)}(Z_i,Z_j)$$

$$= \|Z\|^2 - \sum_{i=1}^{n} |(Z_i,Z)|^2 - \sum_{i=1}^{n} |(Z_i,Z)|^2 + \sum_{i=1}^{n} |(Z,Z_i)|^2$$

$$= \|Z\|^2 - \sum_{i=1}^{n} |(Z,Z_i)|^2.$$

This proves the assertion. □

Corollary 5.1.1. *If* $\{Z_i\}_1^\infty \subset L_2$ *is orthonormal, then*

$$\sum_1^\infty |(Z,Z_i)|^2 \leq \|Z\|^2. \tag{5.1.9}$$

This is so because, according to the previous proposition, (5.1.8) holds for every partial series of this sequence.

The following result is known as the Cauchy inequality.

Proposition 5.1.5. *If* $Z_1, Z_2 \in L_2$, *then*

$$\|Z_1 + Z_2\| \leq \|Z_1\| + \|Z_2\|. \tag{5.1.10}$$

PROOF.

$$\|Z_1 + Z_2\|^2 = (Z_1 + Z_2, Z_1 + Z_2) = \|Z_1\|^2 + (Z_1,Z_2) + (Z_2,Z_1) + \|Z_2\|^2.$$

However,

$$\begin{aligned}(Z_1,Z_2) + (Z_2,Z_1) &= (Z_1,Z_2) + \overline{(Z_1,Z_2)}\\ &\leq 2|(Z_1,Z_2)|\\ &\leq 2\|Z_1\|\cdot\|Z_2\|.\end{aligned}$$

Therefore,

$$\begin{aligned}\|Z_1 + Z_2\|^2 &\leq \|Z_1\|^2 + 2\|Z_1\|\cdot\|Z_2\| + \|Z_2\|^2\\ &= (\|Z_1\| + \|Z_2\|)^2,\end{aligned}$$

which proves (5.1.10). □

From (5.1.10) we readily deduce the "triangle inequality":

$$\|Z_1 - Z_3\| \leq \|Z_1 - Z_2\| + \|Z_2 - Z_3\|. \tag{5.1.11}$$

Remark 5.1.2. The concept of a norm permits us to measure the "distance" $d(Z_1,Z_2)$ between any two members $Z_1, Z_2 \in L_2$ as

$$d(Z_1,Z_2) = \|Z_1 - Z_2\|. \tag{5.1.12}$$

Now, we can prove the following result.

Proposition 5.1.6. *The next four functions are uniformly continuous* [(a) *and* (c) *on* L_2 *and* (b) *and* (d) *on* $L_2 \times L_2$].

$$\text{(a) } \alpha Z; \qquad \text{(b) } (Z_1, Z_2); \qquad \text{(c) } \|Z\|; \qquad \text{(d) } Z_1 + Z_2.$$

PROOF. From (5.1.10), it follows that

(a) $d(\alpha Z_1, \alpha Z_2) = |\alpha| d(Z_1, Z_2)$,
(b) $|(Z_1, Z) - (Z_2, Z)| \leq \|Z\| \cdot \|Z_1 - Z_2\| = \|Z\| d(Z_1, Z_2)$,
(c) $| \|Z_1\| - \|Z_2\| | \leq d(Z_1, Z_2)$,
(d) $\|(Z_1 + Z_2) - (U_1 + U_2)\| \leq \|Z_1 - U_1\| + \|Z_2 - U_2\|$

$$= d(Z_1, U_1) + d(Z_2, U_2).$$

5.2. Convergence in Quadratic Mean

Because $L_2 = L_2\{\Omega, \mathscr{B}, P\}$ is a metric space, with the distance defined by (5.1.12), we can now introduce a notion of convergence.

Definition 5.2.1. A sequence $\{Z_n\}_1^\infty \subset L_2$ is said to converge to an element $Z \in L_2$ if

$$\|Z_n - Z\| = d(Z_n, Z) \to 0 \quad \text{as } n \to \infty. \tag{5.2.1}$$

For this kind of convergence, it is not necessary that the values $Z_n(\omega)$ converge to $Z(\omega)$ pointwise or a.s. We shall call this kind of convergence, "convergence in quadratic mean" or "convergence in the mean square" and write

$$Z_n \xrightarrow{\text{q.m.}} Z \quad \text{or} \quad Z = \text{l.i.m}\, Z_n.$$

Proposition 5.2.1. *If* $\{Z_n\}_1^\infty \subset L_2$ *converges in quadratic mean* (q.m.), *it may have at most one limit.*

PROOF. Suppose that

$$Z_n \xrightarrow{\text{q.m.}} U \quad \text{and} \quad Z_n \xrightarrow{\text{q.m.}} V,$$

where $U, V \subset L_2$; then, for all $n = 1, 2, \ldots$,

$$\|U - V\| \leq \|Z_n - U\| + \|Z_n - V\| \to 0 \quad \text{as } n \to \infty.$$

Hence, $\|U - V\| = 0$, which implies that $U = V$ (a.s.). □

Proposition 5.2.2. *If* $\{Z_n\}_1^\infty \subset L_2$ *converges in q.m. to* $Z \in L_2$, *then*

$$\|Z_n\| \to \|Z\|.$$

PROOF. The obvious inequalities

$$\|Z_n\| \leq \|Z\| + \|Z_n - Z\|,$$
$$\|Z\| \leq \|Z_n\| + \|Z_n - Z\|$$

imply that

$$|\|Z_n\| - \|Z\|| \leq \|Z_n - Z\|,$$

which proves the assertion. □

Definition 5.2.2. A sequence $\{Z_n\} \subset L_2$ is said to be "Cauchy" or "fundamental," if for every $\varepsilon > 0$ there exists a natural number N such that

$$\|Z_n - Z_m\| \leq \varepsilon$$

if $m, n > N$.

Proposition 5.2.3. *If* $\{Z_n\}_1^\infty \subset L_2$ *has a limit* $Z \in L_2$, *then it is a Cauchy sequence.*

PROOF. Let $\varepsilon > 0$ be arbitrary; then there exists a natural number N, such that for any $n > N$

$$\|Z_n - Z\| < \varepsilon/2.$$

Now, if $m, n > N$, then

$$\|Z_n - Z_m\| \leq \|Z_n - Z\| + \|Z_m - Z\| < \varepsilon,$$

which proves the assertion. □

The converse statement is much deeper.

Proposition 5.2.4 (Riesz–Fischer). *If* $\{Z_n\}_1^\infty$ *is a Cauchy sequence, there exists* $Z \in L_2$ *such that*

$$Z_n \xrightarrow{\text{q.m.}} Z.$$

PROOF. Consider the convergent series $\sum_{k=1}^\infty 2^{-k}$. Because $\{Z_n\}_1^\infty$ is fundamental, for every k there is an n_k such that

$$\|Z_n - Z_m\| < 2^{-k} \quad \text{if } m, n \geq n_k.$$

We may suppose, without loss of generality, that $n_1 < n_2 < \cdots$, so that

$$\|Z_{n_{k+1}} - Z_{n_k}\| < 2^{-k}.$$

Therefore,

$$\sum_{k=1}^\infty \|Z_{n_{k+1}} - Z_{n_k}\| < \infty.$$

On the other hand, due to convexity (Jensen's inequality)

$$\|Z_{n_{k+1}} - Z_{n_k}\|^2 = E|Z_{n_{k+1}} - Z_{n_k}|^2 \geq (E|Z_{n_{k+1}} - Z_{n_k}|)^2,$$

so that

$$E|Z_{n_{k+1}} - Z_{n_k}| \leq \|Z_{n_{k+1}} - Z_{n_k}\|.$$

From this, it follows that the series

$$\sum_{k=1}^{\infty} E|Z_{n_{k+1}} - Z_{n_k}|$$

converges. Then, according to the Beppo–Levi theorem, the series

$$|Z_{n_1}| + \sum_{k=1}^{\infty} |Z_{n_{k+1}} - Z_{n_k}|$$

converges (a.s.). Consequently, the series

$$Z_{n_1} + \sum_{k=1}^{\infty} (Z_{n_{k+1}} - Z_{n_k})$$

converges (a.s.). However,

$$Z_{n_1} + \sum_{k=1}^{\infty} (Z_{n_{k+1}} - Z_{n_k}) = \lim_{k\to\infty} Z_{n_k}.$$

Set

$$Z = \lim_{k\to\infty} Z_{n_k},$$

where this limit exists and $Z = 0$ on Ω where this limit does not exist.

Next, we have to show that $Z \in L_2$ and that

$$Z = \text{l.i.m}\, Z_n.$$

To this end, take an $\varepsilon > 0$ and a natural number N such that, for $m, n > N$, $\|Z_n - Z_m\| < \varepsilon$. If $k_0 > N$, then

$$\|Z_n - Z_{n_k}\|^2 < \varepsilon^2$$

for any $n > N$ and $k > k_0$. By applying Fatou's lemma to the sequence $\{\|Z_n - Z_{n_k}\|\}_{k>k_0}$, we obtain that

$$\|Z_n - Z\|^2 \leq \liminf_{k\to\infty} \|Z_n - Z_{n_k}\|^2 \leq \varepsilon^2.$$

In other words, for all $n > N$, $\|Z_n - Z\| < \varepsilon$. From this, it follows that $Z_n - Z \in L_2$ and, consequently, $Z \in L_2$, and Z is the mean square limit of $\{Z_n\}_1^{\infty}$. This completes the proof of the Riesz–Fischer theorem. □

Corollary 5.2.1. *Because* $|EZ_n - EZ|^2 \leq \|Z_n - Z\|^2 \to 0$ *as* $n \to \infty$, *it follows that*

$$EZ_n \to EZ.$$

The following is a criterion for the mean square convergence due to Loève.

Proposition 5.2.5. *A sequence* $\{Z_n\}_1^\infty \subset L_2$ *converges in quadratic mean if and only if*

$$EZ_n\bar{Z}_m \to C \quad (\textit{a finite constant}) \tag{5.2.2}$$

as n and m tend independently to infinity.

PROOF. Assume that (5.2.2) holds; then, as $m, n \to \infty$,

$$\begin{aligned}\|Z_n - Z_m\|^2 = E|Z_n - Z_m|^2 &= E(Z_n - Z_m)(\bar{Z}_n - \bar{Z}_m)\\ &= EZ_n\bar{Z}_n - EZ_n\cdot\bar{Z}_m - EZ_m\bar{Z}_n + EZ_m\bar{Z}_m\\ &\to C - C - C + C = 0.\end{aligned}$$

On the other hand, if $Z_n \xrightarrow{\text{q.m.}} Z$, then

$$EZ_n\bar{Z}_m \to EZ\bar{Z} = \|Z\|^2.$$

This follows from the fact that if $\{Z_n\}_1^\infty \subset L_2$ and $\{U_n\}_1^\infty \subset L_2$ converge in q.m. to Z and U, respectively, then from Corollary 5.2.1, it follows that

$$EZ_n\bar{U}_n \to EZ\bar{U}.$$ □

Remark 5.2.1. The property of the space L_2 established by the Riesz–Fischer theorem is called "completeness." With this property, $L_2 = L_2\{\Omega, \mathscr{B}, P\}$ becomes a Hilbert space.

5.3. Remarks on the Structure of L_2

The structure of the Hilbert space $L_2 = L_2\{\Omega, \mathscr{B}, P\}$ depends clearly on the underlying probability space $\{\Omega, \mathscr{B}, P\}$. In this section, we discuss the problem of approximation of a r.v. $Z \in L_2$ by means of a certain type of bounded r.v. from L_2. We also discuss the problem of separability of the space L_2. To this end, the following definition is needed.

Definition 5.3.1. A (real or complex) r.v. Z on a probability space $\{\Omega, \mathscr{B}, P\}$ is said to be "essentially bounded" if there is a constant $0 < c < \infty$ such that

$$|Z| \le c \quad \text{(a.s.)}.$$

The subset of all essentially bounded r.v.'s of $L_2\{\Omega, \mathscr{B}, P\}$ is denoted by $L_\infty = L_\infty\{\Omega, \mathscr{B}, P\}$. On L_∞, we define a norm by

$$\|Z\| = \|Z\|_\infty = \inf\{c; |Z| \le c \quad \text{(a.s.)}\}. \tag{5.3.1}$$

With this norm, L_∞ becomes a normed linear space. The norm $\|Z\|_\infty$ is often called the "essential supremum" of $|Z|$ and is denoted by

$$\|Z\|_\infty = \text{ess sup}|Z|. \tag{5.3.2}$$

Next, we prove the following result.

Proposition 5.3.1. *Let U be an arbitrary complex r.v. on $\{\Omega, \mathscr{B}, P\}$, then for any $\varepsilon > 0$ there exists a r.v. $Z \in L_\infty\{\Omega, \mathscr{B}, P\}$ such that*

$$P\{U \neq Z\} < \varepsilon.$$

PROOF. Set $A_k = \{|U| > k\}$, $k = 1, 2, \ldots$, and write $G = \{|U| = \infty\}$. We clearly have that $P(G) = 0$,

$$A_1 \supset A_2 \supset \cdots, \qquad \text{and that } G = \bigcap_{k=1}^{\infty} A_k.$$

From this, it follows that $P(A_k) \to 0$ as $k \to \infty$, so that there exists n_0 such that $P(A_{n_0}) < \varepsilon$. Now define the r.v. Z as

$$Z = \begin{cases} U & \text{on } A_{n_0}^c \\ 0 & \text{on } A_{n_0}. \end{cases}$$

Clearly, $|Z| \leq n_0$ and $P\{Z \neq U\} < \varepsilon$. This proves the assertion. □

Definition 5.3.2. A subset $D \subset L_2$ is said to be everywhere "dense" in L_2 if any element of L_2 is the mean square limit of a sequence of elements from D.

It is not difficult to see that a necessary and sufficient condition for a subset $D \subset L_2$ to be everywhere dense in L_2 is that, for any $Z \subset L_2$ and any $\varepsilon > 0$, it is possible to find $Z_0 \in D$ such that

$$\|Z - Z_0\| < \varepsilon.$$

Proposition 5.3.2. *The class $L_\infty \subset L_2$ of all bounded r.v.'s is everywhere dense in L_2.*

PROOF. Let $Z \in L_2$ and consider

$$\varphi(A) = \int_A Z^2 \, dP, \quad A \in \mathscr{B}.$$

Clearly, $\varphi(\cdot)$ is a bounded non-negative countably additive set function on $\mathscr{B}$ such that $\varphi \ll P$. Therefore, for any $\varepsilon > 0$, there exists a $\delta > 0$ such that if

$$P(A) < \delta,$$

then

$$\varphi(A) < \varepsilon^2.$$

Now, according to Proposition 5.3.1, there exists a bounded r.v. Z_0 such that

$$P\{Z \neq Z_0\} < \delta.$$

Without loss of generality, we may assume that $Z_0 = 0$ on the set $\{Z \neq Z_0\}$. Then

$$\|Z - Z_0\|^2 = \int |Z - Z_0|^2 \, dP = \int_{\{Z \neq Z_0\}} |Z|^2 \, dP < \varepsilon^2,$$

which proves the proposition. □

Next, we define two important concepts in L_2 theory. The first one is that of a "linear manifold."

Definition 5.3.3. A subset $M \subset L_2$ is called a linear manifold if, for any Z_1, $Z_2 \in M$ and complex numbers α and β, $\alpha Z_1 + \beta Z_2 \in M$.

Definition 5.3.4. A closed linear manifold $H \subset L_2$ is called a subspace.

Let $G \subset L_2$ be an arbitrary subset. There always exists at least one linear manifold that contains G. The intersection of all linear manifolds containing G is denoted by $\langle G \rangle$ and is called the linear manifold spanned by G. The closed linear manifold is denoted by $\overline{\langle G \rangle}$.

One of the fundamental questions in the theory of L_2 spaces is that of their dimensionality. In this respect, of particular interest are conditions on the probability space $\{\Omega, \mathcal{B}, P\}$ under which $L_2\{\Omega, \mathcal{B}, P\}$ is countably infinite dimensional. This question is closely related to the one of whether there exists a countable everywhere dense subset of L_2. In other words, is the space L_2 separable? In general, this is not the case unless the probability space $\{\Omega, \mathcal{B}, P\}$ has a particular structure.

Remark 5.3.1. A linear inner product space L_2 with the norm $\|h\| = (h, h)^{1/2}$ for $h \in L_2$ is often called unitary. This term, however, is not standard in the literature.

5.4. Orthogonal Projection

To investigate the geometry of the space $L_2\{\Omega, \mathcal{B}, P\}$, we only need the concept of orthogonality. According to Definition 5.1.5, elements $Z_1, Z_2 \in L_2$ are said to be orthogonal, written $Z_1 \perp Z_2$, if $(Z_1, Z_2) = 0$. We say that an element $Z \in L_2$ is orthogonal to a subset $M \subset L_2$, and write

$$Z \perp M$$

if $(Z, U) = 0$ for all $U \in M$.

In this section, we discuss an important and useful result in the theory of Hilbert spaces, the so-called projection theorem, which deals with decomposition of an element $Z \in L_2$ into orthogonal components. Roughly speaking, the content of this theorem can be described as follows. Given a subspace $H \subset L_2$ and an arbitrary $Z \subset L_2$, there exists a unique decomposition of the form

$$Z = Z_H + Z_0,$$

where $Z_H \in H$ and $Z_0 \perp H$. The Z_H is called the orthogonal projection of Z on H.

To this end, let us show that there is always a minimal distance from an element $Z \in L_2$ to a given subspace.

Proposition 5.4.1. *Let $H \subset L_2$ be a subspace and $Z \in L_2$ arbitrary and write*

$$\delta = \inf\{\|Z - U\|; U \in H\}.$$

Then there is $Z_0 \in H$ such that

$$\|Z - Z_0\| = \delta.$$

Proof. Choose any $\{Z_k\}_1^\infty \subset H$ such that $\|Z - Z_k\| \to \delta$. On the other hand, from (5.1.6), we have

$$\|(Z_n - Z) + (Z_k - Z)\|^2 + \|(Z_n - Z) - (Z_k - Z)\|^2$$
$$= 2\|Z_n - Z\|^2 + 2\|Z_k - Z\|^2$$

or

$$\|Z_n - Z_k\|^2 = 2\|Z_n - Z\|^2 + 2\|Z_k - Z\|^2 - 4\|\tfrac{1}{2}(Z_n + Z_k) - Z\|^2.$$

Because $\frac{1}{2}(Z_n + Z_k) \in H$, it follows that

$$\|\tfrac{1}{2}(Z_n + Z_k) - Z\|^2 \geq \delta^2.$$

Therefore,

$$\|Z_n - Z_k\|^2 \leq 2\|Z_n - Z\|^2 + 2\|Z_k - Z\|^2 - 4\delta^2. \tag{5.4.1}$$

By letting n and $k \to \infty$, the right-hand side in (5.4.1) converges to zero. This implies that $\{Z_n\}_1^\infty$ is a Cauchy sequence. Therefore, $Z_n \xrightarrow{\text{q.m.}} Z_0 \in H$. From the continuity of the norm, it follows that

$$\|Z - Z_0\| = \lim_{n\to\infty} \|Z - Z_n\| = \delta. \qquad \square$$

Proposition 5.4.2. *Let $H \subset L_2$ be a subspace and $Z \in L_2$ arbitrary. Let $Z_0 \in H$ be such that*

$$\|Z - Z_0\| = \inf\{\|Z - U\|; U \in H\};$$

then $(Z - Z_0) \perp H$.

Proof. If $U \in H$, so is $Z_0 - \alpha U$ for any complex α. Therefore,

$$\|Z - Z_0\| \leq \|Z - Z_0 + \alpha U\|.$$

Set $Z^* = Z - Z_0$ and $\bar{\beta} = (Z^*, U)$. This yields

$$\|Z^*\|^2 \le \|Z^* + \alpha U\|^2 = \|Z^*\|^2 + \overline{\alpha\beta} + \alpha\beta + |\alpha|^2 \|U\|^2$$

or

$$0 \le \overline{\alpha\beta} + \alpha\beta + |\alpha|^2 \|U\|^2.$$

By successively assigning α values t, $-t$, it, and $-it$ where $t > 0$, one obtains

$$-t\|U\| \le \beta + \bar{\beta} \le t\|U\|,$$

$$-t\|U\| \le i(\beta - \bar{\beta}) \le t\|U\|.$$

Because $t > 0$ is arbitrary, it follows that $\beta + \bar{\beta} = \beta - \bar{\beta} = 0$. Therefore, $(Z^*, U) = \beta = 0$ for all $U \in H$, which proves the proposition. □

Definition 5.4.1. The orthogonal complement of a subspace $H \subset L_2$ is the subset $H^\perp \subset L_2$ of all $Z \in L_2$ such that

$$Z \perp H.$$

Proposition 5.4.3. *The orthogonal complement $H^\perp$ of a subspace H is also a subspace such that $H \cap H^\perp = \{Z_0\}$,* where $Z_0 = 0$ (a.s.).

PROOF. If $Z_1, Z_2 \in H^\perp$, then $(Z_1, Z) = (Z_2, Z) = 0$ for any $Z \in H$. Therefore,

$$0 = \alpha(Z_1, Z) + \beta(Z_2, Z) = (\alpha Z_1 + \beta Z_2, Z)$$

for any α and β. This shows that $\alpha Z_1 + \beta Z_2 \in H^\perp$, so that $H^\perp$ is a linear manifold. To see that it is closed, let $\{U_k\}_1^\infty \subset H^\perp$ converge in q.m. to Z^0. Then, by continuity of the inner product, we have that for any $Z \in H$

$$(Z^0, Z) = \lim_{n\to\infty} (U_n, Z) = 0.$$

Therefore, $Z^0 \perp Z$, so that $Z^0 \in H^\perp$. Finally, if $Z_0 \in H \cap H^\perp$, then $(Z_0, Z_0) = 0$, so that $Z_0 = 0$ (a.s.). □

Definition 5.4.2. Let A and B be two arbitrary subsets of L_2; then

$$A + B = \{Z + U; Z \in A, U \in B\}. \tag{5.4.2}$$

If H_1 and H_2 are two orthogonal subspaces of L_2 one writes $H_1 \oplus H_2$ for $H_1 + H_2$. Thus, the use of the symbol $\oplus$ entails orthogonality of the summands.

Proposition 5.4.4. *If H is a subspace of L_2, then $H \oplus H^\perp = L_2$.*

PROOF. Clearly, $H \oplus H^\perp \subset L_2$. If $Z \in L_2$, then by Proposition 5.4.1 and 5.4.2, there exists $Z_0 \in H$ such that $(Z - Z_0) \perp H$, which implies that $(Z - Z_0) \in H^\perp$. From this, the assertion follows because $Z = Z_0 + (Z - Z_0)$.

Proposition 5.4.5. *If $Z_1 + U_1 = Z_2 + U_2$, where $Z_1, Z_2 \in H$ and $U_1, U_2 \in H^\perp$, then $Z_1 = Z_2$ and $U_1 = U_2$* (a.s.).

PROOF. Because $Z_1 - Z_2 \in H$ and $U_1 - U_2 \in H^\perp$, it follows that $Z_1 - Z_2 = U_2 - U_1 \in H \cap H^\perp$ and the assertion follows because $H \cap H^\perp = \{0\}$. □

Corollary 5.4.1.

$$(H^\perp)^\perp = H.$$

Remark 5.4.1. If $H \subset L_2$ is a subspace, it follows from Propositions 5.4.4 and 5.4.5 that any $Z \in L_2$ can be expressed uniquely as a sum

$$Z = Z_H + Z_{H^\perp}, \tag{5.4.3}$$

where $Z_H \in H$ and $Z_{H^\perp} \in H^\perp$. We call Z_H the orthogonal projection of Z on H. It is clear that $Z = Z_H$ if and only if $Z \in H$.

5.5. Orthogonal Basis

At the end of the Section 5.3 we made a brief commentary concerning the dimensionality of L_2 spaces, although the notion by itself has not been formally defined. This concept, whose formal definition will be given at the end of this section, is closely related to the concepts of "orthogonality" and "completeness" which will be discussed in some detail in this section. We begin with the following definition (see Definition 5.1.6):

Definition 5.5.1. An orthogonal system $\mathcal{O} \subset L_2$ is called "complete" if no other orthogonal collection of elements of L_2 contains $\mathcal{O}$ as its proper subset.

Next, we shall show that the closed linear manifold spanned by a complete orthonormal collection $\mathcal{O} \subset L_2$ is equal to L_2, $\langle \bar{\mathcal{O}} \rangle = L_2$. For this reason a complete orthonormal system is often called a "basis" for the space L_2.

Definition 5.5.2. Let $\{U_k\}_1^\infty \subset L_2$ be an arbitrary orthonormal system and $Z \in L_2$. The complex numbers

$$c_k = (Z, U_k), \quad k = 1, 2, \ldots, \tag{5.5.1}$$

are called the "Fourier coefficients" of Z with respect to the system $\{U_k\}_1^\infty$ and the sum

$$\sum_{k=1}^{\infty} c_k U_k \tag{5.5.2}$$

is called a "Fourier series" of Z with respect to $\{U_k\}_1^\infty$.

From (5.5.1) and the Bessel inequality (5.1.9), it follows that

$$\sum_{k=1}^{\infty} |c_k|^2 \leq \|Z\|^2. \tag{5.5.3}$$

This then clearly implies that

$$\left\| \sum_{k=m+1}^{n} c_k U_k \right\|^2 = \sum_{k=m+1}^{n} |c_k|^2 \to 0 \quad \text{as } m, n \to \infty. \tag{5.5.4}$$

Therefore,

$$\|S_n - S_m\| \to 0 \quad \text{as } m, n \to \infty,$$

where $S_n = \sum_{k=1}^{n} c_k U_k$. Hence, by the Riesz–Fischer theorem,

$$S_n \xrightarrow{\text{q.m.}} S \in L_2.$$

Now we can prove the following result.

Proposition 5.5.2. *Let $\{Z_k\}_1^\infty \subset L_2$ be a complete orthonormal system; then every element $Z \in L_2$ admits an expansion convergent in the mean square*

$$Z = \sum_{k=1}^{\infty} c_k Z_k, \tag{5.5.5}$$

where $c_k = (Z, Z_k)$, i.e.,

$$\left\| Z - \sum_{k=1}^{n} c_k Z_k \right\| \to 0 \quad \text{as } n \to \infty. \tag{5.5.6}$$

Furthermore, we have that

$$\|Z\|^2 = \sum_{k=1}^{n} |c_k|^2, \tag{5.5.7}$$

which is called Parseval's formula.

PROOF. From (5.5.4), it follows that

$$\sum_{k=1}^{n} c_k Z_k \xrightarrow{\text{q.m.}} Z_0 \quad \text{as } n \to \infty. \tag{5.5.8}$$

Now, for any fixed $k = 1, 2, \ldots$ and $n \geq k$,

$$\begin{aligned} |(Z_0, Z_k) - c_k| &= \left| \left(Z_0 - \sum_{j=1}^{n} c_j Z_j, Z_k \right) \right| \\ &\leq \left\| Z_0 - \sum_{j=1}^{n} c_j Z_j \right\| \cdot \|Z_k\| \\ &= \left\| Z_0 - \sum_{j=1}^{n} c_j Z_j \right\|. \end{aligned}$$

As $n \to \infty$ the right-hand side tends to zero due to (5.5.8), which implies that

$$(Z_0, Z_k) = c_k.$$

Hence, for any $k = 1, 2, \ldots,$

$$(Z_0 - Z, Z_k) = (Z_0, Z_k) - (Z, Z_k) = c_k - c_k = 0,$$

so that $Z_0 - Z$ is orthogonal to every element of the complete orthonormal system $\{Z_k\}_1^\infty$. This then implies that $Z_0 - Z = 0$ (a.s.) so that

$$Z_0 = Z \text{ (a.s.)}. \tag{5.5.9}$$

Finally, because

$$\left\| Z - \sum_{k=1}^{n} c_k Z_k \right\|^2 = \|Z\|^2 - \sum_{k=1}^{n} |c_k|^2, \tag{5.5.10}$$

we obtain the Parseval formula by letting $n \to \infty$ in (5.5.10) and invoking (5.5.9). □

Remark 5.5.1. The expansion (5.5.5) is called the generalized Fourier series of $Z \in L_2$, where $c_k = (Z, Z_k)$ are the generalized Fourier coefficients of Z with respect to the orthonormal sequence of $\{Z_k\}_1^\infty$. When $\{Z_k\}_1^\infty$ is a complete orthonormal system, it follows from Proposition 5.5.2 that

$$\overline{\langle \{Z_k\}_1^\infty \rangle} = L_2. \tag{5.5.11}$$

Remark 5.5.2. From Proposition 5.5.2, it also follows that for a given complete orthonormal $\{Z_k\}_1^\infty \subset L_2$ and $Z \in L_2$, we have that (5.5.5) holds where $c_k = (Z, Z_k)$, $k = 1, 2, \ldots$, and (5.5.7) holds. Does the converse also hold? In other words, given an arbitrary sequence of complex numbers $\{c_k\}_1^\infty$ and a complete orthonormal system $\{Z_k\}_1^\infty$, does there exist a $Z \in L_2$ such that $c_k = (Z, Z_k)$ for all $k = 1, 2, \ldots$ and such that (5.5.5) holds? From Bessel's inequality, we know that only those sequences come into consideration for which $\sum_{k=1}^\infty |c_k|^2 < \infty$. This condition is also sufficient. In fact, from (5.5.4) and using the Riesz–Fischer theorem, we have that the series

$$\sum_{k=1}^{\infty} c_k Z_k$$

converges in the mean to a $Z \in L_2$ if $\sum_{k=1}^\infty |c_k|^2 < \infty$, and

$$(Z, Z_k) = c_k, \quad k = 1, 2, \ldots.$$

In other words, if $\{Z_k\}_1^\infty$ is a complete orthonormal system and $\{c_k\}_1^\infty$ a sequence of complex numbers for which $\sum_{k=1}^\infty |c_k|^2 < \infty$, then $\sum_{k=1}^n c_k Z_k \xrightarrow{\text{q.m.}} Z \in L_2$, such that

$$(Z, Z_k) = c_k, \quad k = 1, 2, \ldots, \tag{5.5.12}$$

and no other solution $Z \in L_2$ of the system of equations (5.5.12) exists.

Therefore, every complete orthonormal sequence $\{Z_k\}_1^\infty \in L_2$ establishes, by means of the formula $Z = \sum_{k=1}^\infty c_k Z_k$, a one-to-one correspondence between the elements $Z \in L_2$ and the sequences of complex numbers $\{c_k\}_1^\infty$ satisfying $\sum_{k=1}^\infty |c_k|^2 < \infty$. Now we can give the following definition.

Definition 5.5.3. The dimension of the space L_2 is the cardinality of its complete orthonormal set. If there is a complete orthonormal sequence $\{Z_k\}_1^\infty \subset L_2$, then L_2 is countably infinite dimensional.

Remark 5.5.3. The following, more general, form of Parseval's formula also holds. For any two Z, $U \in L_2$,

$$(Z, U) = \sum_{k=1}^{\infty} (Z, Z_k)\overline{(U, Z_k)}. \tag{5.5.13}$$

This follows from the identity (which is easy to verify)

$$4(Z, U) = \|Z + U\|^2 - \|Z - U\|^2 + i\|Z + iU\| - i\|Z - iU\| \tag{5.5.14}$$

and from (5.5.7).

5.6. Existence of a Complete Orthonormal Sequence in L_2

Definition 5.5.3 specifies that a Hilbert space L_2 is countably infinite dimensional if there exists a complete orthonormal sequence $\{Z_k\}_1^\infty \subset L_2$ (often called its base). In general, such a sequence does not exist. If, however, there exists a countable everywhere dense subset $D_0 \subset L_2$ (in other words, such that any element $Z \in L_2$ is the mean square limit of a sequence of elements from D_0), then there exists a complete orthonormal sequence in L_2.

To show this, write elements of D_0 as a sequence $\{U_k\}_1^\infty$, i.e., $D_0 = \{U_k\}_1^\infty$. From this, by means of the well-known Gram–Schmid orthogonalization procedure, one can always construct an orthonormal family $\{Z_k\}_1^\infty$ as follows. Put

$$Z_1 = U_1/\|U_1\|. \tag{5.6.1}$$

To obtain Z_2, set

$$W_2 = U_2 - (U_2, Z_1)Z_1.$$

Clearly, $(W_2, Z_1) = 0$. Thus,

$$Z_2 = W_2/\|W_2\|.$$

Continuing this processs, we obtain

$$W_k = U_k - \sum_{i=1}^{k-1} (U_k, Z_i)Z_i, \tag{5.6.2}$$

so that

$$Z_k = W_k/\|W_k\|, \tag{5.6.3}$$

and so on. It is easy to verify that the sequence $\{Z_k\}_1^\infty$ is orthonormal.

Let us now show that $\{Z_n\}_1^\infty$ is also complete. If $Z_0 \in L_2$ is orthogonal to every Z_k, it follows from (5.6.1) and (5.6.2) that Z_0 is also orthogonal to every U_i. Consequently,

$$\|Z_0 - U_k\|^2 = (Z_0 - U_k, Z_0 - U_k) = (Z_0, Z_0) + (U_k, U_k) \geq \|Z_0\|^2.$$

But because $\{U_k\}_1^\infty$ is everywhere dense in L_2, the left-hand side can be made arbitrarily small by suitable choice of k. This implies that $\|Z_0\| = 0$, which proves the assertion.

We conclude this section with some remarks on second-order stochastic processes. Let $\{\xi(t); t \in T\}$ be such a process, and denote by $\langle\overline{\xi(T)}\rangle$ the closed linear manifold spanned by this family. Clearly, $\langle\overline{\xi(T)}\rangle$ is a subspace of L_2 and, as such, it is a Hilbert space.

Assume now that $\{\xi(t); t \in T\}$ is a q.m. continuous process and that T is an interval. In other words, at every $t \in T$.

$$\|\xi(t+h) - \xi(t)\|^2 \to 0 \quad \text{as } h \to 0.$$

Denote now by $Q \subset R$ the set of rationals of R. Then, for every $t \in T$, there exists $\{t_k\} \subset Q \cap T$ such that $t_k \to t$ as $k \to \infty$. Therefore, because $\{\xi(t); t \in T\}$ is continuous in quadratic mean, we have, for every $t \in T$,

$$\xi(t) = \operatorname*{l.i.m}_{n\to\infty} \xi(t_n).$$

In other words, the countable family $\{\xi(t); t \in Q \cap T\}$ is everywhere dense in $\langle\overline{\xi(T)}\rangle$. Therefore, according to the previous discussion, there exists a complete countable orthonormal family in $\langle\overline{\xi(T)}\rangle$.

5.7. Linear Operators in a Hilbert Space

In this section, we discuss the concept of a continuous linear mapping of a Hilbert space into itself.

Definition 5.7.1. Let $\mathscr{H}$ be a Hilbert space. A mapping $T: \mathscr{H} \to \mathscr{H}$ is called a "linear operator" if

$$T(c_1X_1 + c_2X_2) = c_1TX_1 + c_2TX_2 \tag{5.7.1}$$

for all $X_1, X_2 \in \mathscr{H}$ and any complex numbers c_1, c_2.

Definition 5.7.2. The operator T is continuous at $X_0 \in \mathscr{H}$ if $X_n \xrightarrow{\text{q.m.}} X_0$ implies that $TX_n \xrightarrow{\text{q.m.}} TX_0$. This means that, for any $\varepsilon > 0$, there is $\delta = \delta(\varepsilon)$ such that

$$\|X_n - X_0\| < \delta \Rightarrow \|TX_n - TX_0\| < \varepsilon.$$

Definition 5.7.3. A linear operator is bounded if there exists a constant $C > 0$ such that

$$\|TX\| \leq C\|X\| \tag{5.7.2}$$

for all $X \in \mathscr{H}$. The least such C is called the "norm" of T and is denoted by $\|T\|$.

From the last definition we have

$$\begin{aligned}\|T\| &= \inf\{C; \|TX\| \leq C\|X\|, X \in \mathscr{H}\}\\ &= \inf\left\{C; \frac{\|TX\|}{\|X\|} \leq C, X \in \mathscr{H}\right\}.\end{aligned}$$

In other words,

$$\|T\| = \sup_{\substack{X \in \mathscr{H}\\ X \neq 0}} \frac{\|TX\|}{\|X\|}. \tag{5.7.3}$$

An alternative formula for the norm of T is

$$\|T\| = \sup_{\|X\|=1} \|TX\|.$$

From (5.7.2), it follows that every bounded linear operator is continuous on $\mathscr{H}$.

Proposition 5.7.1. *If T is a linear and continuous operator on $\mathscr{H}$, there exists a unique linear continuous operator T^* on $\mathscr{H}$ such that $(TX, Y) = (X, T^*Y)$ for all X, $Y \in \mathscr{H}$ and*

$$\|T\| = \|T^*\|.$$

Definition 5.7.4. The operator T^* is called the "adjoint" of T. We now list some simple properties of adjoint operators.

(i) $(\lambda T)^* = \bar{\lambda}T^*$.
(ii) $(T^*Y, X) = \overline{(X, T^*Y)} = \overline{(TX, Y)} = (Y, TX)$.
(iii) $(T^*)^* = T$ [it follows from (ii) and

$$((T^*)^*X, Y) = (X, T^*Y) = (TX, Y).$$

(iv) If T_i: $\mathscr{H} \to \mathscr{H}$ are continuous and linear, $i = 1, 2$, then

$$(T_1 + T_2)^* = T_1^* + T_2^*.$$

[This follows from (iii) and

$$\begin{aligned}((T_1 + T_2)^*X, Y) &= (X, (T_1 + T_2)Y) = (X, T_1Y + T_2Y)\\ &= (X, T_1Y) + (X, T_2Y) = \overline{(T_1Y, X)} + \overline{(T_2Y, X)}\\ &= \overline{(Y, T_1^*X)} + \overline{(Y, T_2^*X)} = (T_1^*X, Y) + (T_2^*X, Y)\\ &= (T_1^*X + T_2^*X, Y).]\end{aligned}$$

(v) $\|T^*T\| = \|TT^*\| = \|T\|^2$. (Clearly, the composites

$$T^*T\colon \mathscr{H} \to \mathscr{H} \quad \text{and} \quad TT^*\colon \mathscr{H} \to \mathscr{H}$$

are well defined.)

(vi) $(T_1 \circ T_2)^* = (T_2^* \circ T_1^*)$.

Definition 5.7.5. An operator $T\colon \mathscr{H} \to \mathscr{H}$ is called

(a) isometric if: $T^*T = I$ (identity operator),
(b) unitary if: $T^*T = TT^* = I$,
(c) self-adjoint if: $T^* = T$,
(d) projection if: $TT = T$ and $T = T^*$,
(e) normal if: $T^*T = TT^*$.

The following result is very useful. Suppose that the space L_2 is separable. Then, according to the Proposition 5.5.4, there exists a complete orthonormal basis $\{Z_k\}_1^\infty \subset L_2$. Therefore, every $Z \in L_2$ has a unique representation

$$Z = \sum_{k=1}^{\infty} c_k Z_k, \qquad c_k = (Z, Z_k).$$

Proposition 5.7.2. *Let $\{b_n\}_1^\infty$ be a bounded sequence of complex numbers and $C = \text{l.u.b}\{|b_k|; k = 1, 2, \dots\}$. There exists a unique operator T such that*

(i) $TZ_k = b_k Z_k$, $k = 1, 2, \dots,$
(ii) $T \sum_{k=1}^\infty c_k Z_k = \sum_{k=1}^\infty c_k TZ_k$,
(iii) $\|T\| = C$,
(iv) $T^*Z_k = \bar{b}_k Z_k$,
(v) $T^*T = TT^*$,
(vi) $T^*(\sum_{k=1}^\infty c_k Z_k) = \sum_{k=1}^\infty c_k T^* Z_k$.

PROOF. Consider $Z \in \mathscr{H}$ and $Z = \sum_{k=1}^\infty c_k Z_k$, where $\{Z_k\}_1^\infty \subset \mathscr{H}$ is complete and orthonormal. The problem is to define TZ. Because

$$\sum_{k=1}^{\infty} |c_k b_k|^2 \le C^2 \sum_{k=1}^{\infty} |c_k|^2 = C^2 \|Z\|^2 \tag{5.7.4}$$

[this follows from (5.5.13) if we put $U = Z$], we can define

$$TZ = \sum_{k=1}^{\infty} b_k c_k Z_k.$$

From (5.7.4), we have that

$$\|TZ\|^2 \le C^2 \|Z\|^2, \tag{5.7.5}$$

which shows that T is continuous and $\|T\| \le C$.

Clearly, $TZ_k = b_k Z_k$; because $\|Z_k\| = 1$,

$$\|T\| = \sup_{\|Z\|=1} \|TZ\| \ge \|TZ_k\| = \|b_k Z_k\| = |b_k|$$

for all $k = 1, 2, \dots$. Hence, $\|T\| \ge C$. This proves (i)–(iii).

To prove (iv) and (v), suppose that

$$Z = \sum_{k=1}^{\infty} c_k Z_k \quad \text{and} \quad T^*Z = \sum_{k=1}^{\infty} \gamma_k Z_k;$$

the problem is to show that $\gamma_k = c_k \bar{b}_k$. For all k,

$$\gamma_k = (T^*Z, Z_k) = (Z, TZ_k) = (Z, b_k Z_k) = \bar{b}_k (Z, Z_k) = \bar{b}_k c_k.$$

Finally, $T^*TZ_k = T^*|b_k|^2 Z_k = TT^*Z_k$; because $\{Z_k\}_1^\infty$ is complete, $T^*T = TT^*$. In addition, if T_0 is another operator such that $T_0 Z_k = c_k Z_k$, then $T_0 Z_k = TZ_k$ for all $k = 1, 2, \ldots$; because $\{Z_k\}_1^\infty$ is complete, $T_0 = T$. □

5.8. Projection Operators

As we have seen in Section 5.4 (Propositions 5.4.1 and 5.4.2), if $H \subset L_2$ is a subspace and $Z \in L_2$ an arbitrary element, there exists a unique decomposition

$$Z = Z_H + Z_{H^\perp},$$

where Z_H is called the orthogonal projection of Z on H and $Z_{H^\perp}$ is orthogonal on $H^\perp$.

Let H be an arbitrary subspace of L_2 and let P be the mapping

$$P: L_2 \to L_2$$

associated with the subspace H such that, for any $Z \in L_2$,

$$PZ = Z_H. \tag{5.8.1}$$

The following properties of the mapping (5.8.1) are not difficult to verify. (Here $0 \in L_2$ is a zero element.)

(i) $$PZ = Z, \quad \forall Z \in H \quad \text{and} \quad PZ = 0 \quad \text{if } Z \in H^\perp; \tag{5.8.2}$$

(ii) $$(PZ, U) = (Z, PU) \quad \text{for all } Z, U \in L_2. \tag{5.8.3}$$

[This can be shown as follows:

$$(PZ, U) = (Z_H, U_H + U_{H^\perp}) = (Z_H, U_H),$$

$$(Z, PU) = (Z_H + Z_{H^\perp}, U_H) = (Z_H, U_H).]$$

It is also easy to see that

(iii) $$P(Z + U) = PZ + PU, \tag{5.8.4}$$

$$Z + U = (Z_H + U_H) + (Z_{H^\perp} + U_{H^\perp}),$$

$$Z_H + U_H \in H, Z_{H^\perp} + U_{H^\perp} \in H_{H^\perp},$$

so that $P(Z + U) = Z_H + U_H = PZ + PU$;

(iv) $$P(cZ) = cPZ; \tag{5.8.5}$$

(v) $$PPZ = PZ;$$

(vi) $$(PZ, Z) = \|PZ\|^2 \leq \|Z\|,$$
$$\|Z\|^2 = \|Z_H\|^2 + \|Z_{H^\perp}\|^2 \geq \|Z_H\|^2 = \|PZ\|^2;$$

(vii) H is the range of P.

Remark 5.8.1. The mapping (5.8.1) is called the projection of L_2 on H. Sometimes, the notation P_H is used to indicate the relationship of P to the subspace $H \subset L_2$.

Proposition 5.8.1. *If T is any projection operator, there is a unique linear subspace $H \subset \mathscr{H}$ such that $T = P_H$.*

Problems and Complements

5.1. If $\{\alpha_i\}_1^n$ and $\{\beta_i\}_1^n$ are sequences of complex numbers, show that (Cauchy's inequality)
$$\left|\sum_{i=1}^n \alpha_i\beta_i\right|^2 \leq \left(\sum_{i=1}^n |\alpha_i|^2\right)\left(\sum_{i=1}^n |\beta_i|^2\right).$$

5.2. Using the Cauchy inequality, show that
$$\left(\sum_{k=1}^n |Z_k|\right)^2 \leq n \sum_{k=1}^n |Z_k|^2.$$

5.3. Show that every metric induced by a norm satisfies
(i) $d(Z_1 + \alpha, Z_2 + \alpha) = d(Z_1, Z_2)$,
(ii) $d(\alpha Z_1, \alpha Z_2) = |\alpha| d(Z_1, Z_2)$,
where α is a complex number. Can every metric be obtained from a norm?

5.4. Can we obtain a norm from a metric?

5.5. Let $Z \in L_2$ and $A \subset L_2$. The distance between Z and A is defined by
$$d(Z, A) = \inf\{d(Z, U); U \in A\}.$$
Show that
$$|d(Z_1, A) - d(Z_2, A)| \leq d(Z_1, Z_2).$$

5.6. Verify the identity
$$\|Z_1 - Z_2\|^2 + \|Z_1 - Z_3\|^2 = \tfrac{1}{2}\|Z_2 - Z_3\|^2 + 2\|Z_1 - \tfrac{1}{2}(Z_2 + Z_3)\|^2.$$

5.7. Let $\{Z_n\}_0^\infty \subset L_2$; show that $Z_n \xrightarrow{\text{q.m.}} Z$ if
$$\|Z_n\| \to \|Z\| \quad \text{and} \quad (Z_n, Z) \to (Z, Z).$$

5.8. Let $Z_1, Z_2 \in L_2\{\Omega, \mathscr{B}, P\}$ be such that $|Z_1| \cdot |Z_2| \geq 1$ (a.s.). Show that
$$(E|Z_1|)(E|Z_2|) \geq 1.$$

5.9. Let C_0 the set of all sequences of complex numbers $\{\alpha_k\}_1^\infty$ such that $\{k; \alpha_k \neq 0\}$ is finite. Define

$$(\mathbf{X}, \mathbf{Y}) = \sum_1^\infty x_k \bar{y}_k,$$

$\mathbf{X}, \mathbf{Y} \in C_0$. Show that C_0 is an inner product space but not a Hilbert space.

5.10. Prove that $L_\infty = L_\infty\{\Omega, \mathscr{B}, P\}$ is a Banach space if $\|Z\| = \text{ess sup}|Z|$.

5.11. Let $L_0 \subset L_2$ be a subspace and $Z \in L_2$. Show that

$$\inf\{\|Z - U\|; U \in L_0\} = \sup\{|(Z, W); W \in L_0^+, \|W\| \leq 1\}.$$

5.12. If $\{Z_n\}_1^\infty$ and $\{U_n\}_1^\infty$ from L_2 are such that

$$Z_n \xrightarrow{\text{q.m.}} Z \quad \text{and} \quad U_n \xrightarrow{\text{q.m.}} U,$$

show that $(Z_n, U_n) \to (Z, U)$.

5.13. Show that in L_2, $Z_1 \perp Z_2$ if and only if

$$\|Z_1 + \alpha Z_2\| = \|Z_1 - \alpha Z_2\|.$$

5.14. Prove that if $\{Z_k\}_1^\infty \subset L_2$ is orthonormal, then, for any $Z \in L_2$,
(i) $\lim_{n\to\infty}(Z, Z_n) = 0$,
(ii) $\|Z_i - Z_j\| = 2$ for all $i \neq j$.

5.15. A subset $M \subset L_2$ is said to be convex if $(0 \leq \alpha \leq 1)$

$$Z_1, Z_2 \in M \Rightarrow A = \{Z \in L_2; Z = \alpha Z_1 + (1 - \alpha)Z_2\} \subset M.$$

If $\{Z_n\} \subset M$ such that $\|Z_n\| \to d = \inf\{\|U\|; U \in M\}$, show that $\{Z_n\}$ is a Cauchy sequence. The set A is called the segment joining Z_1 and Z_2.

5.16. Let $\{Z_n\}_1^\infty \subset L_2$ be an orthonormal sequence. Show that for any U, $V \in L_2$

$$\sum_{k=1}^\infty |(U, Z_k)(V, Z_k)| \leq \|U\| \cdot \|V\|.$$

5.17. Let $\{Z_n\} \subset L_2$ be such that $Z_n \xrightarrow{\text{q.m.}} Z$. If

$$\mathbf{Y}_n = \frac{1}{n} \sum_{i=1}^n Z_i,$$

then show that $\mathbf{Y}_n \xrightarrow{\text{q.m.}} Z$.

5.18. Let $\{Z_k\}_1^n$ be an orthonormal collection. Prove that

$$\inf\left\{\left\| Z - \sum_1^n \alpha_k Z_k \right\|\right\}$$

is attained if $\alpha_k = (Z, Z_k)$.

5.19. If $\{Z_n\} \subset L_2$ is such that $\sup_n \|Z_n\| \leq K < \infty$, show that $Z_n/n \to 0$ (a.e.).

5.20. Let l_p, $1 \leq p < \infty$, be the set of all sequences of real numbers $(\alpha_1, \alpha_2, \ldots)$ such that

$$\sum_{i=1}^{\infty} |\alpha_i|^p < \alpha.$$

Show that l_p is separable.

5.21. Let $\mathscr{H}$ be Hilbert space and φ a linear functional on $\mathscr{H}$. Show that there exists a unique element $x_0 \in \mathscr{H}$ such that

$$\varphi(x) = (x, x_0) \quad \text{for all } x \in \mathscr{H}.$$

(This is the Riesz representation theorem.)

5.22. Prove Proposition 5.7.1.

CHAPTER 6

Second-Order Processes

6.1. Covariance Function $C(s, t)$

There exists a large class of engineering and physics problems whose solutions require only the knowledge of the first two moments and some very general properties of a second-order random process (see Definition 1.5.8). This chapter is concerned with some key properties of complex-valued second-order random processes.

Let $\{\xi(t); t \in T\}$ be a complex-valued second-order random process on $\{\Omega, \mathscr{B}, P\}$, i.e.,

$$\{\xi(t); t \in T\} \subset L_2\{\Omega, \mathscr{B}, P\}. \tag{6.1.1}$$

Second-order processes are often called "Hilbert processes." Separating the real and imaginary parts of the process, we can write

$$\xi(t) = X(t) + iY(t), \tag{6.1.2}$$

where $X(t)$ and $Y(t)$ are two second-order real-valued random processes.

In the sequel, unless otherwise stated, we will always assume that $E\{\xi(t)\} = 0$ for all $t \in T$.

Definition 6.1.1. The covariance function $C(s, t)$ of the process (6.1.1) is by definition the second mixed moment, i.e.,

$$C(s, t) = E\{\xi(s)\overline{\xi(t)}\} \tag{6.1.3}$$

or, according to the definition (5.1.3) of the inner product,

$$C(s, t) = (\xi(s), \xi(t)). \tag{6.1.4}$$

From (6.1.3), it readily follows that

$$C(s,t) = \overline{C(t,s)}, \tag{6.1.5}$$

which is known as the "Hermitian symmetry." We also have that

$$C(t,t) = \operatorname{Var}\{\xi(t)\} = \|\xi(t)\|^2. \tag{6.1.6}$$

The covariance function of a second-order process is always finite-valued. This follows from the Schwarz inequality (see 5.1.1)

$$|C(s,t)|^2 \leq E|\xi(s)|^2 \cdot E|\xi(t)|^2 = \|\xi(s)\|^2 \cdot \|\xi(t)\|^2.$$

A covariance function possesses a number of interesting features, some of which are listed below. The following property imposes definite restriction on the form of such a function.

(i) Every covariance function is non-negative definite. In other words, for any $\{t_1, \ldots, t_n\} \subset T$ and any complex numbers $z_1, \ldots, z_n$, $n = 1, 2, 3, \ldots$,

$$\begin{aligned} \sum_{i=1}^{n} \sum_{j=1}^{n} z_i \bar{z}_j C(t_i, t_j) &= \sum_{i=1}^{n} \sum_{j=1}^{n} z_i \bar{z}_j E\xi(t_i)\overline{\xi(t_j)} \\ &= E\left\{\sum_{i=1}^{n} z_i \xi(t_i) \sum_{j=1}^{n} \overline{z_i \xi(t_j)}\right\} \\ &= \left\| \sum_{i=1}^{n} z_i \xi(t_i) \right\|^2 \geq 0. \end{aligned} \tag{6.1.7}$$

(ii) Any complex-valued function on $T \times T$ which is non-negative definite is Hermitian. To show this, let $R(s,t)$ be such a function and consider

$$\sum_{i=1}^{n} \sum_{j=1}^{n} R(t_i, t_j) z_i \bar{z}_j \geq 0. \tag{6.1.8}$$

From this, for $n = 1$ we have that

$$R(t,t) \geq 0, \quad \forall t \in T.$$

Next, for $n = 2$, (6.1.8) yields

$$R(t_1,t_1)z_1\bar{z}_1 + R(t_1,t_2)z_1\bar{z}_2 + R(t_2,t_1)z_2\bar{z}_1 + R(t_2,t_2)z_2\bar{z}_2 \geq 0.$$

This implies that, for all complex numbers z_1, z_2,

$$R(t_1,t_2)z_1\bar{z}_2 + R(t_2,t_1)\bar{z}_1 z_2 \quad \text{is real.} \tag{6.1.9}$$

For $z_1 = 1$, $z_2 = i$, (6.1.9) becomes

$$(R(t_2,t_1) - R(t_1,t_2))i \quad \text{is real,}$$

so that

$$R(t_2,t_1) - R(t_1,t_2) \quad \text{is pure imaginary.} \tag{6.1.10}$$

Finally, if we set $z_1 = z_2 = 1$ in (6.1.9), we conclude that

$$R(t_1,t_2) + R(t_2,t_1) \quad \text{is real,} \tag{6.1.11}$$

which together with (6.1.10) clearly implies that

$$R(t_1, t_2) = \overline{R(t_2, t_1)}.$$

It seems that the last property of the function $R(s, t)$ implies that any non-negative definite function on $T \times T$ is a covariance function of a second-order stochastic process $\{\xi(t); t \in T\}$.

(iii) For any non-negative definite function $R(s, t)$ on $T \times T$ (real or complex), there exists a second-order process $\{\xi(t); t \in T\}$ whose covariance function is precisely $R(s, t)$.

This has already been established in the case when $R(s, t)$ was real (see Chapter 4; Section 4.4). To show that the statement holds when $R(s, t)$ is complex, consider the Hermitian form

$$H(t_1, \ldots, t_n) = \sum_{i=1}^{n} \sum_{j=1}^{n} R(t_i, t_j) z_i \bar{z}_j, \tag{6.1.12}$$

where $\{t_1, \ldots, t_n\} \subset T$ $n = 1, 2, \ldots$. Let $R_1 = \mathrm{Re}\{R\}$ and $R_2 = \mathrm{Im}\{R\}$; then we can write

$$R(s, t) = R_1(s, t) + iR_2(s, t). \tag{6.1.13}$$

If we set $z_j = u_j - iv_j$, we readily obtain that

$$\begin{aligned} H(t_1, \ldots, t_n) = &\sum_{i=1}^{n} \sum_{j=1}^{n} R_1(t_i, t_j)(u_i u_j + v_i v_j) \\ &- \sum_{i=1}^{n} \sum_{j=1}^{n} R_2(t_i, t_j)(u_i v_j - u_j v_i) \end{aligned} \tag{6.1.14}$$

(there is no imaginary part because, by assumption, $R(s, t)$ is non-negative definite). According to Equation (4.3.3),

$$e^{-1/2H}$$

is the characteristic function of $2n$-dimensional Gaussian distribution of a system of $2n$ real r.v.'s, say

$$(X(t_1), \ldots, X(t_n), Y(t_1), \ldots, Y(t_n)) \tag{6.1.15}$$

with $E(X(t_i)) = E(Y(t_i)) = 0$, $i = 1, \ldots, n$, and

$$E\{X(t_i)X(t_j)\} = E\{Y(t_i)Y(t_j)\} = R_1(t_i, t_j), \tag{6.1.16}$$

$$E\{X(t_i)Y(t_j)\} = -R_2(t_i, t_j). \tag{6.1.17}$$

Set

$$\xi(t) = (X(t) + iY(t))/\sqrt{2}. \tag{6.1.18}$$

We see that

$$E\{\xi(s)\} = 0, \qquad R(s, t) = E\{\xi(s)\overline{\xi(t)}\} = R_1(s, t) + iR_2(s, t). \tag{6.1.19}$$

Therefore, the system $(\xi(t_1), \ldots, \xi(t_n))$ has a Gaussian distribution for all $\{t_1, \ldots, t_n\} \subset T$ and $n = 1, 2, \ldots$. These distributions satisfy the Kolmogorov consistency conditions, so there exists a complex-valued process $\xi(t)$ having $R(s,t)$ as its covariance function.

Next we shall list a few basic properties of the class of covariance functions.

(iv) The class of covariance functions is closed under additions, multiplications, and passages to the limit. In other words, if C_1 and C_2 are two covariance functions of random process with a common parameter set T, then so are $\alpha_1 C_1 + \alpha_2 C_2$ and $C_1 \cdot C_2$ when α_1, $\alpha_2 > 0$. In addition, if $\{C_n\}_1^\infty$ is a sequence of covariance functions and

$$C = \lim_{n\to\infty} C_n,$$

then C is also a covariance function.

It is apparent that non-negative definiteness is preserved under positive linear combinations or under passage to the limit. In view of what was said in (iii), this proves the first and third statement.

Let $\xi_1(t)$ and $\xi_2(t)$ be independent; then

$$\begin{aligned} E\{\xi_1(s)\xi_2(s)\cdot\overline{\xi_1(t)\xi_2(t)}\} &= E\{\xi_1(s)\overline{\xi_1(t)}\xi_2(s)\overline{\xi_2(t)}\} \\ &= E\{\xi_1(s)\overline{\xi_1(t)}\}E\{\xi_2(s)\overline{\xi_2(t)}\} \\ &= C_1(s,t)C_2(s,t). \end{aligned}$$

Because the first member is a covariance function, the second statement is proved. To see that two such processes, $\xi_1(t)$ and $\xi_2(t)$, exist, it suffices to assume that $\xi_1(t)$ is Gaussian on a probability space $\{\Omega_1, \mathscr{B}_1, P_1\}$ and, similarly, that $\xi_2(t)$ is normal on $\{\Omega_2, \mathscr{B}_2, P_2\}$, and then form the product space.

6.2. Quadratic Mean Continuity and Differentiability

Let $\{\xi(t); t \in T\}$ be an L_2 stochastic process with $T \subset R$ an interval. In general, its covariance function $C(s,t)$ does not provide any direct information about properties of sample functions of $\xi(t)$ such as continuity, differentiability, etc. In this section, we will define analogous concepts which make sense in Hilbert space and give criteria for L_2 continuity and differentiability in terms of C. What is needed for this purpose is the notion of L_2 convergence, i.e., convergence in L_2 norm, which is specified by Definition 5.2.1.

Definition 6.2.1. A second-order process $\{\xi(t); t \in T\}$ is said to be L_2 continuous [or continuous in quadratic mean (q.m.)] at a point $t \in T$ if and only if

$$\xi(t+h) \xrightarrow{\text{q.m.}} \xi(t) \quad \text{as } h \to 0. \tag{6.2.1}$$

According to the definition (5.2.1), this is equivalent to

$$\|\xi(t+h) - \xi(t)\|^2 = E|\xi(t+h) - \xi(t)|^2 \to 0$$

as $h \to 0$.

If a process is q.m. continuous at every $t \in T$, we will say it is a q.m. continuous process.

The following two propositions establish a relation between q.m. continuity of a stochastic process and continuity of its covariance function.

Proposition 6.2.1. *Let $\{\xi(t); t \in T\}$ be an L_2 process with $C(s,t) = E\{\xi(s)\overline{\xi(t)}\}$. The process is q.m. continuous at a point $t \in T$ if and only if $C(\cdot,\cdot)$ is continuous at (t,t).*

Proof. Set

$$\begin{aligned}\|\xi(t+h) - \xi(t)\|^2 &= C(t+h,t+h) - C(t+h,t) - C(t,t+h) + C(t,t)\\ &= C(t+h,t+h) - C(t,t) - (C(t+h,t) - C(t,t))\\ &\quad - (C(t,t+h) - C(t,t)).\end{aligned}$$

From this, it is clear that the process is q.m. continuous at the point $t \in T$ if $C(\cdot,\cdot)$ is continuous at (t,t).

Conversely, if $\{\xi(t); t \in T\}$ is q.m. continuous at point $t \in T$, then

$$\begin{aligned}|C(t+h,t+h') - C(t,t)| &= |E\xi(t+h)\overline{\xi(t+h')} - E\xi(t)\overline{\xi(t)}|\\ &= |E(\xi(t+h) - \xi(t))\overline{\xi(t+h')}\\ &\quad + E(\overline{\xi(t+h') - \xi(t)})\xi(t)|\\ &\le |E(\xi(t+h) - \xi(t))\overline{\xi(t+h')}|\\ &\quad + |E(\overline{\xi(t+h') - \xi(t)})\xi(t)|.\end{aligned}$$

From this and the Schwarz inequality (5.1.1), the assertion follows. □

The next proposition shows that q.m. continuity of $\xi(t)$ on T implies continuity of $C(s,t)$ on $T \times T$.

Proposition 6.2.2. *If $C(t,t) = R(t)$ is continuous at every $t \in T$, $C(\cdot,\cdot)$ is continuous on $T \times T$.*

Proof. Consider

$$\begin{aligned}|C(s+h,t+h') - C(s,t)| &= |E\xi(s+h)\overline{\xi(t+h')} - E\xi(s)\overline{\xi(t)}|\\ &= |E(\xi(s+h) - \xi(s))\overline{\xi(t+h')}\\ &\quad + E(\overline{\xi(t+h') - \xi(t)})\xi(s)|\\ &\le |E(\xi(s+h) - \xi(s))\overline{\xi(t+h')}|\\ &\quad + |E(\overline{\xi(t+h') - \xi(t)})\xi(s)|.\end{aligned}$$

Again applying the Schwartz inequality and taking into account the previous proposition, the assertion follows. □

Remark 6.2.1. Continuity in q.m. of a second-order process does not imply sample function continuity. As an example, consider a time-homogeneous Poisson process $N(t)$ (see Remark 2.4.2) with $E\{N(t)\} = \lambda t$. As is known, this process has independent increments so that, for any $0 < s < t$,

$$\begin{aligned} E\{N(s)N(t)\} &= E(N(t) - N(s))N(s) + EN^2(s) \\ &= \lambda(t - s)\lambda s + \lambda s(1 + \lambda s), \end{aligned}$$

which yields

$$C(s, t) = \lambda s.$$

Because $C(s, t)$ is a continuous function, the Poisson process $N(t)$ is q.m. continuous. However, its sample functions are step functions with probability 1.

Definition 6.2.2. A second-order random process $\{\xi(t); t \in T\}$ is said to have a derivative $\xi'(t)$ in q.m. at a point $t \in T$ if

$$\frac{\xi(t + h) - \xi(t)}{h} \xrightarrow{\text{q.m.}} \xi'(t) \quad \text{when } h \to 0.$$

The r.v.

$$\xi'(t) = \frac{d\xi(t)}{dt}$$

is called the q.m. derivative of the random process $\xi(t)$ at the point $t \in T$.

In the sequel, we will need one more definition.

Definition 6.2.3. The second generalized derivative of a covariance function $C(s, t)$ is defined as the limit (if it exists) of the quotient

$$\frac{1}{hh'}\{C(s + h, t + h') - C(s + h, t) - C(s, t + h') + C(s, t)\}$$

as $h, h' \to 0$, which is denoted by

$$\frac{\partial^2 C(s, t)}{\partial s \, \partial t}. \tag{6.2.2}$$

Proposition 6.2.3. *Let $\{\xi(t); t \in T\}$ be a second-order process. A necessary and sufficient condition for q.m. differentiability of $\xi(t)$ at $t \in T$ is that the generalized derivative* (6.2.2) *exists.*

Proof. Write

$$\begin{aligned} &E\left\{\frac{\xi(t + h) - \xi(t)}{h} \, \overline{\frac{\xi(t + h') - \xi(t)}{h'}}\right\} \\ &\quad = \frac{1}{hh'}\{C(t + h, t + h') - C(t, t + h') - C(t + h, t) + C(t, t)\} \end{aligned}$$

and the assertion follows from the Loève criterion (see Proposition 5.2.5) □

Corollary 6.2.1. *If $\{\xi(t); t \in T\}$ is q.m. differentiable at a point $t \in T$, $dE\{\xi(t)\}/dt$ exists and*

$$E\{\xi'(t)\} = \frac{d}{dt}E\{\xi(t)\}. \tag{6.2.3}$$

As a matter of fact, this is implied by the q.m. differentiability of the process at the point $t \in T$ and the inequality

$$\left|E\left\{\xi'(t) - \frac{\xi(t+h) - \xi(t)}{h}\right\}\right| \leq \left(E\left|\xi'(t) - \frac{\xi(t+h) - \xi(t)}{h}\right|^2\right)^{1/2}.$$

If a second-order process $\{\xi(t); t \in T\}$ is q.m. differentiable at every $t \in T$, then $\{\xi'(t); t \in T\}$ is also a second-order random process.

Proposition 6.2.4. *Let $\{\xi(t); t \in T\}$ be a second-order process with covariance function $C(s, t)$. If the generalized derivative*

$$\frac{\partial^2 C(s,t)}{\partial s\, \partial t} \tag{6.2.4}$$

exists for every $s = t$ and $t \in T$, then $\xi(t)$ is q.m. *differentiable on T. In addition,*

$$E\{\xi'(s)\overline{\xi(t)}\} = \frac{\partial C(s,t)}{\partial s} \tag{6.2.5}$$

and

$$E\{\xi'(s)\overline{\xi'(t)}\} = \frac{\partial^2 C(s,t)}{\partial s\, \partial t}. \tag{6.2.6}$$

Proof. Only formulas (6.2.5) and (6.2.6) require a proof. Write

$$\begin{aligned} E\{\xi'(s)\overline{\xi(t)}\} &= \lim_{h\to 0} E\left(\overline{\xi(t)}\frac{\xi(s+h) - \xi(s)}{h}\right) \\ &= \lim_{h\to 0} \frac{C(s+h,t) - C(s,t)}{h} \\ &= \frac{\partial C(s,t)}{\partial s}. \end{aligned}$$

Similarly,

$$\begin{aligned} E\xi'(s)\overline{\xi'(t)} &= \lim_{h,h'\to 0} E\left(\frac{\xi(s+h) - \xi(s)}{h}\,\frac{\overline{\xi(t+h') - \xi(t)}}{h'}\right) \\ &= \lim_{h,h'\to 0} \frac{C(s+h,t+h') - C(s,t+h') - C(s+h,t) + C(s,t)}{hh'} \\ &= \frac{\partial^2 C(s,t)}{\partial s\, \partial t}. \end{aligned}$$

The last result implies that the second generalized derivative exists everywhere on $T \times T$. This proves the proposition. □

Remark 6.2.1. The concept of a stochastic integral in quadratic mean was discussed in some detail in Chapter 3, Section 3.8, but only in the case of real second-order processes. The same concept and results hold in the case of complex-valued second-order processes.

6.3. Eigenvalues and Eigenfunctions of $C(s,t)$

Let $\{\xi(t); t \in T\}$ be a complex-valued second-order stochastic process on a probability space $\{\Omega, \mathscr{B}, P\}$, such that $E\{\xi(t)\} = 0$ for each $t \in T$. Let $C(s,t)$ be its covariance function. In this section, we will give a brief review of some basic properties of eigenvalues and eigenfunctions of $C(s,t)$. This is required for the problem of orthogonal expansions of second-order stochastic processes which will be discussed in the next section.

Denote, as before, by $\langle\overline{\xi(T)}\rangle$ the closed linear manifold spanned by $\{\xi(t); t \in T\}$ (see Definition 5.3.4). Clearly, $\langle\overline{\xi(T)}\rangle$ is a (Hilbert) subspace of $L_2\{\Omega, \mathscr{B}, P\}$. As we have established in Chapter 5, Section 5.6, if the process $\xi(t)$ is q.m. continuous on T, the subspace $\langle\overline{\xi(T)}\rangle$ is separable. In other words, there exists a countable everywhere dense subset in $\langle\overline{\xi(T)}\rangle$. From this, by means of the well-known Gram–Schmidt orthogonalization procedure (see Section 5.6 of Chapter 5) we can always construct an orthonormal family $\{Z_k\}_1^\infty$, which is complete in $\langle\overline{\xi(T)}\rangle$. In such a case, according to Proposition 5.5.2, for every $t \in T$, $\xi(t)$ admits an expansion of the form

$$\xi(t) = \sum_{k=1}^{\infty} \beta_k(t) Z_k, \tag{6.3.1}$$

which is convergent in q.m., where

$$\beta_k(t) = (\xi(t), Z_k), \quad k = 1, 2, \ldots. \tag{6.3.2}$$

From (6.3.1), it seems to follow that under certain conditions we may write

$$\begin{aligned} C(s,t) &= (\xi(s), \xi(t)) = E\{\xi(s)\overline{\xi(t)}\} \\ &= E\left\{\sum_{i=1}^{\infty}\sum_{j=1}^{\infty} \beta_i(s) Z_i \overline{\beta_j(t) Z_j}\right\} \\ &= \sum_{j=1}^{\infty} \beta_j(s)\overline{\beta_j(t)}. \end{aligned} \tag{6.3.3}$$

As we shall see later, $\{\beta_k(t)\}_1^\infty$ is an orthogonal sequence of functions such that

$$(\beta_i(t), \beta_j(t)) = \int_a^b \beta_i(t)\overline{\beta_j(t)}\,dt = \begin{cases} 0, & i \neq j \\ 1/\lambda_i, & i = j \end{cases} \tag{6.3.4}$$

(when $T = [a, b]$). In such a case, if term-by-term integration of (6.3.3) is permitted, we obtain readily that, for all $k = 1, 2, \ldots$,

$$\beta_k(s) - \lambda_k \int_a^b C(s, t)\beta_k(t)\,dt = 0. \tag{6.3.5}$$

This is a Fredholm linear integral equation of the second kind. Therefore, the Fourier coefficients in expansion (6.3.1) are solutions of the integral equation (6.3.5). For this reason, it seems appropriate to list some basic properties of integral equations of the form (6.3.5)

Consider the integral equation

$$\varphi(x) - \lambda \int_a^b K(x, y)\varphi(y)\,dy = 0, \tag{6.3.6}$$

where the kernel $K(x, y)$ is a given (real or complex) function, $\varphi(\cdot)$ is unknown, and λ is a parameter. Equations of this type are called Fredholm homogeneous linear integral equations of the second kind. In general, Equation (6.3.6) has only the trivial solution $\varphi(x) \equiv 0$. For certain critical values of λ, however, there may exist nontrivial solutions.

The values $\lambda_0, \lambda_1, \ldots$ for which nontrivial solutions of (6.3.6) exist are called "eigenvalues" of the kernel $K(x, y)$, and the corresponding φ's are eigenfunctions. The eigenvalues and eigenfunctions depend upon the kernel $K(x, y)$.

Of particular interest here are kernels which are "Hermitian," i.e.,

$$K(x, y) = \overline{K(y, x)}. \tag{6.3.7}$$

As we shall see, such kernels possess a number of important properties, which collectively permit a thorough analysis of the integral equations in question. We now discuss some basic properties of eigenvalues of a Hermitian kernel. The natural context for our discussions will be the Hilbert space $L_2[a, b]$ of square integrable complex-valued functions defined on the interval $[a, b]$. The symbol $\|\cdot\|$ will be used to denote the norm and

$$(f, g) = \int_a^b f(x)\bar{g}(x)\,dx$$

is the inner product. In the case of square integrable complex-valued functions of two independent variables, we will write

$$\|K\| = \left\{\int_a^b \int_a^b |K(x, y)|^2\,dx\,dy\right\}^{1/2}. \tag{6.3.8}$$

Finally, we shall often write the linear transformation (6.3.6) as

$$\varphi = \lambda K\varphi. \tag{6.3.9}$$

Proposition 6.3.1. *Any non-null Hermitian kernel $K(x, y)$ satisfying $\|K\| < \infty$ must have at least one eigenvalue λ.*

In other words, if the kernel $K(x, y)$ is Hermitian, Equation (6.3.6) must have nontrivial solutions. The next proposition is easy to prove.

Proposition 6.3.2. *The eigenvalues of Hermitian kernels are real.*

PROOF. Assume that $\|\varphi\| \neq 0$; then

$$\lambda(K\varphi, \varphi) = (\lambda K\varphi, \varphi) = (\varphi, \varphi) = \|\varphi\|^2,$$

so that $\lambda(K\varphi, \varphi)$ is positive. On the other hand, because K is Hermitian,

$$(K\varphi, \varphi) = (\varphi, \overline{K}\varphi) = (\varphi, K\varphi) = \overline{(K\varphi, \varphi)},$$

which, with the above, implies that λ must be real. □

Corollary 6.3.1. *The eigenvalues of a Hermitian kernel $K(x, y)$ and its conjugate $\overline{K(x, y)}$ are identical.*

Corollary 6.3.2. *If the Hermitian kernel $K(x, y)$ is positive definite, its eigenvalues are positive. As a matter of fact,*

$$\int_a^b \int_a^b K(x, y)\overline{\varphi(x)}\varphi(y)\, dx\, dy > 0$$

and

$$\left(\frac{1}{\lambda}\int_a^b \overline{\varphi}(x)\, dx\right)\lambda \int_a^b K(x, y)\varphi(y)\, dy > 0.$$

Consequently,

$$(1/\lambda)\|\varphi\|^2 > 0,$$

which implies that $\lambda > 0$.

Proposition 6.3.3. *To every eigenvalue λ of a Hermitian kernel $K(x, y)$, there corresponds at least one eigenfunction. The number of linearly independent eigenfunctions corresponding to a given eigenvalue is finite.*

Remark 6.3.1. A kernel of the type

$$K(x, y) = \sum_{i=1}^{n} h_i(x) f_i(y), \tag{6.3.10}$$

where $\{h_i\}_1^n$ and $\{f_i\}_1^n \subset L_2[a, b]$ are two families of linearly independent functions, is called a Pincherle–Goursat kernel.

Proposition 6.3.4. *Every nonzero Hermitian kernel either has a countably infinite number of eigenvalues or is a Pincherle–Goursat kernel.*

We are not going to prove this proposition here. However, we shall prove the following result.

Proposition 6.3.5. *Two eigenfunctions,* $\varphi(\cdot)$ and $\psi(\cdot)$, *of a Hermitian kernel* $K(x, y)$, *corresponding to two different eigenvalues* λ_1 *and* λ_2, *are orthogonal to one another.*

PROOF. Note that

$$(\varphi, \psi) = (\lambda_1 K\varphi, \psi) = \lambda_1 (K\varphi, \psi) = \lambda_1 (\varphi, \bar{K}\psi)$$
$$= \frac{\lambda_1}{\lambda_2}(\varphi, \lambda_2 \bar{K}\psi) = \frac{\lambda_1}{\lambda_2}(\varphi, \lambda_2 K\psi) = \frac{\lambda_1}{\lambda_2}(\varphi, \psi),$$

which proves the assertion. □

6.4. Karhunen–Loève Expansion

Let $\{\xi(t); t \in [a, b]\}$ be second-order complex-valued q.m. continuous process defined on a probability space $\{\Omega, \mathscr{B}, P\}$ with $E\{\xi(t)\} = 0$. Its covariance function $C(s, t)$ is non-negative definite Hermitian and, according to Proposition 2.2, continuous on $[a, b] \times [a, b]$. Unless $C(s, t)$ is a Pincherle–Goursat kernel, the covariance function has a countably infinite number of eigenvalues $\{\lambda_k\}_1^\infty$ (see Proposition 6.3.4) such that $\lambda_k > 0$ for all $k = 1, 2, \ldots$ (see Corollary 6.3.2). The corresponding eigenfunctions $\{\varphi_k(t)\}_1^\infty$ are continuous on $[a, b]$. In addition, we will assume that

$$(\varphi_i, \varphi_j) = \begin{cases} 0 & \text{if } i \neq j \\ 1 & \text{if } i = j \end{cases}$$

for all $i, j = 1, 2, \ldots$ (see Proposition 6.3.5).

Assume that $C(s, t)$ is square integrable; then, for any fixed $s \in [a, b]$, the Fourier series of $C(s, t)$ with respect to s is

$$C(s, t) \sim \sum_{k=1}^{\infty} c_k(t)\varphi_k(s),$$

where

$$c_k(t) = \int_a^b C(s, t)\bar{\varphi}_k(s)\,dt = \frac{1}{\lambda_k}\bar{\varphi}_k(t).$$

Consequently,

$$C(s, t) \sim \sum_{k=1}^{\infty} \frac{1}{\lambda_k}\bar{\varphi}_k(t)\varphi_k(s).$$

If the system of eigenfunctions $\{\varphi_k(t)\}_1^\infty$ is complete, we would have that

$$C(s, t) = \sum_{k=1}^{\infty} \frac{1}{\lambda_k}\bar{\varphi}_k(t)\varphi_k(s)$$

in mean square (see Proposition 5.5.2). In other words,

$$C(s,t) = \underset{n\to\infty}{\text{l.i.m.}} \sum_{k=1}^{n} \frac{1}{\lambda_k} \overline{\varphi_k(t)} \varphi_k(s).$$

The following proposition due to *Mercer* (1909) holds for all square integrable continuous Hermitian kernels whose eigenvalues are all of one sign.

Proposition 6.4.1. *Let $\{\xi(t); t \in [a,b]\}$ be a second-order q.m. continuous stochastic process with $E\{\xi(t)\} = 0$ and square integrable covariance function $C(s,t)$ with eigenvalues $\{\lambda_k\}_1^\infty$ and eigenfunctions $\{\varphi_k(t)\}_1^\infty$, which form an orthonormal sequence. Then,*

$$C(s,t) = \sum_{k=1}^{\infty} \frac{1}{\lambda_k} \overline{\varphi_k(s)} \varphi_k(t), \tag{6.4.1}$$

where the infinite series converges absolutely and uniformly on $[a,b] \times [a,b]$.

A proof of this proposition can be found in Riesz and Sz.-Nagy (1955).

We now embark on a proof of a result known as the Karhunen–Loève orthogonal expansion. As before, let

$$\{\xi(t); t \in [a,b]\} \tag{6.4.2}$$

be a second-order q.m. continuous process with $E\{\xi(t)\} = 0$ and covariance function $C(s,t)$. According to Proposition 3.8.1, the integrals

$$\xi_k = (\lambda_k)^{1/2} \int_a^b \xi(t) \overline{\varphi_k(t)}\, dt \tag{6.4.3}$$

exist for all $k = 1, 2, \ldots$ and represent r.v.'s. Clearly,

$$E\{\xi_k\} = 0, \quad k = 1, 2, \ldots$$

and

$$\begin{aligned} E\{\xi_i \bar{\xi}_j\} &= (\lambda_i \lambda_j)^{1/2} E\left(\int_a^b \int_a^b \xi(s) \overline{\varphi_i(s)}\, \overline{\xi(t)} \varphi_j(t)\, ds\, dt \right) \\ &= (\lambda_i \lambda_j)^{1/2} \int_a^b \int_a^b C(s,t) \varphi_j(t) \overline{\varphi_i(s)}\, dt\, ds. \end{aligned}$$

From the fact that

$$\varphi_j(s) = \lambda_j \int_a^b C(s,t) \varphi_j(t)\, dt,$$

we obtain

$$E\{\xi_i \bar{\xi}_j\} = \left(\frac{\lambda_i}{\lambda_j}\right)^{1/2} \int_a^b \varphi_j(s) \overline{\varphi_i(s)}\, ds = \begin{cases} (\lambda_i/\lambda_j)^{1/2} & \text{if } i = j \\ 0 & \text{if } i \neq j. \end{cases} \tag{6.4.4}$$

This clearly implies that the r.v.'s $\{\xi_k\}_1^\infty$ are uncorrelated.

Proposition 6.4.2 (Karhunen–Loève). *On $[a,b]$, the stochastic process (6.4.2) admits an orthogonal expansion of the form*

$$\xi(t) = \sum_{k=1}^{\infty} \frac{1}{(\lambda_k)^{1/2}} \varphi_k(t)\xi_k, \tag{6.4.5}$$

where the infinite series in (6.4.5) converges in q.m. to $\xi(t)$ uniformly in t.

PROOF. Consider

$$\begin{aligned} E\left|\xi(t) - \sum_{k=1}^{n} \frac{1}{(\lambda_k)^{1/2}} \varphi_k(t)\xi_k\right|^2 \\ &= E\left(\xi(t) - \sum_{k=1}^{n} \frac{1}{(\lambda_k)^{1/2}} \varphi_k(t)\xi_k\right)\left(\overline{\xi(t)} - \sum_{k=1}^{n} \frac{1}{(\lambda_k)^{1/2}} \overline{\varphi_k(t)\xi_k}\right) \\ &= C(t,t) - \sum_{k=1}^{n} \frac{1}{(\lambda_k)^{1/2}} \varphi_k(t) E\{\overline{\xi(t)}\xi_k\} - \sum_{k=1}^{n} \frac{1}{(\lambda_k)^{1/2}} \overline{\varphi_k(t)} E\{\xi(t)\overline{\xi_k}\} \\ &\quad + \sum_{k=1}^{n} \frac{1}{(\lambda_k)^{1/2}} \varphi_k(t)\overline{\varphi_k(t)}. \end{aligned} \tag{6.4.6}$$

Using (6.4.3), it is not difficult to verify that

$$\begin{aligned} E\{\overline{\xi(t)}\xi_k\} &= (\lambda_k)^{1/2} E\left\{\int_a^b \xi(s)\overline{\xi(t)}\overline{\varphi_k(s)}\, ds\right\} \\ &= (\lambda_k)^{1/2} \int_a^b C(s,t)\overline{\varphi_k(s)}\, ds = \frac{1}{(\lambda_k)^{1/2}} \overline{\varphi_k(t)}. \end{aligned}$$

Similarly,

$$E\{\xi(t)\xi_k\} = \frac{1}{(\lambda_k)^{1/2}} \varphi_k(t).$$

This and (6.4.6) then yield

$$E\left|\xi(t) - \sum_{k=1}^{n} \frac{1}{(\lambda_k)^{1/2}} \varphi_k(t)\xi_k\right|^2 = C(t,t) - \sum_{k=1}^{n} \frac{1}{\lambda_k} \varphi_k(t)\overline{\varphi_k(t)}.$$

By letting $n \to \infty$ and invoking the Mercer theorem, we see that the series in (6.4.5) converges in q.m. to $\xi(t)$ for every $t \in [a,b]$. Finally, because $C(t,t)$ is continuous, the convergence is uniform due to Dini's theorem. This proves the assertion. □

EXAMPLE 6.4.1. Suppose that $\xi(t)$ is a real Gaussian process. Then, the r.v.'s ξ_n defined by (6.4.3) are normal and, as such, due to (6.4.4), they are independent. In addition, because

$$\sum_{k=1}^{\infty} E\left\{\frac{1}{(\lambda_k)^{1/2}} \varphi_k(t)\xi_k\right\}^2 = \sum_{k=1}^{\infty} \frac{1}{\lambda_k} \varphi_k^2(t) = C(t,t),$$

the series in (6.4.5) converges (a.s.).

Example 6.4.2. Let $\{\xi(t); t \in [0,1]\}$ be a standard Brownian motion process (see Definition 3.1.1). As we know [see (3.1.4)], its covariance function is

$$C(s,t) = \min\{s,t\}. \tag{6.4.7}$$

To determine its eigenvalues and eigenfunctions, consider the integral equation

$$\varphi_k(s) - \lambda_k \int_0^1 C(s,t)\varphi_k(t)\,dt = 0, \tag{6.4.8}$$

where $C(s,t)$ is specified by (6.4.7). Now write this equation as

$$\lambda_k \int_0^s t\varphi_k(t)\,dt + \lambda_k s \int_s^1 \varphi_k(t)\,dt = \varphi_k(s). \tag{6.4.9}$$

Differentiating both sides with respect to s, we obtain

$$\lambda_k \left\{ s\varphi_k(s) + \int_s^1 \varphi_k(t)\,dt - s\varphi_k(s) \right\} = \varphi_k'(s)$$

or

$$\varphi_k'(s) = \lambda_k \int_s^1 \varphi_k(t)\,dt. \tag{6.4.10}$$

Differentiating once more, we have

$$\varphi_k'' = -\lambda_k \varphi_k(s). \tag{6.4.11}$$

As is well known, the general solution of this second-order linear differential equation is

$$\varphi_k(s) = C_1 \cos \lambda_k s + C_2 \sin(\lambda_k)^{1/2} s. \tag{6.4.12}$$

From Equation (6.4.9), it clearly follows that $\varphi_k(0) = 0$. On the other hand, from Equation (6.4.10), we deduce that

$$\varphi_k'(1) = 0. \tag{6.4.13}$$

From the first initial condition and (6.4.13), we have

$$C_2(\lambda_k)^{1/2} \cos((\lambda_k)^{1/2}) = 0,$$

which implies that $(\lambda_k)^{1/2} = \pi(2k-1)/2$, so that

$$\varphi_k(s) = C_2 \sin\left(\frac{\pi}{2}\right)(2k-1)s, \quad k = 1, 2, \ldots. \tag{6.4.14}$$

Finally, to determine the constant C_2, we use the fact that $\{\varphi_k(t)\}_1^\infty$ are orthonormal functions, so that

$$1 = \int_0^1 \varphi_k^2(t)\,dt = C_2^2 \int_0^1 \sin^2\left(\frac{\pi}{2}\right)(2k-1)t\,dt.$$

From this, we deduce that

$$C_2 = (2)^{1/2}$$

and, thus,

$$\varphi_k(t) = (2)^{1/2} \sin\left(\frac{\pi}{2}\right)(2k-1)t. \tag{6.4.15}$$

Finally, we have

$$\lambda_k = (k - \tfrac{1}{2})^2\pi^2, \quad k = 1, 2, \ldots. \tag{6.4.16}$$

Therefore, according to (6.4.5),

$$\xi(t) = (2)^{1/2} \sum_{k=1}^{\infty} \frac{\sin \pi(k-\frac{1}{2})t}{(k-\frac{1}{2})\pi} \xi_k, \tag{6.4.17}$$

where $\{\xi_k\}_1^\infty$ is a sequence of independent $N(0,1)$ r.v.'s.

6.5. Stationary Stochastic Processes

Let

$$\{\xi(t); t \in R\} \tag{6.5.1}$$

be a second-order complex-valued stochastic process on a probability space $\{\Omega, \mathscr{B}, P\}$. The stochastic process is called "wide sense stationary" (see Definition 1.7.9) if, for all $s, t, h \in R$,

$$E\{\xi(t+h)\} = E\{\xi(t)\} \quad \text{and} \quad E\{\xi(s+h)\overline{\xi(t+h)}\} = E\{\xi(s)\overline{\xi(t)}\}. \tag{6.5.2}$$

In this case, $E\{\xi(t)\}$ is clearly a constant, say μ, and the covariance function $C(s,t)$ is a function of $t - s$, i.e.,

$$\mu = E\{\xi(t)\}, \qquad E\{\xi(s)\overline{\xi(s+\tau)}\} - |\mu|^2 = C(\tau). \tag{6.5.3}$$

From (6.5.3), we deduce that $C(-\tau) = \overline{C(\tau)}$. In what follows, we will assume, without loss of generality, that

$$E\{\xi(t)\} = 0 \quad \text{and} \quad \operatorname{Var}\{\xi(t)\} = C(0) = 1.$$

The stochastic process (6.5.1), regarded as a family of complex-valued r.v.'s is a subset of $L_2\{\Omega, \mathscr{B}, P\}$. Because for every $t \in R$, $\xi(t)$ is a point in $L_2\{\Omega, \mathscr{B}, P\}$, the process (6.5.1) represents a curve in $L_2\{\Omega, \mathscr{B}, P\}$. Because

$$E\{\xi(t)\} = (\xi(t), 1) = 0 \quad \text{and} \quad (\xi(s), \xi(t)) = C(t-s), \tag{6.5.4}$$

this curve lies in the subspace which is the orthogonal complement (see Definition, 5.4.1) to $\{1\}$ in $L_2\{\Omega, \mathscr{B}, P\}$.

According to the theory of second-order processes, the stochastic process (6.5.1) is said to be "continuous" if it is q.m. continuous (see Definition 6.2.1), that is, if

$$E\{\xi|(t+h)-\xi(t)|^2\}\to 0$$

as $h\to 0$ for all $t\in(-\infty,\infty)$. From Chebyshev's inequality, it follows that (6.5.1) is also stochastically continuous. Indeed,

$$P\{|\xi(t+h)-\xi(t)|>\varepsilon\}\le\frac{1}{\varepsilon^2}E\{|\xi(t+h)-\xi(t)|^2\}\to 0$$

for all $t\in R$, as $h\to 0$. We now have the following result.

Proposition 6.5.1. *A wide sense stationary stochastic process is q.m. continuous if and only if the real part of its covariance function is continuous at* 0.

PROOF. For the process (6.5.1),

$$\begin{aligned}E\{|\xi(t+h)-\xi(t)|^2\} &= E(\xi(t+h)-\xi(t))(\overline{\xi(t+h)-\xi(t)})\\ &= E|\xi(t+h)|^2+E|\xi(t)|^2\\ &\quad - E\{\xi(t)\overline{\xi(t+h)}+\xi(t+h)\overline{\xi(t)}\}\\ &= 2C(0)-C(h)-\overline{C(h)}\\ &= 2(C(0)-\mathrm{Re}\{C(h)\})\\ &= 2\,\mathrm{Re}\{1-C(h)\}.\end{aligned}$$

The proof now follows directly. □

As an example consider the random telegraph process.

EXAMPLE 6.5.1. Let $\{\xi(t); t\in R\}$ be a real-valued random process where, for each $t\in R$, the r.v. $\xi(t)$ may assume only two values, -1 and $+1$, with

$$P\{\xi(t)=-1\}=\tfrac{1}{2},\qquad P\{\xi(t)=+1\}=\tfrac{1}{2}.$$

The sequence of transition times $\{\tau_k\}_{-\infty}^{\infty}$ forms a time-homogeneous Poisson process with a parameter $\lambda>0$. More specifically, if $p_k(u)$ is the probability of k transitions in $(t,t+u)$, then

$$p_k(u)=e^{-\lambda u}\left(\frac{(\lambda u)^k}{k!}\right),\quad k=0,1,\ldots.$$

Show that $\xi(t)$ is wide sense stationary and determine its covariance function.

Demonstration:

First we have, for all $t\in R$,

$$E\{\xi(t)\}=1\cdot P\{\xi(t)=1\}+(-1)P\{\xi(t)=-1\}=0.$$

On the other hand, after some straightforward calculations, we obtain

$$P\{\xi(t)=-1,\xi(t+\tau)=-1\}=P\{\xi(t)=1,\xi(t+\tau)=1\},$$

$$P\{\xi(t)=-1,\xi(t+\tau)=1\}=P\{\xi(t)=1,\xi(t+\tau)=-1\}.$$

Next, for $\tau > 0$,

$$P\{\xi(t+\tau) = 1 | \xi(t) = 1\} = \sum_{k=0}^{\infty} p_{2k}(\tau) = e^{-\lambda t}\,\mathrm{ch}(\lambda\tau).$$

Similarly,

$$P\{\xi(t+\tau) = 1 | \xi(t) = -1\} = \sum_{k=0}^{\infty} p_{2k+1}(\tau) = e^{-\lambda\tau}\,\mathrm{sh}(\lambda\tau),$$

where

$$\mathrm{ch}(x) = \tfrac{1}{2}(e^x + e^{-x}), \qquad \mathrm{sh}(x) = \tfrac{1}{2}(e^x - e^{-x}).$$

From this, we obtain

$$\begin{aligned} C(t, t+\tau) &= E\{\xi(t)\xi(t+\tau)\} \\ &= P\{\xi(t) = 1, \xi(t+\tau) = 1\} - P\{\xi(t) = -1, \xi(t+\tau) = 1\} \\ &\quad - P\{\xi(t) = 1, \xi(t+\tau) = -1\} + P\{\xi(t) = -1, \xi(t+\tau) = -1\} \\ &= e^{-\lambda\tau}\{\mathrm{ch}(\lambda\tau) - \mathrm{sh}(\lambda\tau)\} = e^{-2\lambda\tau}. \end{aligned}$$

If, however, $\tau < 0$,

$$C(t+\tau, t) = e^{2\lambda\tau}.$$

Therefore,

$$C(t, t+\tau) = C(\tau) = e^{-2\lambda|\tau|}.$$

This shows that the stochastic process $\xi(t)$ is wide sense stationary. Figure 6.1(a) is a graphical depiction of a sample function and Figure 6.1(b) a depiction of the covariance function $C(\tau)$ of the process $\xi(t)$.

6.6. Remarks on the Ergodicity Property

Consider an arbitrary random process $\xi(t)$ and suppose that we want to determine some characteristics of the process, such as its mean, variance, and covariance function. For the first two characteristics, one needs one-

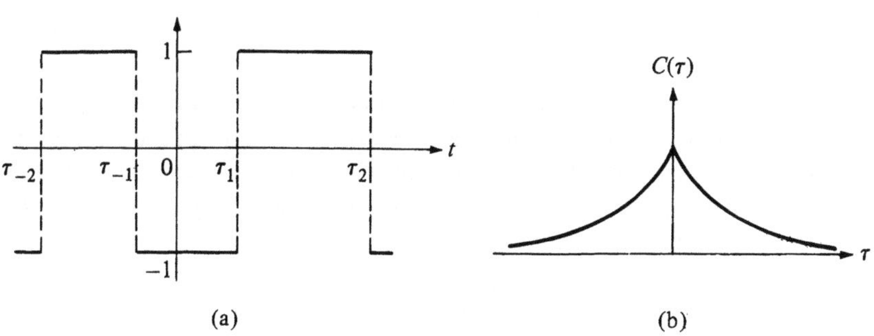

Figure 6.1

dimensional marginal distributions of the process. The evaluation of the covariance function requires bivariate marginal distributions of $\xi(t)$.

In the case when these distributions are not available, estimates of these characteristics can be achieved if enough sample functions of the process $\xi(t)$ are available.

Large classes of stationary random processes have an interesting property known as "ergodicity." This feature, roughly speaking, means that on the basis of a single realization over a sufficiently long time interval, we can obtain an estimate of the mean of the process and, possibly, of some other parameters. More specifically, if $\xi(t)$ is an ergodic stationary process, then the time average is equal to the ensemble average, namely,

$$\lim_{T\to\infty} \frac{1}{2T}\int_{-T}^{T} \xi(t)\,dt = E\{\xi(s)\} = \alpha, \tag{6.6.1}$$

where the convergence is in the q.m. sense. In general, if

$$\lim_{T\to\infty} \frac{1}{2T}\int_{-T}^{T} f(\xi(t))\,dt = E\{f(\xi(s))\} \tag{6.6.2}$$

holds for every Borel function $f(\cdot)$ such that

$$E|f(\xi(t))| < \infty,$$

we say that the process is ergodic. From this, we also obtain that

$$\begin{aligned} C(\tau) &= E\{\xi(t)\xi(t+\tau)\} \\ &= \lim_{T\to\infty} \frac{1}{2T}\int_{-T}^{T} \xi(t)\xi(t+\tau)\,dt. \end{aligned} \tag{6.6.3}$$

In general, it is not easy to give a simple sufficient condition for ergodicity. For Gaussian processes, however, we have the following result. Let $\{\xi(t); -\infty < t < \infty\}$ be Gaussian and stationary with

$$E\{\xi(t)\} = \alpha, \qquad R(\tau) = C(\tau) - \alpha^2.$$

Then $\xi(t)$ is ergodic if

$$\int_{-\infty}^{\infty} |R(\tau)|\,d\tau < \infty. \tag{6.6.4}$$

Sometimes it may happen that (6.6.2) holds for some function $f(\cdot)$. For instance, if it holds for linear functions only, we say that the process is ergodic with respect to its mean. This question can also be asked in a somewhat more direct way. Again, let $\{\xi(t); -\infty < t < \infty\}$ be a stationary process; we are interested in a sufficient condition for

$$\frac{1}{2T}\int_{-T}^{T} \xi(t)\,dt \xrightarrow{\text{q.m.}} \alpha, \tag{6.6.5}$$

where $\alpha = E\{\xi(t)\}$. In other words, we want to know when

$$\lim_{T\to\infty} E\left\{\frac{1}{2T}\int_{-T}^{T}\xi(t)\,dt - \alpha\right\}^2 = 0. \tag{6.6.6}$$

Clearly,

$$E\left\{\frac{1}{2T}\int_{-T}^{T}\xi(t)\,dt - \alpha\right\}^2 = \frac{1}{(2T)^2}\int_{-T}^{T}\int_{-T}^{T}(E\{\xi(t)\xi(s)\} - \alpha^2)\,dt\,ds$$
$$= \frac{1}{(2T)^2}\int_{-T}^{T}\int_{-T}^{T}R(t-s)\,dt\,ds.$$

We change the variables of integration by putting

$$\tau = t - s, \qquad z = t + s,$$
$$t = \tfrac{1}{2}(z+\tau), \qquad s = \tfrac{1}{2}(z-\tau).$$

The Jacobian of the transformation is

$$\begin{vmatrix} \dfrac{\partial t}{\partial \tau} & \dfrac{\partial t}{\partial z} \\ \dfrac{\partial s}{\partial \tau} & \dfrac{\partial s}{\partial z} \end{vmatrix} = \begin{vmatrix} \dfrac{1}{2} & \dfrac{1}{2} \\ \dfrac{1}{2} & -\dfrac{1}{2} \end{vmatrix} = -\frac{1}{2}.$$

On the other hand, the domains of integration are shown in Figure 6.2.

Therefore, we have

$$\frac{1}{(2T)^2}\int_{-T}^{T}\int_{-T}^{T}R(t-s)\,dt\,ds$$
$$= \frac{1}{2(2T)^2}\left(\int_{-2T}^{0}\int_{-\tau-2T}^{\tau+2T}R(\tau)\,dz\,d\tau + \int_{0}^{2T}\int_{\tau-2T}^{-\tau+2T}R(\tau)\,dz\,d\tau\right)$$

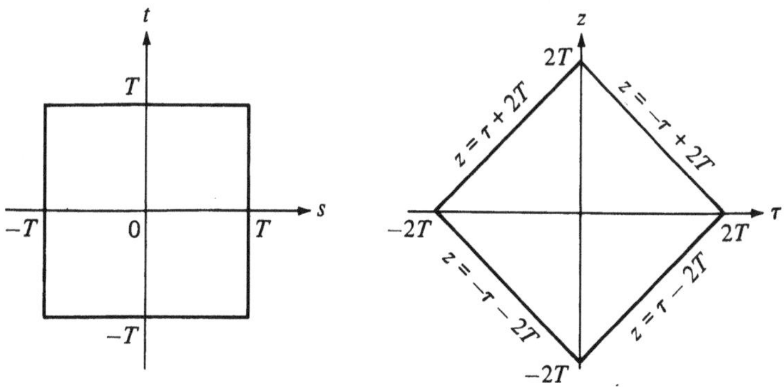

Figure 6.2

$$= \frac{1}{2(2T)^2}\left(\int_{-2T}^{0} R(\tau)(4T)\,d\tau + \int_{0}^{2T} R(\tau)(4T - 2\tau)\,d\tau\right)$$

$$= \frac{1}{2T}\left[\int_{-2T}^{0} R(\tau)\,d\tau + \int_{0}^{2T} R(\tau)\left(1 - \frac{\tau}{2T}\right)d\tau\right]$$

$$\leq \frac{1}{2T}\left[\int_{-2T}^{0} |R(\tau)|\,d\tau + \int_{0}^{2T} |R(\tau)|\left(1 - \frac{\tau}{2T}\right)d\tau\right]$$

$$\leq \frac{1}{2T}\int_{-2T}^{2T} |R(\tau)|\,d\tau \leq \frac{1}{2T}\int_{-\infty}^{\infty} |R(\tau)|\,d\tau.$$

Thus, (6.6.5) holds if

$$\int_{-\infty}^{\infty} |R(\tau)|\,d\tau < \infty,$$

which is a simple sufficient condition for ergodicity.

Problems and Complements

6.1. Show that the following functions are non-negative definite:

(i) $$R_1(s,t) = \min(s,t), \quad s, t \geq 0;$$

(ii) $$R_2(s,t) = \begin{cases} 1 - |t-s|, & |t-s| \leq 1 \\ 0, & |t-s) > 1 \end{cases} \quad s, t \in R;$$

(iii) $$R_3(s,t) = \min(s,t) - st, \quad s, t \in [0,1];$$

(iv) $$R_4(s,t) = \exp\{-|t-s|\}, \quad s, t \in R.$$

6.2. Let $C(s,t)$ be a covariance function and $P_n(\cdot)$ a polynomial with positive coefficients. Show that

$$C_1(s,t) = P_n(C(s,t))$$

is also a covariance function.

6.3. Consider the stochastic process $\{\xi(t); t \geq 0\}$, where

$$\xi(t) = X\cos(\eta t + \theta)$$

and $X > 0$, $\eta > 0$ are r.v.'s independent of θ with finite second moments, whereas θ is uniform in $[0, 2\pi]$. Show that $\xi(t)$ is stationary.

6.4. Let $h_1(t), \ldots, h_n(t)$ be real functions and $a_1, \ldots, a_n$ positive constants. Show that

$$C(s,t) = \sum_{k=1}^{n} a_k h_k(s) h_k(t)$$

is a covariance function.

6.5. Let $\{\xi(t); t \geq 0\}$ be a stochastic process defined by

$$\xi(t) = \alpha\sin(\beta t + X),$$

where $\alpha > 0$ and $\beta > 0$ are constants and $X \sim N(0,1)$. Determine $E\{\xi(t)\}$ and $E\{\xi(s)\xi(t)\}$. Is the process wide sense stationary?

6.6. Let $\{X_n\} \subset L_2$ and $X_n \xrightarrow{\text{q.m.}} X$. Is $X \in L_2$? Show that

(i) $EX_n \to EX$,
(ii) $E|X_n|^2 \to E|X|^2$,
(iii) $(X_n, X) \to (X, X)$.

6.7. Let $\{Z_n\}$, $\{U_n\} \subset L_2$ be such that

$$Z_n \xrightarrow{\text{q.m.}} Z, \qquad U_n \xrightarrow{\text{q.m.}} U.$$

Show that $EZ_n\overline{U}_n \to EZ\overline{U}$.

6.8. Let $\{X_n\} \subset L_2$ and $X_n \xrightarrow{\text{q.m.}} X$. Find a condition under which $X_n \xrightarrow{\text{q.m.}} X \Rightarrow X_n \to X$ (a.s.).

6.9. Let $\{\xi(t); t \in T\} \subset L_2$ be a random process and $t_0 \in T$. Show that $\xi(t) \xrightarrow{\text{q.m.}} Z \in L_2$ as $t \to t_0$ if and only if for all $\{t_n\}_1^\infty$, $\{s_n\}_1^\infty$ such that $t_n \to t_0$, $s_n \to t_0$, we have

$$E\xi(t_n)\overline{\xi(s_n)} \to C_0 \quad \text{(constant)}.$$

6.10. If $\{\xi(t); t \in [a,b]\} \subset L_2$ is L_2-continuous, show that, in L_2,

$$\frac{d}{dt}\int_a^t \xi(s)\,ds = \xi(t), \quad a \le t \le b.$$

6.11. Find the Karhunen–Loève expansion on the interval $[0,1]$ of an L_2 process with covariance function

$$C(s,t) = st.$$

6.12. Determine eigenvalues and eigenfunctions of the Fredholm homogeneous integral equation

$$\varphi(x) - \lambda \int_0^\pi C(x,t)\varphi(t)\,dt = 0,$$

where

$$C(x,t) = \begin{cases} \cos x \sin t, & 0 \le x \le t \\ \cos t \sin x, & t \le x \le \pi. \end{cases}$$

6.13. Let $C(s,t)$ be given by

$$C(s,t) = \begin{cases} s(t-1), & 0 \le s \le t \\ t(s-1), & t \le s \le 1. \end{cases}$$

Find the eigenvalues and complete orthonormal eigenfunctions.

6.14. Consider $(0 \le x, y \le 1)$

$$K(x,y) = \min(x,y) - xy.$$

Show that the kernel is non-negative definite and find its eigenvalues and eigenfunctions.

6.15. Determine the eigenvalues and eigenfunctions of the symmetric kernel

$$K(s,t) = \begin{cases} t(s+1), & t \le s \le 1 \\ s(t+1), & t \le s \le 1. \end{cases}$$

CHAPTER 7

Spectral Analysis of Stationary Processes

7.1. Preliminaries

Let $\{\xi(t); t \in R\}$ be a wide sense stationary, complex-valued random process with $E\{\xi(t)\} = 0$ and

$$C(t) = E\{\xi(s)\overline{\xi(s+t)}\}. \tag{7.1.1}$$

In this chapter, we will continue to build a theory of this particular class of stochastic processes based on the covariance function alone, using methods discussed in Chapter 5.

As we have established in Chapter 6, the covariance $C(t)$ is a non-negative definite complex function of real argument. In this section, we will see that the class of such functions coincides with the class of complex functions of real argument $\phi(t)$ which can be written as

$$\phi(t) = K\phi_0(t), \quad K > 0, \tag{7.1.2}$$

where $\phi_0(t)$ is the characteristic function of a real-valued r.v. Consequently, every covariance function should have a representation of the form

$$\phi(t) = \int_{-\infty}^{\infty} e^{itx}\, dF(x), \tag{7.1.3}$$

where $F(\cdot) \geq 0$ is a nondecreasing bounded function on R. Let us first prove the following result.

Proposition 7.1.1. *The covariance function $C(t)$ is continuous on R if it is continuous at $t = 0$.*

PROOF. Because $C(t)$ is non-negative definite,

$$\sum_{i=1}^{n}\sum_{j=1}^{n} C(t_i - t_j)z_i\bar{z}_j \geq 0. \tag{7.1.4}$$

Set $n = 3$, $z_1 = 1$, $z_2 = z$, $z_3 = -z$, $t_1 = 0$, $t_2 = u$, $t_3 = v$ in (7.1.4) to obtain

$$\begin{aligned} &C(0) + \bar{z}C(-u) - \bar{z}C(-v) + zC(u) + |z|^2C(u) - |z|^2C(u - v) \\ &\quad - zC(v) - |z|^2C(u - v) + |z|^2C(0) \\ &= C(0) + 2\,\mathrm{Re}\{z[C(u) - C(v)]\} + 2|z|^2[C(0) - \mathrm{Re}\{C(u - v)\}] \geq 0. \end{aligned}$$

Writing

$$C(u) - C(v) = |C(u) - C(v)|e^{i\theta},$$

$$z = xe^{-\theta i}, \quad x \text{ real},$$

the last inequality becomes

$$C(0) + 2x|C(u) - C(v)| + 2x^2[C(0) - \mathrm{Re}\{C(u - v)\}] \geq 0.$$

Because this holds for all $x \in R$, the discriminant cannot be positive, so that

$$|C(u) - C(v)|^2 \leq 2C(0)[C(0) - \mathrm{Re}\{C(u - v)\}]. \tag{7.1.5}$$

Now, by assumption, $C(\cdot)$ is continuous at 0. Thus, because in the last inequality the right-hand side tends to zero as $u \to v$, so does the left-hand side. This proves the assertion. □

Remark 7.1.1. From (7.1.5) we see that $C(t)$ is uniformly continuous on R if it is continuous at zero.

The following is the celebrated Bochner–Khinchin theorem.

Proposition 7.1.2. *A complex-valued function $C(t)$ defined on R continuous at zero is the covariance function of a wide sense stationary stochastic process if and only if it can be written in the form*

$$C(t) = \int_{-\infty}^{\infty} e^{itx}\,dF(x), \tag{7.1.6}$$

where $F(\cdot)$ is a real nondecreasing bounded function on R.

The function $F(\cdot)$ is referred to as the spectral distribution of the process $\xi(t)$. It is uniquely defined up to an additive constant and we can always suppose that

$$F(-\infty) = 0, \qquad F(+\infty) = C(0). \tag{7.1.7}$$

In addition, we assume that $F(\cdot)$ is right-continuous. If $F(\cdot)$ is absolutely continuous, the derivative

$$f(\cdot) = F'(\cdot)$$

exists and is called the spectral density of the process.

Remark 7.1.2. Let $\varphi(\cdot)$: $R \to R_+ = [0, \infty)$ be a continuous symmetric function such that $\varphi(0) = 1$ and $\varphi(t) \to 0$ as $t \to \infty$. In addition, if $\varphi(\cdot)$ is convex on R_+, then $\varphi(\cdot)$ is a characteristic function. This result is due to Pólya. It is useful in establishing if a certain function is non-negative definite.

EXAMPLE 7.1.1. Let us see if the following functions are non-negative definite:

(i) $$B_0(t) = \begin{cases} 1 - |t| & \text{if } |t| \le 1 \\ 0 & \text{if } |t| > 1; \end{cases}$$

(ii) $$B_1(t) = e^{-|t|};$$

(iii) $$B_2(t) = e^{|t|}.$$

Clearly, $B_0(t)$ satisfies all the conditions of the Pólya theorem and, therefore, represents a characteristic function. The same conclusion holds for $B_1(t)$. However, $B_2(t)$ is not non-negative definite.

EXAMPLE 7.1.2. Let $\{Z_k\}_1^n \subset L_2\{\Omega, \mathscr{B}, P\}$ be an orthonormal family of complex-valued r.v.'s with $E\{Z_k\} = 0$, $k = 1, \ldots, n$; then

$$\xi(t) = \sum_{k=1}^{n} Z_k \exp(i\lambda_k t), \quad \lambda_k \text{ real},$$

is a wide sense stationary stochastic process. Its covariance function is

$$E\{\xi(s)\overline{\xi(s+t)}\} = (\xi(s), \xi(s+t)) = \sum_{k=1}^{n} \exp(i\lambda_k t).$$

The next proposition gives the spectral representation of a covariance function in the discrete case.

Proposition 7.1.3 (Herglotz's Theorem). *Let $C(n)$ be the covariance function of a wide sense stationary sequence $\{\xi_n\}_1^\infty$ with $E\{\xi_n\} = 0$, then*

$$C(n) = \int_{-\pi}^{\pi} e^{i\lambda n}\, dF(\lambda), \tag{7.1.8}$$

where $F(\cdot)$ is a bounded nondecreasing function with support $[-\pi, \pi]$.

The last two propositions will be proved in the next section.

Remark 7.1.3. If the covariance function $C(t)$ is absolutely integrable on $(-\infty, \infty)$, i.e., if

$$\int_{-\infty}^{\infty} |C(t)|\, dt < \infty, \tag{7.1.9}$$

then, as is known, $F(\cdot)$ is absolutely continuous and

$$f(x) = \frac{1}{2\pi} \int_{-\infty}^{\infty} e^{-ixt} C(t)\, dt. \tag{7.1.10}$$

If (7.1.9) does not hold, we have, for any $x_1 < x_2$,

$$F(x_2) - F(x_1) = \frac{1}{2\pi} \lim_{T\to\infty} \int_{-T}^{T} \frac{\exp(itx_1) - \exp(-itx_2)}{it} C(t)\,dt. \quad (7.1.11)$$

Remark 7.1.4. If $F(\cdot)$ is a symmetric function, i.e., if

$$F(x) = C(0) - F(-x+0), \quad (7.1.12)$$

the covariance function is real and

$$C(t) = \int_{-\infty}^{\infty} (\cos tx)\,dF(x). \quad (7.1.13)$$

In addition, if (7.1.9) holds,

$$f(x) = \frac{1}{\pi} \int_0^{\infty} C(t)(\cos tx)\,dt. \quad (7.1.14)$$

A covariance function is real if and only if (7.1.12) holds.

Example 7.1.3. For the random telegraph process (see Example 6.5.1), we have established that

$$C(t) = e^{-2\lambda|t|}, \quad -\infty < t < \infty.$$

This is clearly an absolutely integrable function so we can use the formula (7.1.14)

$$f(x) = \frac{1}{\pi} \int_0^{\infty} e^{-2\lambda t} \cos(xt)\,dt.$$

Integrating by parts twice, we obtain

$$\frac{1}{\pi} \int_0^{\infty} e^{-2\lambda t} \cos(xt)\,dt = \frac{1}{2\pi\lambda}\left(1 - \frac{x^2}{2\lambda} \int_0^{\infty} e^{-2\lambda t} \cos(xt)\,dt\right).$$

Therefore,

$$f(x) = \frac{2\lambda}{(4\lambda^2 + x^2)\pi}.$$

7.2. Proof of the Bochner–Khinchin and Herglotz Theorems

We shall first prove the Bochner–Khinchin theorem. For this purpose we need the following lemma.

Lemma 7.2.1. *Let $\varphi(\cdot)$ be bounded and integrable on $[-T, T]$ and such that, for all $x \in R$,*

$$f_T(x) = \int_{-T}^{T} e^{-itx}\varphi(t)\,dt \geq 0. \tag{7.2.1}$$

Then, $f_T(x)$ is integrable on $R = (-\infty, \infty)$.

PROOF. From (7.2.1), it is clear that $f_T(\cdot)$ is continuous and, therefore, integrable on every finite interval. Next, set

$$G_T(x) = \int_{-x}^{x} f_T(u)\,du. \tag{7.2.2}$$

Because $f_T(\cdot) \geq 0$, $G_T(x_1) \leq G_T(x_2)$ for every $x_1 \leq x_2$. We must show that $G_T(\cdot)$ is bounded. To this end, set ($s > 0$)

$$\phi_T(s) = \frac{1}{s}\int_{s}^{2s} G_T(x)\,dx \geq G_T(s).$$

Consequently, if $\phi_T(\cdot)$ is bounded, so is $G_T(\cdot)$. From (7.2.1) and (7.2.2), we have

$$G_T(x) = \int_{-x}^{x} du \int_{-T}^{T} e^{itu}\varphi(t)\,dt = 2\int_{-T}^{T} \frac{\sin tx}{t}\varphi(t)\,dt.$$

Therefore,

$$\begin{aligned}\phi_T(s) &= \frac{2}{s}\int_{s}^{2s} dx \int_{-T}^{T} \frac{\sin tx}{t}\varphi(t)\,dt \\ &= \frac{2}{s}\int_{-T}^{T} \frac{1}{t^2}\{\cos st - \cos 2st\}\varphi(t)\,dt \\ &= \frac{2}{s}\int_{-T}^{T} \frac{1}{t^2}\left\{1 - 2\sin^2\left(\frac{st}{2}\right) - 1 + 2\sin^2 st\right\}\varphi(t)\,dt.\end{aligned}$$

Set $M = \sup_t |\varphi(t)|$; then, clearly,

$$|\phi_T(s)| \leq 6M \int_{-\infty}^{\infty} \left(\frac{\sin x}{x}\right)^2 dx,$$

which proves the lemma. □

We now prove the Bochner–Khinchin theorem. First, let us show that the function $C(t)$ defined by (7.1.6) is non-negative definite. Indeed,

$$\begin{aligned}\sum_{j=1}^{n}\sum_{k=1}^{n} C(t_j - t_k) z_j \bar{z}_k &= \sum_{j=1}^{n}\sum_{k=1}^{n}\left(\int_{-\infty}^{\infty} \exp[i(t_j - t_k)x]\,dF(x)\right) z_j \bar{z}_k \\ &= \int_{-\infty}^{\infty}\left(\sum_{j=1}^{n}\exp(it_j x) z_j \overline{\sum_{k=1}^{n}\exp(it_k x) z_k}\right) dF \\ &= \int_{-\infty}^{\infty}\left|\sum_{j=1}^{n}\exp(it_j x) z_j\right|^2 dF(x) \geq 0.\end{aligned}$$

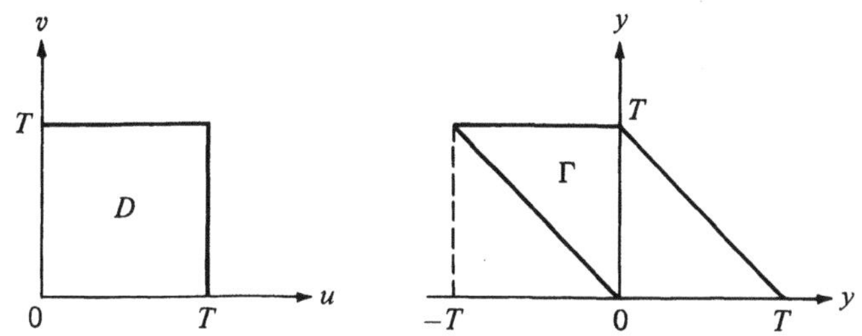

Figure 7.1

Next, let us show that any non-negative definite function which is continuous at zero has a representation in the form (7.1.6). To this end, consider

$$f_T(x) = \frac{1}{2\pi T}\int_0^T\int_0^T C(u-v)\exp[-i(u-v)x]\,du\,dv. \tag{7.2.3}$$

Clearly, $f_T(x) \geq 0$ because

$$C(u-v)e^{-iu}e^{iv} \geq 0.$$

Let us now make the change of variables in the double integral (7.2.3) as follows:

$$t = u, \qquad y = u - v, \qquad du\,dv = |J|\,dt\,dy = dt\,dy,$$

where J is the Jacobian of the transformation. From this we obtain (see Figure 7.1)

$$\begin{aligned} f_T(x) &= \frac{1}{2\pi T}\left(\int_{-T}^{0} dy \int_{-y}^{T} C(y)e^{-iyx}\,dt + \int_0^T dy \int_0^{T-y} C(y)e^{-iyx}\,dt\right) \\ &= \frac{1}{2\pi T}\left(\int_{-T}^{0} C(y)e^{iyx}(T+y)\,dy + \int_0^T C(y)e^{-iyx}(T-y)\,dy\right) \\ &= \frac{1}{2\pi}\int_{-T}^{T} C(y)\left(1-\frac{|y|}{T}\right)e^{-iyx}\,dy. \end{aligned}$$

According to Lemma 7.2.1, $f_T(x)$ is integrable over $(-\infty, \infty)$. Therefore, the inverse Fourier transform of $f_T(x)$ exists and is equal to

$$C(y)\left(1-\frac{|y|}{T}\right) = \int_{-\infty}^{\infty} f_T(x)e^{ixy}\,dx \tag{7.2.4}$$

for all $|y| \leq T$.

From (7.2.4), we have

$$C(0) = \int_{-\infty}^{\infty} f_T(x)\,dx$$

for all $T > 0$. Because

$$C(y)\left(1 - \frac{|y|}{T}\right) \to C(y) \quad \text{as } T \to \infty$$

uniformly on every finite interval and

$$f_T(x) \to f(x) \quad \text{as } T \to \infty,$$

we obtain that

$$C(y) = \int_{-\infty}^{\infty} f(x) e^{ixy}\, dx,$$

which proves the proposition when $F(x)$ in (7.1.6) is absolutely continuous. In a similar fashion, one can show it holds in the general case.

Proof of the Herglotz Theorem. Clearly, $C(n)$ given by Equation (7.1.8) is non-negative definite. For each $n = 1, 2, \ldots$ and $x \in [-\pi, \pi]$, define $f_n(x)$ by

$$f_n(x) = \frac{1}{2n\pi} \sum_{k=1}^{n} \sum_{v=1}^{n} C(k - v) e^{-ixk} e^{ixv};$$

then $f_n(\cdot) \geq 0$. Because there are $n - |m|$ couples (k, v) for which $k - v = m$, we have

$$f_n(x) = \frac{1}{2\pi} \sum_{r=-n+1}^{n-1} \left(1 - \frac{|r|}{n}\right) C(r) e^{-ixr}. \tag{7.2.5}$$

Set

$$F_n(x) = \frac{1}{2\pi} \int_{-\pi}^{x} f_n(u)\, du.$$

From (7.2.5) and the fact that

$$\int_{-\pi}^{\pi} e^{ixk}\, dx = 0 \qquad \text{if } k \neq 0,$$

it follows that

$$\frac{1}{2\pi} \int_{-\pi}^{\pi} f_n(u)\, du = C(0).$$

Therefore, $F_n(\cdot)$ is a nondecreasing continuous function. Also from (7.2.5),

$$\int_{-\pi}^{\pi} e^{itx}\, dF_n(x) = \frac{1}{2\pi} \int_{-\pi}^{\pi} e^{itx} f_n(x)\, dx = \left(1 - \frac{|t|}{n}\right) C(t). \tag{7.2.6}$$

From the Helly compactness theorem, there is a subsequence of $\{F_n\}$ which converges weakly to a bounded distribution F and, hence, converges completely because $F_n(-\pi) = 0$ and $F_n(\pi) = C(0)$ for all n. Equation (7.1.8) then holds as a consequence of the second Helly theorem and (7.2.6). □

7.3. Random Measures

In this section, we define the concept of an orthogonal random measure and of a stochastic integral with respect to it. We begin with some notation and definitions needed for this purpose. Let $\{\Omega, \mathscr{B}, P\}$ be a probability space and $\{S, \mathscr{S}\}$ an arbitrary measurable space. Let $\mathscr{S}_0$ be an algebra which generates the σ-algebra $\mathscr{S}$.

Definition 7.3.1. A mapping

$$\eta: \mathscr{S}_0 \to L_2\{\Omega, \mathscr{B}, P\}. \tag{7.3.1}$$

satisfying the conditions

$$\begin{aligned} &\text{(i)}\ \ \eta(\theta) = 0, \text{ where } \theta \text{ is the empty set,} \\ &\text{(ii)}\ \ \eta(A \cup B) = \eta(A) + \eta(B) \text{ (a.s.) for any disjoint } A, B \in \mathscr{S}_0 \end{aligned} \tag{7.3.2}$$

is called an elementary random (or stochastic) measure.

From the definition it is clear that

$$m(A) = \|\eta(A)\|^2 < \infty \quad \text{for any } A \in \mathscr{S}_0. \tag{7.3.3}$$

Definition 7.3.2. An elementary random measure is said to be orthogonal if

$$(\eta(A), \eta(B)) = 0 \quad \text{for all disjoint } A, B \in \mathscr{S}_0. \tag{7.3.4}$$

For an elementary orthogonal random measure, the set function $m(\cdot)$ on $\mathscr{S}_0$ defined by (7.3.3) is finitely additive. Indeed, if A_1, $A_2 \in \mathscr{S}_0$ are such that $A_1 \cap A_2 = \theta$, then

$$m(A_1 \cup A_2) = \|\eta(A_1 \cup A_2)\|^2 = \|\eta(A_1) + \eta(A_2)\|^2$$

(due to (7.3.2.ii). Now, due to (7.3.4),

$$\|\eta(A_1) + \eta(A_2)\|^2 = \|\eta(A_1)\|^2 + \|\eta(A_2)\|^2 = m(A_1) + m(A_2).$$

Assume now that $m(\cdot)$ is subadditive; then it can be extended to a measure on $\{S, \mathscr{S}\}$. This extension, which is unique, will again be denoted by m. Then $m(\cdot)$ is called the measure associated with $\eta(\cdot)$. In the following, we will denote by

$$L_2(m) = L_2\{S, \mathscr{S}, m\} \tag{7.3.5}$$

the Hilbert space of complex-valued functions on S which are square integrable with respect to m.

Let $\{B_k\}_1^n \subset \mathscr{S}_0$ be disjoint and consider

$$h(s) = \sum_{k=1}^{n} c_k I_{B_k}(s), \tag{7.3.6}$$

where $c_1, \ldots, c_n$ are complex numbers. Define

$$\psi(h) = \int_S h(s)\eta(ds) = \sum_{k=1}^{n} c_k \eta(B_k). \tag{7.3.7}$$

This integral is well defined and, clearly, $\psi(h) \in L_2\{\Omega, \mathscr{B}, P\}$. If

$$f(s) = \sum_{i=1}^{n} \alpha_i I_{B_i}(s),$$

where $\alpha_1, \ldots, \alpha_n$ are complex numbers,

$$\begin{aligned}(\psi(h), \psi(f)) &= \sum_{k=1}^{n} \sum_{i=1}^{n} c_k \bar{\alpha}_i (\eta(B_k), \eta(B_i)) \\ &= \sum_{j=1}^{n} c_j \bar{\alpha}_j m(B_j) = \int_S h(s)\overline{f(s)} m(ds) = (h, f).\end{aligned} \tag{7.3.8}$$

The last inner product is defined in $L_2(m)$ [see (7.3.5)]. Thus, the mapping ψ preserves the inner product.

Denote by $L_2^* \subset L_2(m)$ the set of all functions on S of type (7.3.6). The integral (7.3.7) represents a linear mapping of L_2^* to $L_2\{\Omega, \mathscr{B}, P\}$. Because the inner product is preserved, the mapping is clearly isometric. Now, let $g \in L_2(m)$ be arbitrary and let $\{h_n\}_1^\infty$ be a sequence of functions of type (7.3.6) such that

$$\|g - h_n\| \to 0 \quad \text{as } n \to \infty. \tag{7.3.9}$$

Then, due to (7.3.8),

$$\|\psi(h_n) - \psi(h_m)\| = \|h_n - h_m\| \to 0 \quad \text{as } m, n \to \infty, \tag{7.3.10}$$

which shows that $\{\psi(h_n)\}_1^\infty$ is a Cauchy sequence. By the Riesz–Fisher theorem (Proposition 5.2.4), there exists a r.v., say $\psi(g) \in L_2\{\Omega, \mathscr{B}, P\}$, such that

$$\|\psi(g) - \psi(h_n)\| \to 0 \quad \text{as } n \to \infty.$$

This r.v. is called the stochastic integral of $g \in L_2(m)$ with respect to the elementary orthogonal random measure η. We denote it by

$$\psi(g) = \int_S g(s)\eta(ds). \tag{7.3.11}$$

Thus, the mapping

$$\psi\colon L_2\{S, \mathscr{S}, m\} \to L_2\{\Omega, \mathscr{B}, P\} \tag{7.3.12}$$

represents an isometric isomorphism. The following properties of ψ are straightforward consequences of its construction: For any $f, g \in L_2(m)$,

$$\begin{aligned}&\text{(i)} && (\psi(f), \psi(g)) = (f, g); \quad \|\psi(f)\|^2 = \|f\|^2 = \int_S |f(s)|^2 m(ds); \\ &\text{(ii)} && \psi(c_1 f + c_2 g) = c_1\psi(f) + c_2\psi(g) \quad \text{(a.s.)}; \\ &\text{(iii)} && \|\psi(f_n) - \psi(f)\| \to 0 \quad \text{if } \|f - f_n\| \to 0 \text{ as } n \to \infty.\end{aligned} \tag{7.3.13}$$

Can an elementary orthogonal stochastic measure η, which is defined on $\mathscr{S}_0$, be extended to $\mathscr{S}$. The answer is affirmative and we will use the definition of the integral (7.3.11) to prove it. For $A \in \mathscr{S}$, define

$$\tilde{\eta}(A) = \int_S I_A(s)\eta(ds). \tag{7.3.14}$$

It follows from (7.3.13.ii) that

$$\tilde{\eta}(A_1 \cap A_2) = \tilde{\eta}(A_1) + \tilde{\eta}(A_2) \quad \text{(a.s.)}$$

for $A_1, A_2 \in \mathscr{S}$ such that $A_1 \cap A_2 = \theta$. From (7.3.13.i), we also have

$$\|\tilde{\eta}(A)\|^2 = m(A), \quad A \in \mathscr{S}. \tag{7.3.15}$$

Let us show that $\tilde{\eta}$ is countably additive in the q.m. sense. Let $\{B_k\}_1^\infty \subset \mathscr{S}$ be disjoint and let

$$B = \bigcup_k B_k.$$

Then,

$$\begin{aligned}\tilde{\eta}(B) - \sum_1^n \tilde{\eta}(B_k) &= \psi(I_B) - \sum_1^n \psi(I_{B_k}) \\ &= \sum_1^\infty \psi(I_{B_k}) - \sum_1^n \psi(I_{B_k}) = \sum_{k=n+1}^\infty \psi(I_{B_k}).\end{aligned}$$

However,

$$\left\| \sum_{k=n+1}^\infty \psi(I_{B_k}) \right\| = \sum_{k=n+1}^\infty \|\psi(I_{B_k})\| = \sum_{k=n+1}^\infty m(B_k) \to 0$$

as $n \to \infty$. Consequently,

$$\left\| \tilde{\eta}(B) - \sum_1^n \tilde{\eta}(B_k) \right\| \to 0 \quad \text{as } n \to \infty.$$

From (7.3.13.i), we also have

$$(\tilde{\eta}(A_1), \tilde{\eta}(A_2)) = 0$$

if $A_1, A_2 \in \mathscr{S}$ are disjoint.

The mapping

$$\tilde{\eta}: \mathscr{S} \to L_2\{\Omega, \mathscr{B}, P\}$$

is countably additive in the q.m. sense and coincides with η on $\mathscr{S}_0$. We will call $\tilde{\eta}$ an orthogonal stochastic measure and write from now on η instead of $\tilde{\eta}$. We also call

$$\psi(g) = \int_S g(s)\eta(ds),$$

defined above, a stochastic integral with respect to η.

7.4. Process with Orthogonal Increments

In the rest of this chapter $\{S, \mathscr{S}, m\} = \{R, \mathscr{R}, m\}$, where R is the real line and $\mathscr{R}$ is the σ-algebra of Borel subsets of R. The purpose of this section is to show that orthogonal random measures are generated by processes with orthogonal increments, and vice versa. Let $\{Z(t); t \in R\} \subset L_2\{\Omega, \mathscr{B}, P\}$, with $E\{Z(t)\} = 0$ for all $t \in R$.

Definition 7.4.1. The stochastic process $\{Z(t); t \in R\}$ is said to have orthogonal increments if, for any $t_0 < t_1 < t_2$,

$$E(Z(t_1) - Z(t_0))\overline{(Z(t_2) - Z(t_1))} = 0. \tag{7.4.1}$$

In the following, we shall say that the process is right-continuous in the q.m. sense if, for every $t \in R$ fixed,

$$\|Z(t) - Z(t_k)\| \to 0 \quad \text{as } t_k \downarrow t. \tag{7.4.2}$$

Let us now show that an orthogonal stochastic measure η generates a process having orthogonal increments. Let m be the finite measure associated with η. Set

$$Z(t) = \eta((-\infty, t]). \tag{7.4.3}$$

Then

$$\|Z(t)\|^2 = m((-\infty, t]) = F(t) \tag{7.4.4}$$

and, for all $s < t$,

$$\|Z(t) - Z(s)\|^2 = \|\eta((s, t])\|^2 = m((s, t]). \tag{7.4.5}$$

From this, we clearly have that $Z(t)$ is right-continuous in the q.m. sense. In addition, the function $F(\cdot)$ defined by (7.4.4) is also right-continuous. Finally, because

$$E(Z(t_1) - Z(t_0))(Z(t_2) - Z(t_1)) = (\eta((t_0, t_1]), \eta((t_1, t_2])) = 0,$$

the process $Z(t)$ defined by (7.4.3) has orthogonal increments.

Conversely, let $\{Z(t); t \in R\}$ be as in Definition 7.4.1 and, for any $s < t$, define

$$F(t) - F(s) = \|Z(t) - Z(s)\|^2. \tag{7.4.6}$$

Because the process has orthogonal increments, for any $s < \tau < t$,

$$\begin{aligned}\|Z(t) - Z(s)\|^2 &= \|Z(t) - Z(\tau) + Z(\tau) - Z(s)\|^2 \\ &= \|Z(t) - Z(\tau)\|^2 + \|Z(\tau) - Z(s)\|^2 \\ &\geq \|Z(\tau) - Z(s)\|^2.\end{aligned}$$

Therefore,

$$F(t) \geq F(\tau), \tag{7.4.7}$$

which implies that $F(\cdot)$ is a nondecreasing function on R. Finally, from (7.4.6), we see that $Z(t)$ is q.m. continuous at $t \in R$ if and only if $F(t)$ is continuous at t. In the following, we assume that

$$F(-\infty) = 0 \quad \text{and} \quad F(+\infty) < \infty. \tag{7.4.8}$$

Let $m(\cdot)$ be the Lebesgue–Stieltjes measure that F generates. We see that, for any interval $(s, t]$,

$$m((s,t]) = F(t) - F(s) = \|Z(t) - Z(s)\|.$$

Therefore, if $\mathscr{L}$ is the algebra of sets

$$A = \bigcup_{k=1}^{n} (a_k, b_k],$$

where $(a_k, b_k]$ are disjoint and

$$\eta(A) = \sum_{k=1}^{n} [Z(b_k) - Z(a_k)], \tag{7.4.9}$$

it is evident that

$$\|\eta(A)\|^2 = m(A),$$

where

$$m(A) = \sum_{k=1}^{n} [F(b_k) - F(a_k)] \tag{7.4.10}$$

and

$$(\eta((a_i, b_i]), \eta((a_j, b_j])) = 0 \quad \text{for all } i \neq j.$$

This clearly implies that $\eta(A)$ defined by (7.4.9) represents an elementary orthogonal stochastic measure. The set function $m(\cdot)$ defined by (7.4.1) has a unique extension to a measure on $\mathscr{R}$. From this and the preceding section, we see that $\eta(\cdot)$ can be also extended to $\mathscr{R}$ and that

$$\|\eta(A)\|^2 = m(A), \quad A \in \mathscr{R}.$$

Consequently, there exists an isomorphism between processes $\{Z(t); t \in R\} \subset L_2\{\Omega, \mathscr{B}, P\}$ with orthogonal increments continuous from the right in the q.m. sense and such that

$$\|Z(t)\|^2 = F(t), \qquad F(-\infty) = 0, \qquad F(+\infty) < \infty,$$

and orthogonal stochastic measures η with $m(\cdot)$, the measure associated with it. The correspondence is given by

$$Z(t) = \eta((-\infty, t]), \qquad F(t) = m((-\infty, t]),$$

and

$$\eta((s,t]) = Z(t) - Z(s), \qquad m((s,t]) = F(t) - F(s).$$

Finally, the stochastic integral

$$\int_{-\infty}^{\infty} h(t)\,dZ(t) \tag{7.4.11}$$

will mean the stochastic integral

$$\int_{-\infty}^{\infty} h(t)\eta(dt).$$

7.5. Spectral Representation

This section is concerned with the spectral representation of a wide sense stationary stochastic process. Mathematically speaking, the spectral representation establishes an isometric isomorphism between the closed linear manifold spanned by the process and a certain Hilbert space $L_2\{R, \mathscr{R}, F\}$ of complex functions. Such a representation is often useful in a large class of scientific and engineering problems, for example, guidance and control, signal detection, image enhancement, and so on.

Let $\{\xi(t); t \in R\} \subset L_2\{\Omega, \mathscr{B}, P\}$ be a wide sense stationary stochastic process with

$$E\{\xi(t)\} = 0, \qquad C(t) = E(\xi(s)\overline{\xi(s+t)}). \tag{7.5.1}$$

If the covariance function $C(t)$ is continuous at zero, then by Proposition 7.1.2, there exists a unique nondecreasing bounded function $F(\cdot)$ on R such that

$$C(t) = \int_{-\infty}^{\infty} e^{itx}\,dF(x). \tag{7.5.2}$$

The function $F(\cdot)$, which we assume to be continuous from the right, is called the "spectral distribution" of the process $\xi(t)$. The set of the discontinuity points of $F(\cdot)$, which is always finite or countably infinite, is called the "spectrum" of the process $\xi(t)$.

Proposition 7.5.1. *Assume that the covariance function $C(t)$ is continuous at zero. There exists unique orthogonal stochastic measure η with values in $L_2\{\Omega, \mathscr{B}, P\}$ such that*

$$\xi(t) = \int_{-\infty}^{\infty} e^{i\lambda t}\eta(d\lambda) \quad \text{(a.s.)}. \tag{7.5.3}$$

Furthermore,

$$\|\eta(A)\|^2 = m(A) = \int_A dF \quad \textit{for all } A \in \mathscr{R}, \tag{7.5.4}$$

where $m(\cdot)$ is the Lebesgue–Stieltjes measure associated with η and generated by the spectral distribution F.

PROOF. There are various ways to prove this proposition. The method adopted here is a version of Stone's spectral representation theorem for a family of unitary operators in a Hilbert space.

Let $L_2(m) = L_2\{R, \mathscr{R}, m\}$ be the Hilbert space of complex square integrable functions on R with the inner product

$$(f, h) = \int_{-\infty}^{\infty} f(\lambda)\overline{h(\lambda)}m(d\lambda). \tag{7.5.5}$$

Consider the system of complex functions

$$\{e^{i\lambda t}; t \in R\} \subset L_2\{R, \mathscr{R}, m\}.$$

Denote by $\langle\{e^{i\lambda t}; t \in R\}\rangle$ the linear manifold spanned by $\{e^{i\lambda t}; t \in R\}$. This linear manifold is dense in $L_2(m)$ (see Proposition 5.3.2). Therefore,

$$\langle\overline{\{e^{i\lambda t}; t \in R\}}\rangle = L_2\{R, \mathscr{R}, m\}. \tag{7.5.6}$$

As usual, we denote by $\overline{\langle\xi(R)\rangle}$ the closed linear manifold spanned by $\{\xi(t); t \in R\}$.

We now establish a one-to-one correspondence between the linear manifolds $\langle\{e^{i\lambda t}; t \in R\}\rangle$ and $\langle\xi(R)\rangle$ by setting

$$e^{i\lambda t} \leftrightarrow \xi(t), \quad t \in R. \tag{7.5.7}$$

Clearly then, for any $\{t_1, \ldots, t_n\} \subset R$, we have

$$\sum_{j=1}^{n} c_j e^{i\lambda t_j} \leftrightarrow \sum_{j=1}^{n} c_j \xi(t_j). \tag{7.5.8}$$

Note that (7.5.7) is a consistent definition in the sense that

$$\sum_{j=1}^{n} c_j e^{i\lambda t_j} = 0 \text{ (a.e.) } m \Leftrightarrow \sum_{j=1}^{n} c_j \xi(t_j) = 0 \text{ (a.s.) } P.$$

The correspondence (7.5.7) is an isometry. As a matter of fact, keeping in mind (7.5.2) and (7.5.7), we have

$$\begin{aligned}(e^{i\lambda t_1}, e^{i\lambda t_2}) &= \int_{-\infty}^{\infty} e^{i\lambda(t_1 - t_2)}\,dF(\lambda) = C(t_1 - t_2)\\ &= E(\xi(t_1)\overline{\xi(t_2)}) = (\xi(t_1), \xi(t_2)).\end{aligned} \tag{7.5.9}$$

Similarly,

$$\left(\sum_j \alpha_j e^{i\lambda t_j}, \sum_k \beta_k e^{i\lambda t_k}\right) = \left(\sum_j \alpha_j \xi(t_j), \sum_k \beta_k \xi(t_k)\right). \tag{7.5.10}$$

Now, consider $\zeta \in \overline{\langle\xi(R)\rangle}$. There exists $\{\zeta_n\}_1^\infty \subset \overline{\langle\xi(R)\rangle}$ such that

$$\|\zeta - \zeta_n\| \to 0 \quad \text{as } n \to \infty.$$

Consequently, $\{\zeta_n\}_1^\infty$ is a Cauchy sequence and, therefore, so is $\{f_n\}_1^\infty \subset L_2\{R, \mathscr{R}, m\}$, where

$$f_n \leftrightarrow \zeta_n.$$

Because $L_2\{R, \mathscr{R}, m\}$ is complete, there exists $f \in L_2(m)$ such that $\|f_n - f\| \to 0$ as $n \to \infty$. The converse also holds, namely, if $f \in L_2(m)$ and $\|f - f_n\| \to 0$, where $\{f_n\} \subset \langle\{e^{it\lambda}; t \in R\}\rangle$, there is $\eta \in \overline{\langle\xi(R)\rangle}$ and $\{\eta_n\} \subset \langle\xi(R)\rangle$ such that

$$\|\eta - \eta_n\| \to 0 \quad \text{as} \quad n \to \infty \quad \text{and} \quad \eta_n \leftrightarrow f_n.$$

This clearly implies that the correspondence can be extended to elements of $\overline{\langle\{e^{it\lambda}; t \in R\}\rangle}$ and $\overline{\langle\xi(R)\rangle}$.

Next, consider

$$f(\lambda) = I_B(\lambda), \quad B \in \mathscr{R},$$

and let $\eta(B) \in \overline{\langle\xi(R)\rangle}$ be such that

$$I_B(\lambda) \leftrightarrow \eta(B).$$

Clearly,

$$\|I_B\|^2 = m(B);$$

so that due to isometry,

$$\|\eta(B)\|^2 = m(B).$$

In addition, if $B_1 \cap B_2 = \theta$, where $B_1, B_2 \in \mathscr{R}$,

$$(\eta(B_1), \eta(B_2)) = 0$$

and

$$\left\| \eta(B) - \sum_{k=1}^{n} \eta(B_k) \right\|^2 \to 0 \quad \text{as } n \to \infty,$$

where $\{B_k\}_1^\infty \subset \mathscr{R}$ are disjoint and $B = \bigcup_k B_k$.

Therefore, the family $\eta(B)$, $B \in \mathscr{R}$, forms an orthonormal stochastic measure with respect to which we can define the stochastic integral

$$\psi(f) = \int_{-\infty}^{\infty} f(\lambda)\eta(d\lambda), \quad f \in L_2(m). \tag{7.5.11}$$

Let $f \in L_2(m)$ and $f \leftrightarrow \zeta$ and set

$$\zeta = \phi(f).$$

Let us show that

$$\psi(f) = \phi(f) \quad \text{(a.s.) } P. \tag{7.5.12}$$

If

$$f(\lambda) = \sum_{k=1}^{n} c_k I_{B_k}(\lambda) \tag{7.5.13}$$

with $B_k = (a_k, b_k]$ disjoint, then from (7.5.11),

$$\psi(f) = \sum_{k=1}^{n} c_k \eta(B_k),$$

which is evidently equal to $\phi(f)$. Hence, (7.5.12) holds for functions of the form (7.5.13). However, if $f \in L_2(m)$ and

$$\|f - f_n\| \to 0 \quad \text{as } n \to \infty,$$

where $f_n(\lambda)$ are of the form (7.5.13), then

$$\|\phi(f) - \phi(f_n)\| \to 0 \quad \text{as } n \to \infty,$$

$$\|\psi(f) - \psi(f_n)\| \to 0 \quad \text{[by (7.3.13.iii)]},$$

which proves (7.5.12).

Finally, consider

$$f(\lambda) = e^{i\lambda t};$$

then, by (7.5.7), $\phi(e^{i\lambda t}) = \xi(t)$. On the other hand, due to (7.5.11),

$$\psi(e^{i\lambda t}) = \int_{-\infty}^{\infty} e^{i\lambda t} \eta(d\lambda).$$

This and (7.5.12) then yield

$$\xi(t) = \int_{-\infty}^{\infty} e^{i\lambda t} \eta(d\lambda) \quad \text{(a.s.)}$$

for all $t \in R$. This proves the proposition. □

Remark 7.5.1. Let $\{Z(t); t \in R\}$ be a stochastic process with orthogonal increments corresponding to an orthogonal stochastic measure $\eta(\cdot)$. Then, from (7.4.11) and (7.4.12), we deduce that

$$\xi(t) = \int_{-\infty}^{\infty} e^{i\lambda t}\, dZ(\lambda). \tag{7.5.14}$$

Remark 7.5.2. If $E\{\xi(t)\} = \alpha$, then we have

$$\xi(t) = \alpha + \int_{-\infty}^{\infty} e^{i\lambda t} \eta(d\lambda).$$

Remark 7.5.3. From the proof of the proposition it is clear that there must be many different representations of a wide sense stationary process. The one that we give here is particularly useful, because it allows application of classical harmonic analysis in the studying of an important class of random processes.

Let $C(n)$ be the covariance function of a wide sense stationary sequence $\{\xi_n\}_0^\infty$ with $E\{\xi_n\} = 0$. According to Herglotz's theorem (Proposition 7.1.3) the covariance function has the spectral representation

$$C(n) = \int_{-\pi}^{\pi} e^{in\lambda}\, dF(\lambda),$$

where $F(\cdot)$ is a real nondecreasing bounded function.

Proposition 7.5.2. *There is an orthogonal random measure* η *on* $\mathcal{R} \cap (\pi, \pi]$ *such that, for every n,*

$$\xi_n = \int_{-\pi}^{\pi} e^{in\lambda} \eta(d\lambda) \tag{7.5.15}$$

and $\|\eta(d\lambda)\|^2 = dF(\lambda)$.

The proof of this proposition is essentially the same as the proof of the previous proposition.

Remark 7.5.4. The spectral representation of a wide sense stationary process implies that, for each t, $\xi(t)$ can be approximated in the q.m. sense by a finite sum of the form

$$\sum_{k=1}^{n} e^{i\lambda_k t} \Delta Z(\lambda_k),$$

where $\Delta Z(\lambda_k)$ are orthogonal r.v.'s from $\overline{\langle \xi(R) \rangle}$.

7.6. Ramifications of Spectral Representation

As we have established (see Remark 7.4.1), the covariance function $C(t)$ of a second-order stationary process is real if and only if its spectral distribution $F(\cdot)$ is symmetric. Something similar holds for stationary processes.

Let $\{\xi(t); t \in R\}$ be a real wide sense stationary process and consider its spectral representation

$$\xi(t) = \int_{-\infty}^{\infty} e^{i\lambda t}\, dZ(\lambda).$$

Because $\xi(t) = \overline{\xi(t)}$, we have

$$\xi(t) = \int_{-\infty}^{\infty} e^{-i\lambda t}\, d\overline{Z(\lambda)} = \int_{-\infty}^{\infty} e^{i\lambda t}\, d\overline{Z(-\lambda)},$$

which implies that

$$dZ(\lambda) = d\overline{Z(-\lambda)}. \tag{7.6.1}$$

From this [see also (7.4.11) and (7.4.12)], one can deduce that, for any $A \in \mathcal{R}$,

$$\eta(A) = \overline{\eta(-A)}, \tag{7.6.2}$$

where $\eta(\cdot)$ is the orthogonal random measure induced by $\{Z(t); t \in R\}$ and $-A = \{x; -x \in A\}$.

Set

$$\eta(\cdot) = \eta_1(\cdot) + i\eta_2(\cdot). \tag{7.6.3}$$

It is readily seen that $\eta_1(\cdot)$ and $\eta_2(\cdot)$ are real measures. According to (7.6.2), we have

$$\eta_1(-A) = \eta_1(A), \qquad \eta_2(-A) = -\eta_2(A). \tag{7.6.4}$$

Again consider the real wide sense stationary stochastic process $\xi(t)$ and write its spectral representation, using (7.6.3), as

$$\begin{aligned}\xi(t) &= \int_{-\infty}^{\infty} (\cos \lambda t + i \sin \lambda t)(\eta_1(d\lambda) + i\eta_2(d\lambda)) \\ &= \int_{-\infty}^{\infty} \cos \lambda t\, \eta_1(d\lambda) - \int_{-\infty}^{\infty} \sin \lambda t\, \eta_2(d\lambda) + i\left\{\int_{-\infty}^{\infty} \cos \lambda t\, \eta_2(d\lambda)\right. \\ &\quad \left. + \int_{-\infty}^{\infty} \sin \lambda t\, \eta_1(d\lambda)\right\}.\end{aligned}$$

Because $\xi(t)$ is real,

$$\int_{-\infty}^{\infty} \cos \lambda t\, \eta_2(d\lambda) + \int_{-\infty}^{\infty} \sin \lambda t\, \eta_1(d\lambda) = 0.$$

This and (7.6.4) yield

$$\begin{aligned}\xi(t) &= \int_{-\infty}^{\infty} \cos \lambda t\, \eta_1(d\lambda) + \int_{-\infty}^{\infty} \sin \lambda t\, \eta_2(d\lambda) \\ &= \int_{-\infty}^{\infty} \cos \lambda t\, dZ_1(\lambda) + \int_{-\infty}^{\infty} \sin \lambda t\, dZ_2(\lambda),\end{aligned} \tag{7.6.5}$$

where

$$Z(\lambda) = Z_1(\lambda) + iZ_2(\lambda). \tag{7.6.6}$$

Now consider the operation of q.m. differentiation of a wide sense stationary stochastic process $\xi(t)$. According to Proposition 6.2.4, the q.m. derivative of a second-order process with covariance function $C(s, t)$ exists if

$$\left.\frac{\partial^2 C(s,t)}{\partial s \partial t}\right|_{s=t} \tag{7.6.7}$$

exists at every $t \in T$. For a wide sense stationary stochastic process $\{\xi(t); t \in R\}$ with covariance function $C(t)$, (7.6.7) is equivalent to the requirement that $C''(0)$ exists. This, on the other hand, is equivalent to

$$C''(0) = \int_{-\infty}^{\infty} \lambda^2\, dF(\lambda) < \infty. \tag{7.6.8}$$

Assume now that (7.6.8) holds. It is not difficult to show that

$$\frac{e^{i\lambda h}-1}{h} \xrightarrow{\text{q.m.}} i\lambda \quad \text{(with respect to } F) \tag{7.6.9}$$

as $h \to 0$. Next, taking into account (7.5.14), we have

$$\frac{\xi(t+h)-\xi(t)}{h} = \int_{-\infty}^{\infty} \frac{e^{i\lambda h}-1}{h} e^{i\lambda t}\eta(d\lambda).$$

The following is clearly permissible due to (7.6.8):

$$\underset{h\to 0}{\text{l.i.m.}} \frac{\xi(t+h)-\xi(t)}{h} = \int_{-\infty}^{\infty} \underset{h\to 0}{\text{l.i.m.}} \left(\frac{e^{i\lambda h}-1}{h}\right) e^{i\lambda t}\eta(d\lambda).$$

From this and (7.6.9), we deduce

$$\xi'(t) = i\int_{-\infty}^{\infty} \lambda e^{i\lambda t}\eta(d\lambda). \tag{7.6.10}$$

Let us show that $\xi'(t)$ is also a wide sense stationary process. Because by assumption $E\{\xi(t)\} = 0$, clearly

$$E\{\xi'(t)\} = 0. \tag{7.6.11}$$

In addition,

$$E\{\xi'(s)\overline{\xi'(s+t)}\} = \int_{-\infty}^{\infty} \lambda^2 e^{i\lambda t}\, dF(\lambda) = C_{\xi''}(t),$$

which proves the assertion.

Let $\{\xi(t); t\in T\}$ bewide sense stationary with spectral representation (7.5.14) and let $\zeta \in \langle\overline{\xi(R)}\rangle$. What is the structure of the r.v. ζ?

Proposition 7.6.1. *Let $\zeta \in \langle\overline{\xi(R)}\rangle$, then there exists $\varphi \in L_2\{R, \mathscr{R}, F\}$ such that* (a.s.)

$$\zeta = \int_{-\infty}^{\infty} \varphi(\lambda)\eta(d\lambda). \tag{7.6.12}$$

PROOF. Set

$$\zeta_n = \sum_{k=1}^{n} \alpha_k \xi(t_k); \tag{7.6.13}$$

then by (7.5.14)

$$\zeta_n = \int_{-\infty}^{\infty} \left(\sum_{k=1}^{n} \alpha_k e^{i\lambda t_k}\right)\eta(d\lambda). \tag{7.6.14}$$

In other words, (7.6.12) holds for

$$\varphi_n(\lambda) = \sum_{k=1}^{n} \alpha_k e^{i\lambda t_k}. \tag{7.6.15}$$

In the general case, for $\zeta \in \langle\overline{\xi(R)}\rangle$, there exists a sequence of r.v.'s of type (7.6.13) such that

$$\|\zeta - \zeta_n\| \to 0 \quad \text{as } n \to \infty.$$

But then

$$\|\varphi_n - \varphi_m\| = \|\zeta_n - \zeta_m\| \to 0 \quad \text{as } n, m \to \infty.$$

Consequently $\{\varphi_n\}$ is Cauchy in $L_2\{R, \mathscr{R}, F\}$, so that there exists an $\varphi \in L_2\{R, \mathscr{R}, F\}$ such that

$$\|\varphi - \varphi_n\| \to 0 \quad \text{as } n \to \infty.$$

But then [see (7.3.13.iii)]

$$\|\psi(\varphi) - \psi(\varphi_n)\| \to 0 \quad \text{as } n \to \infty$$

and because

$$\zeta_n = \psi(\varphi_n),$$

we have

$$\zeta = \psi(\varphi) \quad \text{(a.s.)},$$

which proves the assertion. □

7.7. Estimation, Prediction, and Filtering

The general estimation problem in $L_2\{\Omega, \mathscr{B}, P\}$ can be formulated as follows. Let $Y \in L_2$ be arbitrary but fixed and let $S \subset L_2$ be a subspace. Find the element $\hat{Y} \in S$ closest to Y. From the definition of $\hat{Y}$, it clearly follows that

$$\|Y - \hat{Y}\| = \inf\{\|Y - U\|; U \in S\}. \tag{7.7.1}$$

In Hilbert space terminology, $\hat{Y}$ is the orthogonal projection of Y on S, characterized as the element of S such that

$$Y - \hat{Y} \perp S. \tag{7.7.2}$$

This implies that

$$(Y, U) = (\hat{Y}, U) \tag{7.7.3}$$

for each $U \in S$. The following two particular cases are of special interest.

(*i*) *Pure Prediction.* Let $\{\xi(s); s \in R\} \subset L_2\{\Omega, \mathscr{B}, P\}$ be a stochastic process. Assume that this process is observed up to time t, i.e., its values $\xi(s)$ are known for all $s \leq t$. It is required to predict its value at some moment of time $t + \tau$, $\tau > 0$. In this case,

$$Y = \xi(t + \tau) \quad \text{and} \quad S = L_2\{\Omega, \mathscr{F}_\xi^t, P\}, \tag{7.7.4}$$

where $\mathscr{F}_\xi^t = \sigma\{\xi(s); s \leq t\}$.

(*ii*) *Filtering and Prediction.* Let $\{X(t); t \in R\}$ and $\{Z(t); t \in R\} \subset L_2\{\Omega, \mathscr{B}, P\}$ be given processes, where $X(t)$ is the one that interests us. Unfortunately, $X(t)$ cannot be observed directly as it is "contaminated" by the unobservable disturbance $Z(t)$. What is observed is the sum

$$Y(t) = X(t) + Z(t). \tag{7.7.5}$$

Here is a simple example of what an engineer would call a "signal" $X(t)$ plus unwanted noise $Z(t)$, giving $Y(t)$, which is what we observe. The problem here is to separate the noise from the signal. More precisely, one must find an estimate $\hat{X}(t)$ of $X(t)$ from observations $Y(s)$ for $s \le t$. This is called filtering. Clearly, $\hat{X}(t)$ is the orthogonal projection of $X(t)$ on $L_2\{\Omega, \mathscr{F}_Y^t, P\}$. If we have to find an estimate $\hat{X}(t + \tau)$ of $X(t + \tau)$, where $\tau > 0$, from values $Y(s)$ for $s \le t$, this is called filtering and prediction.

We now outline a general solution of the estimation problem. Let $\mathscr{F} \subset \mathscr{B}$ be a sub-σ-algebra. We shall denote by $L_2(\mathscr{F}) = L_2\{\Omega, \mathscr{F}, P\}$, the Hilbert subspace of L_2, which consists of those elements which are $\mathscr{F}$-measurable. We want to show that the orthogonal projection of Y on $L_2(\mathscr{F})$ (see Section 10.1 of Chapter 10) is

$$\hat{Y} = E\{Y|\mathscr{F}\}. \tag{7.7.6}$$

Clearly, $E\{Y|\mathscr{F}\} \in L_2(\mathscr{F})$. To prove the assertion, it suffices to show that (7.7.2) holds. Take any $Z \in L_2(\mathscr{F})$; then,

$$\begin{aligned}(Y - E\{Y|\mathscr{F}\}, Z) &= (Y, Z) - (E\{Y|\mathscr{F}\}, Z)\\ &= E(Y\bar{Z}) - E(\bar{Z}E\{Y|\mathscr{F}\})\\ &= E(Y\bar{Z}) - E(E\{Y\bar{Z}|\mathscr{F}\}) = 0.\end{aligned}$$

Therefore,

$$Y - E\{Y|\mathscr{F}\} \perp L_2(\mathscr{F}). \tag{7.7.7}$$

Remark 7.7.1. The estimate (7.7.6) is clearly unbiased; that is,

$$E\{\hat{Y}\} = E(E\{Y|\mathscr{F}\}) = E\{Y\}.$$

Remark 7.7.2. The quantity

$$\delta^2 = \|Y - \hat{Y}\|^2 \tag{7.7.8}$$

is called the mean square error of the estimate (7.7.6).

The next simple example contains the basic idea used in the following to resolve an estimation problem.

Example 7.7.1. In regression analysis, the following problem is of some interest. Let $X_1, X_2 \in L_2\{\Omega, \mathscr{B}, P\}$; from the family of Borel functions,

$$h: C \to C$$

such that $h(X_1) \in L_2$, (C is the set of complex numbers) find one that minimizes the norm

$$\|X_2 - h(X_1)\|.$$

Clearly, $h(X_1) \in L_2(\sigma\{X_1\})$; therefore, the norm will achieve its minimum if $h(\cdot)$ is such that

$$X_2 - h(X_1) \perp L_2(\sigma\{X\}),$$

where $L_2(\sigma\{X\}) = L_2\{\Omega, \sigma\{X\}, P\}$. From this and (7.7.6), it then follows that

$$h(X_1) = E\{X_2|X_1\} \quad \text{(a.s.).} \tag{7.7.9}$$

From this example we see that the best approximation to X_2 in the class of all Borel functions $h(\cdot)$ of X_1 [which, of course, must be such that $h(X_1) \in L_2$] is $E\{X_2|X_1\}$. This solution, unfortunately, has very limited practical applicability because the expectation is not easy to determine. A more useful solution can be obtained by seeking an approximation, not in the class of all Borel functions but in a subclass consisting of the linear functions

$$h(X_1) = \alpha X_1.$$

In such a case, the condition $X_2 - \alpha X_1 \perp \langle \overline{X_1} \rangle$ would yield

$$(X_2, X_1) = \alpha \|X_1\|^2$$

from which we obtain α.

The next example is slightly more general.

Example 7.7.2. Let $\{Z_k\}_1^n \subset L_2\{\Omega, \mathscr{B}, P\}$ be an orthonormal sequence with $E\{Z_k\} = 0$, $k = 1, \ldots, n$. Write

$$\mathscr{F}_n = \sigma\{Z_1, \ldots, Z_n\}$$

and consider $L_2(\mathscr{F}_n)$. Let $Y \in L_2$ be arbitrary; then, according to (7.7.6), the element in $L_2(\mathscr{F}_n)$ closest to Y is $\hat{Y}$, defined by

$$\hat{Y} = E\{Y|\mathscr{F}_n\}. \tag{7.7.10}$$

Now, consider the linear manifold $\langle\{Z_k\}_1^n\rangle$ spanned by $\{Z_k\}_1^n$. Clearly, $\langle\{Z_k\}_1^n\rangle$ is the subspace (because $\langle\{Z_k\}_1^n\rangle$ is closed) consisting of all the linear combinations $\sum_{k=1}^n c_k Z_k$, where c_k, $k = 1, \ldots, n$, are complex numbers. Therefore, the element $\hat{Y}_0 \in \langle\{Z_k\}_1^n\rangle$ closest to Y must be of the form

$$\hat{Y}_0 = \sum_{k=1}^n \alpha_k Z_k.$$

To determine the coefficients α_k, observe that

$$Y - \sum_{k=1}^n \alpha_k Z_k \perp \langle\{Z_k\}_1^n\rangle$$

which, due to orthonormality of $\{Z_k\}_1^n$, yields

$$(Y, Z_i) = \alpha_i, \quad i = 1, \ldots, n.$$

Therefore,

$$\hat{Y}_0 = \sum_{k=1}^{n} (Y, Z_k) Z_k. \tag{7.7.11}$$

Note that $\hat{Y}$ given by (7.7.10) is, in general, a better estimate of Y than $\hat{Y}_0$. As matter of fact, from (7.7.7), we have

$$\|Y - \hat{Y}_0\|^2 = \|Y - \hat{Y}\|^2 + \|\hat{Y} - \hat{Y}_0\|^2. \tag{7.7.12}$$

This fact also follows from $\langle \{Z_k\}_1^n \rangle \subset L_2(\mathscr{F}_n)$.

Remark 7.7.3. If $Y, Z_1, \ldots, Z_n$ form a Gaussian system,

$$E\{Y | Z_1, \ldots, Z_n\} = \sum_{k=1}^{n} c_k Z_k \in \langle \{Z_k\}_1^n \rangle,$$

where, clearly, $c_k = (Y, Z_k)$. Therefore, in this case

$$\hat{Y} = \hat{Y}_0.$$

To summarize, using the L_2 norm as our criterion of closeness, we have shown that the closest element in the subspace $L_2(\mathscr{F}_n)$ to an arbitrary $Y \in L_2$ is simply the conditional expectation $E\{Y | \mathscr{F}_n\}$. On the other hand, the closest element to Y in the subspace $\langle \{Z_k\}_1^n \rangle$ is $\hat{Y}_0$ given by (7.7.11). From (7.7.12), we see that $\hat{Y}$ represents a better approximation of Y than $\hat{Y}_0$.

7.8. An Application

Consider the simple model

$$X_n = \xi_n + Z_n, \tag{7.8.1}$$

where $\{\xi_n\}_{-\infty}^{\infty}$ and $\{Z_n\}_{-\infty}^{\infty} \subset L_2\{\Omega, \mathscr{B}, P\}$. Here, we interpret ξ_n as a signal and Z_n as noise. Neither ξ_n nor Z_n can be observed directly. It is the superposition X_n of these two that can be observed.

For the signal $\{\xi_n\}_{-\infty}^{\infty}$ and the noise $\{Z_n\}_{-\infty}^{\infty}$, the following conditions are assumed to hold:

(i) $\{\xi_n\}_{-\infty}^{\infty}$ and $\{Z_n\}_{-\infty}^{\infty}$ are both wide sense stationary processes with

$$E\{\xi_n\} = E\{Z_n\} = 0, \qquad \|\xi_n\|^2 = \|Z_n\|^2 = 1. \tag{7.8.2}$$

(ii) $\{\xi_n\}_{-\infty}^{\infty}$ and $\{Z_n\}_{-\infty}^{\infty}$ are uncorrelated. In other words,

$$(\xi_j, Z_k) = 0 \quad \text{for all } k, j: \ldots -1, 0, 1, \ldots. \tag{7.8.3}$$

Denote by

$$C_\xi(k) = (\xi_i, \xi_{i+k}), \qquad C_Z(k) = (Z_i, Z_{i+k}), \tag{7.8.4}$$

$k = \cdots -1, 0, 1, \ldots$. Clearly,

$$C_\xi(-k) = \overline{C_\xi(k)}, \qquad C_Z(-k) = \overline{C_Z(k)}. \tag{7.8.5}$$

In addition, after some straightforward calculations, we have

$$C_X(k) = C_\xi(k) + C_Z(k). \tag{7.8.6}$$

Let us define the problem of interest here. We want to find a linear estimate of a r.v. $Y \in L_2$ on the basis of observations

$$X_m, X_{m+1}, \ldots, X_{m+r}. \tag{7.8.7}$$

This is equivalent to the problem of finding the orthogonal projection $\hat{Y}_0$ of Y on the linear manifold $\langle\{X_{m+i}\}_0^r\rangle$ (which is a closed subspace of L_2).

The best linear predictor $\hat{Y}_0 \in \langle\{X_{m+i}\}_0^r\rangle$. This clearly implies that it must be of the form

$$\hat{Y}_0 = \sum_{i=0}^{r} h_i X_{m+i}, \tag{7.8.8}$$

where $h_0, \ldots, h_r$ are some (complex) constants. Because

$$Y - \hat{Y}_0 \perp \langle\{X_{m+i}\}_0^r\rangle,$$

it follows that

$$(Y, X_{m+j}) = (\hat{Y}_0, X_{m+j})$$

for any $j = 0, \ldots, r$. In other words,

$$\begin{aligned}(Y, X_{m+j}) &= \sum_{i=0}^{r} h_i (X_{m+i}, X_{m+j}) \\ &= \sum_{i=0}^{r} h_i C_X(i-j).\end{aligned}$$

Now consider the case when

$$Y = \xi_{m+r+k}$$

(prediction with filtration). In such a case,

$$\begin{aligned}(Y, X_{m+j}) &= (\xi_{m+r+k}, \xi_{m+j} + Z_{m+j}) \\ &= (\xi_{m+r+k}, \xi_{m+j}) \\ &= C_\xi(j - r - k).\end{aligned}$$

Thus, the system of equations characterizing $\hat{Y}_0$ becomes

$$\sum_{i=0}^{r} h_i C_X(i-j) = C_\xi(j-r-k), \quad j = 0, \ldots, r,$$

where $j = 0, \ldots, r$. A solution of this system yields $h_0, \ldots, h_r$.

7.9. Linear Transformations

Let $\{\xi(t); t \in R\} \subset L_2\{\Omega, \mathscr{B}, P\}$ be a wide sense stationary q.m. continuous process with $E\{\xi(t)\} = 0$, spectral distribution $F(\cdot)$, and orthogonal random measure η. Then (see Proposition 7.5.1), $\xi(t)$ has the spectral representation

$$\xi(t) = \int_{-\infty}^{\infty} e^{i\lambda t}\eta(d\lambda) \tag{7.9.1}$$

where $\|\eta(d\lambda)\|^2 = dF(\lambda)$.

Let $\{\zeta(t); t \in R\} \subset L_2\{\Omega, \mathscr{B}, P\}$ be another wide sense stationary process such that

$$\zeta(t) = \int_{-\infty}^{\infty} e^{i\lambda t}\varphi(\lambda)\eta(d\lambda) \tag{7.9.2}$$

where $\varphi(\cdot) \in L_2\{R, \mathscr{R}, F\}$. We then say that the stationary process $\zeta(t)$ was obtained from $\xi(t)$ by a linear transformation. The function $\varphi(\cdot)$ in (7.9.2) is called the "spectral characteristic" of the transformation.

Being a stationary process, $\zeta(t)$ has its own spectral representation

$$\zeta(t) = \int_{-\infty}^{\infty} e^{i\lambda t}\eta_0(d\lambda).$$

It then follows from (7.9.2) that

$$\eta_0\, d\lambda = \varphi(\lambda)\eta(d\lambda). \tag{7.9.3}$$

In addition, because

$$E\{\zeta(s)\overline{\zeta(s + t)}\} = \int_{-\infty}^{\infty} e^{i\lambda t}|\varphi(\lambda)|^2\, dF(\lambda),$$

the spectral distribution $F_0(\cdot)$ of $\zeta(t)$ is given by

$$dF_0(\lambda) = |\varphi(\lambda)|^2\, dF(\lambda). \tag{7.9.4}$$

EXAMPLE 7.9.1. One of the simplest linear transformations of a stationary stochastic process $\xi(t)$ is the "shift" operator T_s defined by

$$T_s\xi(t) = \xi(t + s). \tag{7.9.5}$$

From this, we have

$$T_s\xi(t) = \int_{-\infty}^{\infty} e^{i\lambda t}e^{i\lambda s}\eta(d\lambda)$$

which is a particular case of (7.9.2), where $\varphi(\lambda) = e^{i\lambda s}$. Clearly, we then have that $F_0 \equiv F$ and $\eta_0\, d\lambda = e^{i\lambda s}\eta\, d\lambda$.

Consider the closed linear manifold $\langle\overline{\xi(R)}\rangle$ spanned by the process $\xi(t)$. According to Proposition 7.6.1, an arbitrary element $\zeta \in \langle\overline{\xi(R)}\rangle$ can be represented as

$$\zeta = \int_{-\infty}^{\infty} \varphi(\lambda)\eta(d\lambda), \quad \varphi \in L_2\{R, \mathscr{R}, F\}.$$

Actually, the following result holds. If $\{\zeta(t); t \in R\} \subset L_2\{\Omega, \mathscr{B}, P\}$ is wide sense stationary and such that

$$\langle\overline{\zeta(R)}\rangle \subset \langle\overline{\xi(R)}\rangle, \tag{7.9.6}$$

then there exists $\varphi(\cdot) \in L_2\{R, \mathscr{R}, F\}$ such that

$$\zeta(t) = \int_{-\infty}^{\infty} e^{i\lambda t}\varphi(\lambda)\eta(d\lambda). \tag{7.9.7}$$

To show this, consider

$$\zeta_n(t) = \sum_{k=1}^{n} \alpha_k \xi(t + t_k) \in \langle\overline{\xi(R)}\rangle.$$

From (7.9.1), we obtain the spectral representation of $\zeta_n(t)$ as

$$\zeta_n(t) = \int_{-\infty}^{\infty} e^{i\lambda t}\varphi_n(\lambda)\eta(d\lambda)$$

where

$$\varphi_n(\lambda) = \sum_{k=1}^{n} \alpha_k e^{i\lambda t_k}.$$

The general case is now obtained for arbitrary $\varphi(\cdot)$ by passage to the limit.

Remark 7.9.1. When (7.9.6) holds, we often say that the process $\zeta(t)$ is subordinated to $\xi(t)$.

Remark 7.9.2. The derivative (7.6.1) of a wide sense stationary process is another particular case of (7.9.2), where $\varphi(\lambda) = i\lambda$.

Consider now another type of linear operation of a wide sense stationary process. The transformation

$$\zeta(t) = \int_{-\infty}^{\infty} h(t - s)\xi(s)\,ds \tag{7.9.8}$$

of the process $\xi(t)$ is called "an admissible linear filter." The complex function $h(\cdot)$ is absolutely integrable, i.e.,

$$\int_{-\infty}^{\infty} |h(s)|\,ds < \infty \tag{7.9.9}$$

and square integrable on every finite interval. The integral (7.9.8) exists if and only if the integral

$$\int_{-\infty}^{\infty}\int_{-\infty}^{\infty} h(t-s_1)C(s_1-s_2)\bar{h}(t-s_2)\,ds_1\,ds_2$$
$$= \int_{-\infty}^{\infty}\int_{-\infty}^{\infty} h(s_1)\bar{h}(s_2)C(s_1-s_2)\,ds_1\,ds_2$$

exists, which clearly holds due to (7.9.9).

Remark 7.9.3. If $h(u) = 0$ for $u < 0$, the filter is said to be physically realizable. The integral (7.9.8) is also known as "a moving average process."

EXAMPLE 7.9.2. Let $\varphi(\cdot)$ be the Fourier transform of some integrable function $h(\cdot)$, namely,

$$\varphi(\lambda) = \int_{-\infty}^{\infty} e^{-i\lambda t}h(t)\,dt. \tag{7.9.10}$$

Then, the linear transformation of the process $\xi(t)$ with the spectral characteristic (7.9.10) is the process

$$\begin{aligned}\zeta(t) &= \int_{-\infty}^{\infty} e^{i\lambda t}\varphi(\lambda)\eta(d\lambda)\\ &= \int_{-\infty}^{\infty} e^{i\lambda t}\left(\int_{-\infty}^{\infty} e^{-i\lambda s}h(s)\,ds\right)\eta(d\lambda)\\ &= \int_{-\infty}^{\infty}\left(\int_{-\infty}^{\infty} e^{i\lambda(t-s)}h(s)\,ds\right)\eta(d\lambda)\\ &= \int_{-\infty}^{\infty}\left(\int_{-\infty}^{\infty} e^{i\lambda u}h(t-u)\,du\right)\eta(d\lambda)\\ &= \int_{-\infty}^{\infty} h(t-u)\xi(u)\,du.\end{aligned}$$

Clearly,

$$\begin{aligned}E\{\zeta(s)\overline{\zeta(s+t)}\} &= E\left\{\int_{-\infty}^{\infty} h(s-u)\xi(u)\,du \int_{-\infty}^{\infty} \bar{h}(s+t-v)\bar{\xi}(v)\,dv\right\}\\ &= \int_{-\infty}^{\infty}\int_{-\infty}^{\infty} h(s_1)h(s_2)C(s_2-s_1)\,ds_1\,ds_2.\end{aligned}$$

7.10. Linear Prediction, General Remarks

Let $\{\xi(t); t \in R\} \subset L_2\{\Omega, \mathscr{B}, P\}$ be a wide sense stationary stochastic process with $E\{\xi(t)\} = 0$. As before, we denote by $\langle\overline{\xi(R)}\rangle$ the closed linear manifold spanned by $\xi(t)$. Clearly, $\langle\overline{\xi(R)}\rangle \subset L_2\{\Omega, \mathscr{B}, P\}$. Because for every $t \in R$, $\xi(t)$

is a "point" in $L_2\{\Omega, \mathscr{B}, P\}$, the whole process represents a curve in the Hilbert space $L_2\{\Omega, \mathscr{B}, P\}$. The closed linear manifold $\langle\overline{\xi(R)}\rangle$ is the smallest subspace of $L_2\{\Omega, \mathscr{B}, P\}$ which contains this curve.

For any $t \in R$, write $R_t = (-\infty, t]$; the $\langle\overline{\xi(R_t)}\rangle$ is defined to be the closed linear manifold spanned by r.v.'s $\xi(s)$, $s \leq t$. The subspace $\langle\overline{\xi(R_t)}\rangle$ is often said to represent the past and present of the stochastic process $\{\xi(s); s \in R\}$, given that t is the present time. Clearly, for any $t_1 \leq t_2$,

$$\langle\overline{\xi(R_{t_1})}\rangle \subseteq \langle\overline{\xi(R_{t_2})}\rangle.$$

Denote by

$$\langle\overline{\xi(-\infty)}\rangle = \bigcap_{t \in R} \langle\overline{\xi(R_t)}\rangle.$$

Therefore, in the terminology just introduced, $\langle\overline{\xi(-\infty)}\rangle$ represents the infinitely remote past.

Definition 7.10.1. If

$$\langle\overline{\xi(-\infty)}\rangle = \langle\overline{\xi(R)}\rangle$$

or, equivalently, if $\langle\overline{\xi(R_t)}\rangle$ is the same for all $t \in R$, the stochastic process $\xi(t)$ is called "deterministic." If, on the other hand,

$$\langle\overline{\xi(-\infty)}\rangle \subseteq \langle\overline{\xi(R)}\rangle,$$

the stochastic process is said to be nondeterministic. Finally, if $\langle\xi(-\infty)\rangle = \{0\}$, $\xi(t)$ is called "purely nondeterministic."

To elucidate somewhat the physical meaning of these notions, we will explain them in terms of linear prediction. As we have seen in Section 7.7, the problem of linear prediction of a process $\xi(t)$ is to find an estimate of $\xi(t + \tau)$ for some $\tau > 0$, given the values of $\xi(s)$ for $s \leq t$. The best linear predictor of $\xi(t + \tau)$ is its orthogonal projection on $\langle\overline{\xi(R_t)}\rangle$, which is denoted by $\tilde{\xi}(t + \tau)$. From Proposition 5.4.2, we know that the deviation

$$\xi(t + \tau) - \tilde{\xi}(t + \tau) \perp \langle\overline{\xi(R)}\rangle. \tag{7.10.1}$$

Its norm

$$\delta(\tau) = \|\xi(t + \tau) - \tilde{\xi}(t + \tau)\| \tag{7.10.2}$$

is called the "error" of the linear prediction. It is independent of t because the process $\xi(t)$ is wide sense stationary. Note also that

$$\tilde{\xi}(t + \tau) \in \langle\overline{\xi(R_t)}\rangle. \tag{7.10.3}$$

Let us show that, for any $0 \leq \tau_1 \leq \tau_2$,

$$\delta(\tau_1) \leq \delta(\tau_2) \quad \text{and} \quad \delta(\tau) = 0 \quad \text{for } \tau \leq 0. \tag{7.10.4}$$

The orthogonal projection of $\xi(t + \tau)$ on $\langle\overline{\xi(R_t)}\rangle$ is the same as the projection of $\xi(u)$ on $\langle\overline{\xi(R_{u-\tau})}\rangle$ with $u = t + \tau$. Now, because $\langle\overline{\xi(R_{u-\tau_2})}\rangle \subset \langle\overline{\xi(R_{u-\tau_1})}\rangle$,

$$\inf\{\|\xi(u) - Z\|; Z \in \langle\overline{\xi(R_{u-\tau_1})}\rangle\} \leq \inf\{\|\xi(u) - Z\|; Z \in \langle\overline{\xi(R_{u-\tau_2})}\rangle\}$$

(see Proposition 5.4.1). This proves (7.10.4).

If the stochastic process $\{\xi(t); t \in R\}$ is deterministic, then $\xi(t + \tau) \in \langle\overline{\xi(R_t)}\rangle$ for all $t, \tau \in R$, so that

$$\xi(t + \tau) = \tilde{\xi}(t + \tau).$$

This implies that $\delta(\tau) = 0$ for all $\tau \leq \tau_0$. Conversely, if $\delta(\tau_0) = 0$ for some $\tau_0 > 0$, the process is deterministic. Indeed, due to (7.10.4) then, $\delta(\tau) = 0$ for all $\tau \leq \tau_0$. Therefore,

$$\xi(t + \tau) = \tilde{\xi}(t + \tau) \in \langle\overline{\xi(R_t)}\rangle$$

for all $t \leq \tau_0$ and all $t \in R$. This then implies that

$$\langle\overline{\xi(R_{t+\tau})}\rangle \subset \langle\overline{\xi(R_t)}\rangle$$

for all $t \in R$ and $\tau > 0$. Consequently, $\langle\overline{\xi(R_{t+\tau})}\rangle = \langle\overline{\xi(R_t)}\rangle$.

In the following, for convenience sake, we will denote by

$$P_t\xi(t + \tau), \quad \tau \geq 0 \tag{7.10.5}$$

the orthogonal projection of $\xi(t + \tau)$ on $\langle\overline{\xi(R_t)}\rangle$. With this notation, the error of the linear prediction (7.10.2) can now be written as

$$\delta(\tau) = \|\xi(t + \tau) - P_t\xi(t + \tau)\|. \tag{7.10.6}$$

Next, we shall prove the following result.

Proposition 7.10.1. *The stochastic process $\xi(t)$ is purely nondeterministic if and only if $\delta(\tau) \to C(0)$ as $\tau \to +\infty$.*

PROOF. Assume that $\delta(\tau) \to C(0)$ as $\tau \to +\infty$. Because $\langle\overline{\xi(-\infty)}\rangle \subset \langle\overline{\xi(R_s)}\rangle$ for all $s \in R$, we have

$$\|\xi(t) - P_s\xi(t)\| \leq \|\xi(t) - P_{-\infty}\xi(t)\|. \tag{7.10.7}$$

Next, because $(\xi(t) - P_s\xi(t)) \perp P_s\xi(t)$,

$$\|\xi(t)\|^2 = \|P_s\xi(t)\|^2 + \|\xi(t) - P_s\xi(t)\|^2. \tag{7.10.8}$$

We also have

$$\|\xi(t)\|^2 = \|P_{-\infty}\xi(t)\|^2 + \|\xi(t) - P_{-\infty}\xi(t)\|^2.$$

From this, (7.10.7) and (7.10.8), we deduce that

$$\|P_s\xi(t)\| \geq \|P_{-\infty}\xi(t)\|.$$

However, $\|\xi(t)\|^2 = C(0)$ for all t, and

$$\|\xi(t) - P_s\xi(t)\|^2 = \delta(t - s) \to C(0) \tag{7.10.9}$$

as $s \to -\infty$. Therefore, by (7.10.8),

$$\|P_s\xi(t)\|^2 \to 0 \quad \text{as } s \to -\infty.$$

Consequently, $\xi(t) \perp \langle\overline{\xi(-\infty)}\rangle$ for all t, so $\langle\overline{\xi(-\infty)}\rangle = \{0\}$.

Conversely, suppose that $\langle\overline{\xi(-\infty)}\rangle = \{0\}$ and let

$$P_{t-s}\xi(t) \xrightarrow{\text{q.m.}} Z \quad \text{as } s \to \infty;$$

then $Z \in \langle\overline{\xi(-\infty)}\rangle = \{0\}$ so that

$$\lim_{\tau\to\infty} \delta(\tau) = \lim_{\tau\to\infty} \|\xi(t+\tau) - P_t\xi(t+\tau)\|^2$$

$$= \lim_{\tau\to\infty} \|\xi(t) - P_{t-\tau}\xi(t)\|^2 = \|\xi(t)\|^2 = C(0).$$

This completes the proof of the assertion. □

7.11. The Wold Decomposition

Let $\{\xi(t); t \in R\} \subset L_2\{\Omega, \mathscr{B}, P\}$ be a wide sense stationary nondeterministic stochastic process. The Wold decomposition, which will be discussed next, shows that such a process can be represented as a sum of two mutually orthogonal second-order processes, one of which is deterministic and the other purely nondeterministic.

Proposition 7.11.1. *Let $\{\xi(t); t \in R\}$ be wide sense stationary with $E\{\xi(t)\} = 0$. Then it can be written as*

$$\xi(t) = \xi_1(t) + \xi_2(t), \tag{7.11.1}$$

where $\{\xi_1(t); t \in R\}$ is purely nondeterministic and $\{\xi_2(t); t \in R\}$ is deterministic. In addition,

$$\xi_1(s) \perp \xi_2(t) \tag{7.11.2}$$

for any $s, t \in R$.

PROOF. Set

$$P_{-\infty}\xi(t) = \xi_2(t) \quad \text{and} \quad \xi_1(t) = \xi(t) - \xi_2(t). \tag{7.11.3}$$

Clearly, $\xi_2(t) \in \langle\overline{\xi(-\infty)}\rangle$ for all t and, according to Proposition 5.4.2,

$$\xi_1(t) \perp \langle\overline{\xi(-\infty)}\rangle, \tag{7.11.4}$$

which implies that $\{\xi_1(t); t \in R\}$ and $\{\xi_2(t); t \in R\}$ are orthogonal processes. Next, because $\xi_1(t) \in \langle\overline{\xi(R_t)}\rangle$ for all t, it follows that

$$\langle\overline{\xi_1(R_t)}\rangle \subset \langle\overline{\xi(R_t)}\rangle \Rightarrow \langle\overline{\xi_1(-\infty)}\rangle \subset \langle\overline{\xi(-\infty)}\rangle. \tag{7.11.5}$$

This and (7.11.4) then imply that $\langle\overline{\xi_1(-\infty)}\rangle = \{0\}$ so that the process $\xi_1(t)$ is purely nondeterministic.

On the other hand, it is clear that

$$\langle\overline{\xi_2(R_t)}\rangle \subset \langle\overline{\xi(-\infty)}\rangle, \quad \forall t \in R, \tag{7.11.6}$$

so that $\langle\overline{\xi_2(R)}\rangle \subset \langle\overline{\xi(-\infty)}\rangle$. Because $\xi(t) = \xi_1(t) + \xi_2(t)$,

$$\langle\overline{\xi(R_t)}\rangle \subset \langle\overline{\xi_1(R_t)}\rangle \oplus \langle\overline{\xi_2(R_t)}\rangle, \quad \forall t \in R. \tag{7.11.7}$$

From this, (7.11.5), and (7.11.6), we deduce that

$$\langle\overline{\xi(-\infty)}\rangle \subset \langle\overline{\xi_1(R_t)}\rangle \oplus \langle\overline{\xi_2(R_t)}\rangle. \tag{7.11.8}$$

Now, (7.11.4) clearly implies that

$$\langle\overline{\xi_1(R_t)}\rangle \perp \langle\overline{\xi(-\infty)}\rangle. \tag{7.11.9}$$

Because [see (7.11.8)] $\langle\overline{\xi(-\infty)}\rangle \subseteq \langle\overline{\xi_1(R_t)}\rangle \oplus \langle\overline{\xi_2(R_t)}\rangle$ for all t, it seems clear that $\langle\overline{\xi(-\infty)}\rangle \subset \langle\overline{\xi_2(R_t)}\rangle$ for every t. This and (7.11.6) yield

$$\langle\overline{\xi(-\infty)}\rangle = \langle\overline{\xi_2(R_t)}\rangle$$

for all t, so that $\langle\overline{\xi_2(-\infty)}\rangle = \langle\overline{\xi_2(R_t)}\rangle$ for all t, from which we conclude that $\{\xi_2(t); t \in R\}$ is deterministic. Using a similar argument we can show that the decomposition (7.11.1) is unique. □

Corollary 7.11.1. *Both processes, $\xi_1(t)$ and $\xi_2(t)$ are wide sense stationary with $E\{\xi_1(t)\} = E\{\xi_2(t)\} = 0$.*

We now give without proof a necessary and sufficient condition for a wide stationary stochastic process $\{\xi(t); t \in R\}$ to be nondeterministic. Let $\{Z(t); t \in R\}$ be the stochastic process with orthogonal increments associated with $\xi(t)$. Then we have the following result.

Proposition 7.11.2. *For the stationary process $\xi(t)$ be nondeterministic, it is necessary and sufficient that it has the representation*

$$\xi(t) = \int_{-\infty}^{t} \alpha(t - s)\, dZ(s), \tag{7.11.10}$$

where

$$\int_{-\infty}^{\infty} |\alpha(s)|^2\, ds < \infty, \qquad \langle\overline{\xi(R_t)}\rangle = \langle\overline{Z(R_t)}\rangle. \tag{7.11.11}$$

Assume that the spectral distribution $F(\cdot)$ of the process $\xi(t)$ is absolutely continuous, i.e.,

$$F(\lambda) = \int_{-\infty}^{\lambda} f(s)\, ds;$$

then the following result holds.

Proposition 7.11.3. *The stochastic process $\xi(t)$ is purely nondeterministic if and only if its spectral density $f(\cdot)$ exists and*

$$\int_{-\infty}^{\infty} \frac{\ln f(\lambda)}{1+\lambda^2} d\lambda > -\infty. \tag{7.11.12}$$

Remark 7.11.1. The stochastic process $\xi(t)$ is deterministic if

$$\int_{-\infty}^{\infty} \frac{\ln f(\lambda)}{1+\lambda^2} d\lambda = -\infty. \tag{7.11.13}$$

If $f(\cdot)$ does not exist, then

$$\xi(t) = \xi_1(t) + \xi_2(t).$$

Remark 7.11.2. The spectral characteristic [see (7.9.2)] $\varphi(\lambda)$ of the linear transformation (7.11.10) is given by

$$\varphi(\lambda) = \int_0^{\infty} e^{-i\lambda t} \alpha(t)\, dt \tag{7.11.14}$$

and represents the boundary value of the analytical function

$$\varphi^*(z) = \int_0^{\infty} e^{-izt} \alpha(t)\, dt$$

on the half-plane $\mathrm{Im}\{z\} < 0$.

Example 7.11.1. Let $\{\xi(t); t \in R\}$ be wide sense stationary with

$$E\{\xi(t)\} = 0 \quad \text{and} \quad C(t) = \alpha e^{-t^2/4}, \quad \alpha > 0.$$

Because the condition (7.1.9) holds, the spectral density $f(\cdot)$ exists [see (7.1.10)] and

$$f(\lambda) = \frac{1}{2\pi} \int_{-\infty}^{\infty} e^{-i\lambda t} \alpha e^{-t^2/4}\, dt = \frac{\alpha}{\pi^{1/2}} e^{-\lambda^2}.$$

From this, we clearly see that (7.11.13) holds, which implies that the process $\xi(t)$ is deterministic.

Example 7.11.2. Let us show that the wide sense stationary process

$$\xi(t) = \sum_{k=1}^{\infty} e^{i\lambda_k t} \eta\{\lambda_k\}, \qquad E\{\xi(t)\} = 0,$$

is deterministic. Indeed, the process has a discrete spectrum [i.e., its spectral distribution $F(\cdot)$ is a step function]. To show this, consider

$$\begin{aligned} C(t) &= E\{\xi(s)\overline{\xi(s+t)}\} \\ &= \sum_{k=1}^{\infty} e^{-i\lambda_k t} E(\eta\{\lambda_k\})^2 \\ &= \int_{-\infty}^{\infty} e^{i\lambda t}\, dF(\lambda), \qquad dF(\lambda) = 0, \qquad \lambda \neq \lambda_k. \end{aligned}$$

If we set $\ln\{dF(\lambda)/d\lambda\} = -\infty$ when $dF(\lambda)/d\lambda = 0$, the condition (7.11.13) holds.

EXAMPLE 7.11.3. The following example is particularly instructive. Let $\{\xi(t); t \in R\}$ be wide sense stationary with $E\{\xi(t)\} = 0$ and covariance $C(t)$, which we assume to be infinitely differentiable at zero. Let $F(\cdot)$ be the corresponding spectral distribution. Then, according to (7.1.6),

$$C(t) = \int_{-\infty}^{\infty} e^{it\lambda}\, dF(\lambda).$$

From this, it readily follows that

$$\int_{-\infty}^{\infty} \lambda^{2n}\, dF(\lambda) < \infty \tag{7.11.15}$$

and vice versa, i.e., if (7.11.15) holds for all $n = 1, 2, \ldots$, the covariance function is infinitely differentiable at zero.

According to (7.6.8), then (7.11.15) implies the existence (in the q.m. sense) of the derivative $\xi'(t)$. Next, because $\xi'(t)$ is also wide sense stationary with covariance function

$$C_{\xi'}(t) = \int_{-\infty}^{\infty} \lambda^2 e^{i\lambda t}\, dF(\lambda),$$

we clearly have

$$C_{\xi}''(0) = -\int_{-\infty}^{\infty} \lambda^4\, dF(\lambda).$$

Therefore, because this integral is finite, the first derivative of $\xi'(t)$ (in the q.m. sense) exists and represents a wide sense stationary process and so on.

In general, if the condition (7.11.15) is met, the nth derivative $\xi^{(n)}(t)$, $n = 1, 2, \ldots$, exists and has the representation

$$\xi^{(n)}(t) = \int_{-\infty}^{\infty} (i\lambda)^n e^{i\lambda t} \eta(d\lambda). \tag{7.11.16}$$

From this, it follows that, for any $\tau > 0$,

$$\xi(t + \tau) = \sum_{k=0}^{\infty} \frac{\tau^k}{k!} \xi^{(k)}(t).$$

This allows an exact prediction of the process for all $\tau > 0$.

7.12. Discrete Parameter Processes

Let $E\{\xi_n\}_{-\infty}^{\infty}$ be a wide sense stationary process with

$$E\{\xi_n\} = 0 \quad \text{and} \quad C(n) = E\{\xi_k \bar{\xi}_{k+n}\}.$$

According to the Wold decomposition, we have

$$\xi_n = \xi_{n1} + \xi_{n2}, \tag{7.12.1}$$

where ξ_{n1} and ξ_{n2} are two wide sense stationary processes such that $\xi_{k1} \perp \xi_{j2}$ for all k, j. The first component in (7.12.1), ξ_{n1} is deterministic and the second component ξ_{n2} represents a purely nondeterministic process. The deterministic component is perfectly predictable, and an explicit formula may be obtained for predicting the purely nondeterministic component as the following proposition shows.

Proposition 7.12.1. *Let $\{X_n\}_{-\infty}^{\infty}$ be a wide sense stationary, purely nondeterministic process such that $E\{X_n\} = 0$. Then,*

$$X_n = \sum_{k=0}^{\infty} \alpha_k \zeta_{n-k}, \tag{7.12.2}$$

where $\{\zeta_m\}_{-\infty}^{\infty}$ is a sequence of pairwise orthogonal r.v.'s with

$$E\{\zeta_m\} = 0, \qquad E|\zeta_m|^2 = 1, \quad \text{and} \quad \langle\overline{\{X_k\}_{-\infty}^{m}}\rangle = \langle\overline{\{\zeta_k\}_{-\infty}^{m}}\rangle \tag{7.12.3}$$

for all m and $\{\alpha_k\}_1^{\infty}$ a sequence of complex numbers such that

$$\sum_{k=1}^{\infty} |\alpha_k|^2 < \infty.$$

Remark 7.12.1. The sequence $\{\zeta_m\}_{-\infty}^{\infty}$ is often called an "innovation sequence" for the process $\{X_n\}_{-\infty}^{\infty}$; the reason for this is that ζ_{n+1} provides "new information" needed to obtain $\langle\overline{\{X_k\}_{-\infty}^{n+1}}\rangle$.

We may drop the requirement that $\langle\overline{\{X_k\}_{-\infty}^{m}}\rangle = \langle\overline{\{\zeta_k\}_{-\infty}^{m}}\rangle$ for all m and (7.12.2) will still hold as the following proposition shows.

Proposition 7.12.2. *A wide sense stationary process $\{X_n\}_{-\infty}^{\infty}$ is purely nondeterministic if and only if it can be represented as*

$$X_n = \sum_{k=0}^{\infty} b_k Z_{n-k} \quad \text{(a.s.)}, \tag{7.12.4}$$

where $\{Z_m\}_{-\infty}^{\infty}$ is an orthonormal system of r.v.'s and

$$\sum_{k=0}^{\infty} |b_k|^2 < \infty.$$

Remark 7.12.1. A process in the form (7.12.4) is called a one-sided moving average process.

An analogous result to Proposition 7.11.3 also holds in the discrete case.

Proposition 7.12.3. *Let $\{\xi_n\}_{-\infty}^{\infty}$ be a nondegenerate wide sense stationary stochastic process with $E\{\xi_n\} = 0$. If the process is purely nondeterministic, there exists a spectral density $f(\lambda)$ such that*

$$\int_{-\pi}^{\pi} \ln f(\lambda)\, d\lambda > -\infty. \tag{7.12.5}$$

Conversely, if the process $\{\xi_n\}_{-\infty}^{\infty}$ *has a spectral density satisfying* (7.12.5), *the process is purely nondeterministic.*

Remark 7.12.2. From (7.12.5), it follows that $f(\cdot) > 0$ (a.e.) with respect to Lebesgue measure.

Remark 7.12.3. If

$$\int_{-\pi}^{\pi} \ln f(\lambda)\, d\lambda = -\infty, \tag{7.12.6}$$

the process $\{\xi_n\}_{-\infty}^{\infty}$ is deterministic.

EXAMPLE 7.12.1. Let $\{Z_n\}_{-\infty}^{\infty}$ be an orthonormal sequence of r.v.'s, i.e., $E\{Z_n\} = 0$, $\|Z_n\| = 1$ for all n and $(Z_j, Z_k) = 0$ for all $j \neq k$. From this, we obtain that the covariance function $C_Z(n)$ is defined by

$$C_Z(n) = \begin{cases} 1, & n = 0 \\ 0, & n \neq 0. \end{cases} \tag{7.12.7}$$

Assume that the process $\{Z_n\}_{-\infty}^{\infty}$ has a spectral density $f(\lambda)$. Taking into account that

$$\{(2\pi)^{-1/2} e^{-in\lambda}\}_{-\infty}^{\infty} \tag{7.12.8}$$

is a complete orthonormal system on $(-\pi, \pi)$ we can write

$$f(\lambda) = \frac{1}{2\pi} \sum_{k=-\infty}^{\infty} c_k e^{-ik\lambda},$$

where

$$c_k = \int_{-\pi}^{\pi} e^{-ik\lambda} f(\lambda)\, d\lambda.$$

But from this, (7.12.7), and Herglotz's theorem [see (7.1.8)], we conclude that

$$c_k = C_Z(k) = \begin{cases} 1, & k = 0 \\ 0, & k \neq 0. \end{cases}$$

Therefore,

$$f(\lambda) = \begin{cases} 1/2\pi, & -\pi \leq \lambda \leq \pi \\ 0, & \text{otherwise.} \end{cases} \tag{7.12.9}$$

The stochastic process $\{Z_n\}_{-\infty}^{\infty}$ is often called "white noise." Because the condition (7.12.5) is clearly satisfied, the process is nondeterministic.

EXAMPLE 7.12.2. Let $\{\xi_n\}_{-\infty}^{\infty}$ be a stochastic process defined by

$$\xi_n = \sum_{k=-\infty}^{\infty} \alpha_k Z_{n-k}, \tag{7.12.10}$$

where $\{Z_n\}_{-\infty}^{\infty}$ is a white noise process (see the previous example) and the sequence of complex numbers $\{\alpha_n\}_{-\infty}^{\infty}$ is such that

$$\sum_{k=-\infty}^{\infty} |\alpha_k|^2 < \infty. \tag{7.12.11}$$

The process $\{\xi_n\}_{-\infty}^{\infty}$ is called a moving average.

Clearly, $E\{\xi_n\} = 0$ for all n; in addition,

$$\begin{aligned} E\{\xi_m \bar{\xi}_{m+n}\} &= \sum_{j=-\infty}^{\infty} \sum_{k=-\infty}^{\infty} \alpha_j \bar{\alpha}_k E\{Z_{m-j} \bar{Z}_{m+n-k}\} \\ &= \sum_{k=-\infty}^{\infty} \alpha_{k-n} \bar{\alpha}_k \\ &= \sum_{k=-\infty}^{\infty} \alpha_k \bar{\alpha}_{k+n}. \end{aligned} \tag{7.12.12}$$

From this and (7.12.1), it follows that the $\{\xi_n\}_{-\infty}^{\infty}$ is wide sense stationary. Write

$$C_\xi(n) = E\{\xi_m \bar{\xi}_{m+n}\}.$$

Then the spectral density $f_\xi(\lambda)$ of the process $\{\xi_n\}_{-\infty}^{\infty}$ is equal to

$$\begin{aligned} f_\xi(\lambda) &= \frac{1}{2\pi} \sum_{n=-\infty}^{\infty} C_\xi(n) e^{-i\lambda n} \\ &= \frac{1}{2\pi} \sum_{n=-\infty}^{\infty} \sum_{k=-\infty}^{\infty} \alpha_k \bar{\alpha}_{k+n} e^{-i\lambda n} \\ &= \frac{1}{2\pi} \sum_{k=-\infty}^{\infty} \alpha_k e^{i\lambda k} \sum_{n=-\infty}^{\infty} \overline{\alpha_{n+k} e^{i\lambda(n+k)}} \\ &= \frac{1}{2\pi} \left| \sum_{r=-\infty}^{\infty} \alpha_r e^{i\lambda r} \right|^2 > 0. \end{aligned}$$

Remark 7.12.3. The last example shows that for a moving average process the spectral density $f(\lambda) > 0$ (a.e.) with respect to the Lebesgue measure. One can show that the converse also holds. In other words, a wide sense stationary process with spectral density $f(\lambda) > 0$ (a.e.) can be represented as a moving average.

7.13. Linear Prediction

A particular case of a general estimation theory is that of (pure) prediction; the basic idea was outlined at the beginning of Section 7.7. In Section 7.10, general ideas of linear prediction were discussed in some detail. This section is concerned with some basic methods and techniques of the theory of linear prediction.

Let $\{\xi(t); t \in R\}$ be a wide sense stationary process with $E\{\xi(t)\} = 0$, covariance function $C(t)$ (continuous at zero), and a spectral distribution $F(\lambda)$. Given values of the process on a set $T_0 \subseteq R_t = (-\infty, t]$, we want to "predict" its value at a time $t + h$, $h > 0$. The prediction $\tilde{\xi}(t + h)$ of $\xi(t + h)$ is clearly a functional of $\{\xi(s); s \in T_0\}$, i.e.,

$$\tilde{\xi}(t + h) = \tilde{\xi}\{\xi(s); s \in T_0\}. \tag{7.13.1}$$

We are concerned with the linear prediction of $\xi(t + h)$. As we have seen in Sections 7.7 and 7.10, the prediction is called "linear" if

$$\tilde{\xi}(t + h) \in \langle\overline{\xi(T_0)}\rangle. \tag{7.13.2}$$

The best possible prediction in such a case would be the element in $\langle\overline{\xi(T_0)}\rangle$ closest (in the L_2 sense) to $\xi(t + h)$. From Proposition 5.4.1, we know that there exists such an element (uniquely determined) which is the orthogonal projection of $\xi(t + h)$ on $\langle\overline{\xi(T_0)}\rangle$. This projection will be denoted by

$$P_{T_0}\xi(t + h). \tag{7.13.3}$$

From Proposition 5.4.2, it follows that

$$\xi(t + h) - P_{T_0}\xi(t + h) \perp \langle\overline{\xi(T_0)}\rangle.$$

From this, it follows that, for any $\zeta \in \langle\overline{\xi(T_0)}\rangle$,

$$(\xi(t + h), \zeta) = (P_{T_0}\xi(t + h), \zeta). \tag{7.13.4}$$

The norm of the deviation $\xi(t + h) - P_{T_0}\xi(t + h)$,

$$\begin{aligned} \delta(h) &= \|\xi(t + h) - P_{T_0}\xi(t + h)\| \\ &= (E|\xi(t + h) - P_{T_0}\xi(t + h)|^2)^{1/2}, \end{aligned} \tag{7.13.5}$$

is called the "error" of the linear prediction.

EXAMPLE 7.13.1. Assume that $T_0 = \{t_0, t_1, \ldots, t_n\}$ with $t_0 < t_1 < \cdots < t_n \leq t$. In this case

$$P_{T_0}\xi(t + h) = \sum_{k=0}^{n} \alpha_k \xi(t_k), \tag{7.13.6}$$

where the sequence of complex constants must be such that (7.13.4) is satisfied. In other words, for any $j = 0, \ldots, n$,

$$(\xi(t + h), \xi(t_j)) = \sum_{k=0}^{n} \alpha_k (\xi(t_k), \xi(t_j)) \tag{7.13.7}$$

or

$$\sum_{k=0}^{n} \alpha_k C(t_j - t_k) = C(t + h - t_j). \tag{7.13.8}$$

The solution of this system of linear equations will give us the values of the coefficients $\alpha_0, \ldots, \alpha_n$, for which the right-hand side of (7.13.6) represents the orthogonal projection of $\xi(t+h)$ on $\langle\overline{\xi(T_0)}\rangle$.

Let us now determine the value of the "error" of the linear prediction. From (7.13.5) and (7.13.6), we have

$$\begin{aligned}\delta^2(h) &= \left\| \xi(t+h) - \sum_{k=0}^{n} \alpha_k \xi(t_k) \right\|^2 \\ &= \left(\xi(t+h) - \sum_{k=0}^{n} \alpha_k \xi(t_k), \xi(t+h) - \sum_{k=0}^{n} \alpha_k \xi(t_k) \right) \\ &= \left(\xi(t+h) - \sum_{k=0}^{n} \alpha_k \xi(t_k), \xi(t+h) \right) \\ &= C(0) - \sum_{k=0}^{n} \alpha_k C(t+h-t_k). \end{aligned} \tag{7.13.9}$$

From this and (7.13.8), we obtain

$$\delta^2(h) = C(0) - \sum_{k=0}^{n} \sum_{j=0}^{n} \alpha_k \alpha_j C(t_k - t_j). \tag{7.13.10}$$

Next, we shall attempt to determine the best linear predictor of $\xi(t+h)$ when

$$T_0 = R_t.$$

As before [see (7.10.5)], the orthogonal projection of $\xi(t+h)$ on the subspace $\langle\overline{\xi(R_t)}\rangle$ will be denoted by $P_t\xi(t+h)$. Because $P_t\xi(t+h) \in \langle\overline{\xi(R_t)}\rangle$, it follows from Proposition 7.6.1 and (7.9.7) that $P_t\xi(t+h)$ has the representation

$$P_t\xi(t+h) = \int_{-\infty}^{\infty} e^{i\lambda t} \varphi(\lambda, h) \eta(d\lambda), \tag{7.13.11}$$

where $\varphi(\lambda, h) \in L_2\{R, \mathscr{R}, F\}$ is the spectral characteristic [see (7.9.2)] of the prediction process.

As an illustration, consider the previous example, and in particular (7.13.6). Using the spectral representation of the process $\xi(t)$, we obtain

$$P_{T_0}\xi(t+h) = \int_{-\infty}^{\infty} \left(\sum_{k=0}^{n} \alpha_k e^{i\lambda t_k} \right) \eta(d\lambda).$$

Here,

$$\varphi(\lambda, h) = \sum_{k=0}^{n} \alpha_k e^{i\lambda t_k} \tag{7.13.12}$$

[the right-hand side of (7.13.12) depends on h because the coefficients are functions of h].

To determine $\varphi(\lambda, h)$ we must devise an analog of the system of linear equations (7.13.8). The following method is due to Yaglom. Starting from the fact that

$$\xi(t+h) - P_t\xi(t+h) \perp \overline{\langle \xi(R_t) \rangle}, \tag{7.13.13}$$

it follows that, for any $s \geq 0$,

$$(\xi(t+h) - P_t\xi(t+h), \xi(t-s)) = 0$$

or, equivalently,

$$E\{\xi(t+h)\bar{\xi}(t-s)\} - E\{P_t\xi(t+h)\bar{\xi}(t-s)\} = 0. \tag{7.13.14}$$

Using the spectral representations

$$\xi(t+h) = \int_{-\infty}^{\infty} e^{i\lambda(t+h)}\eta(d\lambda),$$

$$\xi(t-s) = \int_{-\infty}^{\infty} e^{i\lambda(t-s)}\eta(d\lambda),$$

and (7.13.11), Equation (7.13.14) becomes

$$\int_{-\infty}^{\infty} e^{i\lambda s}[e^{i\lambda h} - \varphi(\lambda, h)]\,dF(\lambda) = 0, \tag{7.13.15}$$

or if $F(\cdot)$ is absolutely continuous,

$$\int_{-\infty}^{\infty} e^{i\lambda s}[e^{i\lambda h} - \varphi(\lambda, h)]f(\lambda)\,d\lambda = 0. \tag{7.13.16}$$

The corresponding prediction error is

$$\begin{aligned}\delta^2(h) &= E|\xi(t+h) - P_t\xi(t+h)|^2 \\ &= E\left|\int_{-\infty}^{\infty} e^{i\lambda(t+h)}\eta(d\lambda) - \int_{-\infty}^{\infty} e^{i\lambda t}\varphi(\lambda, h)\eta(d\lambda)\right|^2 \\ &= E\left|\int_{-\infty}^{\infty} e^{i\lambda t}[e^{i\lambda h} - \varphi(\lambda, h)]\eta(d\lambda)\right|^2 \\ &= \int_{-\infty}^{\infty} |e^{i\lambda t} - \varphi(\lambda, h)|^2 f(\lambda)\,d\lambda \\ &= C(0) - \int_{-\infty}^{\infty} |\varphi(\lambda, h)|^2 f(\lambda)\,d\lambda. \end{aligned} \tag{7.13.7}$$

7.14. Evaluation of the Spectral Characteristic $\varphi(\lambda, h)$

According to (7.13.11), the best linear prediction in the L_2 sense, of $\xi(t+h)$, $h > 0$, given all the values of $\xi(s)$ for $s \leq t$, is completely characterized by its spectral characteristic $\varphi(\lambda, h)$. In some special cases, the determination of

$\varphi(\lambda, h)$ is simple enough (see Example 7.13.1). However, in general, the situation is far more complex. First, the spectral characteristic $\varphi(\lambda, h)$ may not even exist (except as a generalized function). Even if it exists, there is generally speaking, no sufficiently simple expression for $P_t\xi(t+h)$. Second, from Proposition 7.6.1, we know that the $\varphi(\lambda, h) \in L_2\{R, \mathscr{R}, F\}$ and must be a q.m. limit of a sequence of linear combinations of $e^{i\lambda s}$, $s \geq 0$. Finally, the condition (7.13.15) must be satisfied. As a convenience, we list these requirements below. A function $\varphi(\lambda, h)$ is a spectral characteristic if the following holds:

(i) $\varphi(\lambda, h) \in L_2\{R, \mathscr{R}, F\}$, where $F(\cdot)$ is the spectral distribution which is assumed to be absolutely continuous so that the spectral density $f(\lambda) = F'(\lambda)$ exists.

(ii) $\varphi(\lambda, h)$ is the q.m. limit of a sequence of linear combinations of $e^{iu\lambda}$.

(iii)
$$\int_{-\infty}^{\infty} e^{is\lambda}[e^{ih\lambda} - \varphi(\lambda, h)]f(\lambda)\,d\lambda = 0 \tag{7.14.1}$$

for all $s \geq 0$.

Write
$$H(\lambda, h) = [e^{ih\lambda} - \varphi(\lambda, h)]f(\lambda) \tag{7.14.2}$$
and consider the function $H(z, h)$ of a complex variable z which is assumed to be single-valued and holomorphic (analytical) in the upper half of the complex plane $\{z; \operatorname{Im}\{z\} \geq 0\}$. The following proposition gives a sufficient condition for (7.14.1) to hold.

Proposition 7.14.1. *If, as* $|z| \to \infty$, $|H(z,h)|$ *vanishes faster than* $1/|z|^{1+\varepsilon}$ *for some* $\varepsilon > 0$, *then* (7.14.1) *holds.*

PROOF. Because $e^{iz\lambda}$ and $H(z, \lambda)$ are both holomorphic in the upper half of the complex plane, their product is also holomorphic. Therefore, according to the Cauchy theorem,
$$\int_C e^{izs} H(z, h)\,dz = 0$$
for any closed contour $C \subset \{z; \operatorname{Im}\{z\} \geq 0\}$. Take $C = C_R$, where C_R is formed by the semicircle $|z| = R$ in the upper half of the complex plane and $[-R, R]$. According to the residue theorem,
$$\int_{-R}^{R} e^{is\lambda} H(\lambda, h)\,d\lambda = \int_{C_R} e^{iz\lambda} H(z, h)\,dz. \tag{7.14.3}$$
However ($M > 0$ is a constant),
$$\left|\int_{C_R} e^{iz\lambda} H(z, \lambda)\,dz\right| \leq \max_{|z|=R} |H(z, \lambda)|\pi R \leq \frac{M}{R^{\varepsilon}} \to 0$$
as $R \to \infty$. From this and (7.14.3), the assertion follows. □

In the following, we shall assume that $f(\cdot)$ is bounded, i.e.,

$$\sup_{\lambda \in R} f(\lambda) < \infty, \tag{7.14.4}$$

and that $\varphi(\lambda, h) \in L_2\{R, \mathscr{R}, F\}$ is such that

$$\varphi(z, h) \text{ is holomorphic in } \{z; \operatorname{Im}\{z\} \geq 0\}, \tag{7.14.5}$$

$$\varphi(z, h) = O(|z|^r) \quad \text{as } |z| \to \infty, \quad z \in \{z; \operatorname{Im}\{z\} \geq 0\} \tag{7.14.6}$$

for some $r \to 0$.

The conditions of Proposition 7.14.1, (7.14.4), (7.14.5), and (7.14.6) are sufficient for (i)–(iii) to hold. We will illustrate with some examples how these conditions are used to construct a spectral characteristic $\varphi(\lambda, h)$. We will confine ourselves to the case where $f(\lambda)$ is a rational function.

EXAMPLE 7.14.1. Let the wide sense stationary process $\{\xi(t); t \in R\}$ have the covariance function

$$C(t) = \sigma^2 e^{-\alpha|t|}, \quad \alpha > 0.$$

As in Example 7.1.3, we can show that the corresponding spectral density is

$$f(\lambda) = \frac{\sigma^2}{\pi} \frac{\alpha}{\lambda^2 + \alpha^2} = \frac{\sigma^2}{\pi} \frac{\alpha}{(\lambda + \alpha i)(\lambda - \alpha i)}.$$

Now, we have [see (7.14.2)]

$$H(z, h) = \frac{\sigma^2}{\pi} \frac{e^{izh} - \varphi(z, \lambda)}{(z + \alpha i)(z - \alpha i)}.$$

The function $f(z)$ possesses a single pole, $z = \alpha i$, in the upper half-plane $\{z; \operatorname{Im}\{z\} \geq 0\}$. Because $H(z, h)$ is assumed holomorphic in this domain, this pole must be canceled by the pole of the numerator at the same point. Therefore, we must have

$$e^{i(\alpha i)h} - \varphi(\alpha i, \lambda) = 0, \qquad \varphi(\alpha i, h) = e^{-\alpha h}. \tag{7.14.7}$$

In addition, the requirement of Proposition 7.14.1 must be satisfied. The only function $\varphi(z, h)$ which satisfies all these conditions is $\varphi(z, h) = e^{-\alpha h}$ (constant). Thus,

$$\varphi(\lambda, h) = e^{-\alpha h}. \tag{7.14.8}$$

From this and (7.13.11), we deduce that the best linear predictor $P_t\xi(t + h)$ of $\xi(t + h)$ is

$$P_t\xi(t + h) = e^{-\alpha h} \int_{-\infty}^{\infty} e^{it\lambda} \eta(d\lambda) = e^{-\alpha h} \xi(t) \tag{7.14.9}$$

which depends only on the value of $\xi(t)$ at the last observed instant of time. The prediction error is easily found to be

$$\delta^2 = \sigma^2(1 - e^{2\alpha h}).$$

The following example is more instructive although slightly more complicated.

EXAMPLE 7.14.2. Let $\{\xi(t); t \in R\}$ be a wide sense stationary stochastic process with covariance function

$$C(t) = \sigma^2(\cos \alpha t + \sin \alpha |t|)e^{-\alpha|t|}, \quad \alpha > 0. \tag{7.14.10}$$

Find the best linear prediction of $\xi(t + h)$ given $\xi(s)$ for all $s \le t$,

Let us determine the spectral distribution of $\xi(t)$. Because condition (7.1.9) holds and $C(t)$ is real, we can use formula (7.1.14) to obtain the spectral density

$$f(\lambda) = \frac{1}{\pi}\int_0^\infty C(t)\cos t\lambda\, dt$$
$$= \frac{\sigma^2}{\pi}\int_0^\infty e^{\alpha t}(\cos \alpha t + \sin \alpha t)\cos \lambda t\, dt.$$

After some calculation, we have

$$f(\lambda) = \frac{4\alpha^3}{\lambda^4 + (\alpha\sqrt{2})^4}. \tag{7.14.11}$$

Because

$$\lambda^4 + (\alpha\sqrt{2})^4 = [\lambda^2 + i(\alpha\sqrt{2})^2][\lambda^2 - i(\alpha\sqrt{2})^2]$$
$$= [\lambda + (1 + i)\alpha][\lambda - (1 + i)\alpha][\lambda + (1 - i)\alpha][\lambda - (1 - i)\alpha],$$

we obtain

$$f(\lambda) = \frac{4\alpha^3}{[\lambda + (1 + i)\alpha][\lambda - (1 + i)\alpha][\lambda + (1 - i)\alpha][\lambda - (1 - i)\alpha]}.$$

Hence,

$$H(\lambda, h) = \frac{4\alpha^3[e^{ih\lambda} - \varphi(\lambda, h)]}{[\lambda + (1 + i)\alpha][\lambda - (1 + i)\alpha][\lambda + (1 - i)\alpha][\lambda - (1 - i)\alpha]}.$$

Becase the poles of the denominator in the upper half-plane are

$$\lambda = (1 + i)\alpha, \qquad \lambda = -(1 - i)\alpha,$$

it is clear that the numerator must also vanish at these points, so that

$$\varphi((1 + i)\alpha, h) = e^{ih(1+i)\alpha} = e^{-h(1-i)\alpha}, \tag{7.14.12}$$

$$\varphi(-(1 - i)\alpha, h) = e^{ih(i-1)\alpha} = e^{-h(1+i)\alpha}. \tag{7.14.13}$$

On the other hand, from condition (i) we see that

$$\int_{-\infty}^{\infty} |\varphi(\lambda, h)|^2 f(\lambda)\, d\lambda < \infty,$$

which clearly holds if

$$\varphi(\lambda, h) = a\lambda + b. \tag{7.14.14}$$

With this spectral characteristic, all the conditions concerning asymptotic behavior of $\varphi(z, h)$ and $H(z, h)$ as $|z| \to \infty$ are also satisfied.

To determine the coefficients a and b in (7.14.14), we use conditions (7.14.12) and (7.14.13) to obtain the system of linear equations

$$a(1 + i)\alpha + b = e^{-h(1-i)\alpha},$$
$$-a(1 - i)\alpha + b = e^{-h(1+i)\alpha}.$$

Solving this system, we obtain

$$a = \frac{ie^{-h\alpha}}{\alpha} \sin h\alpha, \qquad b = e^{-h\alpha}(\cos h\alpha + \sin h\alpha).$$

From this and (7.14.14), it follows that

$$\varphi(\lambda, h) = \frac{\lambda ie^{-h\alpha}}{\alpha} \sin h\alpha + e^{-h\alpha}(\cos h\alpha + \sin h\alpha). \tag{7.14.15}$$

We now substitute (7.14.15) in (7.13.11) to obtain

$$\begin{aligned} P_t\xi(t + h) &= \int_{-\infty}^{\infty} e^{it\lambda}\left(\frac{\lambda ie^{-h\alpha}}{\alpha} \sin h\alpha + e^{-h\alpha}(\cos h\alpha + \sin h\alpha)\right)\eta(d\lambda) \\ &= \frac{e^{-h\alpha}}{\alpha} \sin h\alpha \int_{-\infty}^{\infty} i\lambda e^{i\lambda t}\eta(d\lambda) + e^{-h\alpha}(\cos h\alpha + \sin h\alpha) \int_{-\infty}^{\infty} e^{i\lambda t}\eta(d\lambda). \end{aligned}$$

From this and (7.11.16), we conclude that

$$P_t\xi(t + h) = e^{-h\alpha}\left(\frac{\sin h\alpha}{\alpha}\xi'(t) + (\cos h\alpha + \sin h\alpha)\xi(t)\right). \tag{7.14.16}$$

7.15. General Form of Rational Spectral Density

In this section we will discuss the problem of evaluation of the spectral characteristics $\varphi(\lambda, h)$ in the case of rational spectral densities of the form

$$f(\lambda) = K\left|\frac{(\lambda - \alpha_1)\ldots(\lambda - \alpha_k)}{(\lambda - \beta_1)\ldots(\lambda - \beta_n)}\right|^2, \tag{7.15.1}$$

where $K > 0$, $k < n$, and $\{\alpha_i\}$ and $\{\beta_r\}$ are the complex numbers such that

$$\operatorname{Im}\{\alpha_i\} > 0, \quad j = 1, \ldots, k, \qquad \operatorname{Im}\{\beta_r\} > 0, \quad r = 1, \ldots, n. \tag{7.15.2}$$

EXAMPLE 7.15.1. The spectral density $f(\lambda)$ given by

$$f(\lambda) = K\frac{\lambda^2 + \beta^2}{\lambda^4 + \alpha^2}, \quad K > 0, \quad \alpha > 0, \quad \beta > 0,$$

can be written in the form (7.15.1) as follows. Clearly,

$$\lambda^2 + \beta^2 = |\lambda - i\beta|^2, \qquad (\lambda^2)^2 + (\alpha^2)^2 = |\lambda^2 - i\alpha^2|^2.$$

Hence,

$$f(\lambda) = \left|\frac{\lambda - i\beta}{(\lambda - \alpha\sqrt{i})(\lambda + \alpha\sqrt{i})}\right|^2.$$

The function $[e^{i\lambda h} - \varphi(\lambda, h)]$ will be holomorphic in the upper part of the complex plane if the spectral characteristic $\varphi(\lambda, h)$ has singularities only at points $\alpha_1, \ldots, \alpha_k$. In such a case, $\varphi(\lambda, h)$ must be in the form

$$\varphi(\lambda, h) = \frac{Q(\lambda, h)}{(\lambda - \alpha_1)\ldots(\lambda - \alpha_k)} = \frac{Q(\lambda, h)}{\prod_1^k (\lambda - \alpha_j)}, \tag{7.15.3}$$

where $Q(\lambda, h)$ is an entire function (i.e., it does not have singularities at finite points). This takes care of condition (7.14.5). It is also required [see condition (i) of Section 7.14] that

$$\int_{-\infty}^{\infty} |\varphi(\lambda, h)|^2 f(\lambda)\, d\lambda < \infty. \tag{7.15.4}$$

In addition [see (7.14.2)],

$$H(\lambda, h) = \frac{K\{e^{i\lambda h}\prod_1^k (\lambda - \alpha_j) - Q(\lambda, h)\}}{(\prod_1^n (\lambda - \beta_j))(\prod_1^n (\lambda - \bar{\beta}_j))} \prod_1^k (\lambda - \bar{\alpha}_j) \tag{7.15.5}$$

must be holomorphic in the upper part of the complex plane and such that condition (7.14.6) holds. All these requirements clearly imply that $Q(\lambda, h)$ must be a polynomial function

$$Q(\lambda, h) = c_0 + c_1\lambda + \cdots + c_{n-1}\lambda^{n-1}. \tag{7.15.6}$$

The coefficients $\{c_j\}_0^{n-1}$ will be determined first under the condition that all $\beta_1, \ldots, \beta_n$ in (7.15.1) are different, i.e.,

$$\beta_1 \neq \beta_2 \neq \cdots \neq \beta_n. \tag{7.15.7}$$

Because, by assumption, $H(\lambda, h)$ is holomorphic in the upper half of the complex plane, it follows readily that

$$e^{i\lambda h} \prod_{j=1}^{k} (\lambda - \alpha_j) - \sum_{j=0}^{n-1} c_j \lambda^j = 0 \tag{7.15.8}$$

for $\lambda = \beta_l$, $l = 1, \ldots, n$. Solution of this system of n linear equations gives the unknown constants $c_0, \ldots, c_{n-1}$.

Example 7.15.2. Determine the best linear predictor for $\xi(t + h)$, $h > 0$, based on $\xi(s)$, $s \leq t$, if the spectral density of the process $\xi(t)$ is given by

$$f(\lambda) = \frac{1}{(\lambda^2 + 2\alpha^2)(\lambda^2 + \alpha^2/2)}, \quad \alpha > 0.$$

We can clearly write

$$f(\lambda) = \frac{1}{|(\lambda - i\alpha\sqrt{2})(\lambda - i\alpha/\sqrt{2})|^2}, \tag{7.15.9}$$

so that $n = 2$, $\beta_1 = i\alpha\sqrt{2}$, $\beta_2 = i\alpha/\sqrt{2}$, and conditions (7.15.2) and (7.15.7) are satisfied. From this it clearly follows that

$$Q(\lambda, h) = c_0 + c_1\lambda,$$

so that Equation (7.15.8) becomes

$$e^{i\lambda h} - (c_0 + c_1\lambda) = 0$$

for $\lambda = i\alpha\sqrt{2}$ and $\lambda = i\alpha/\sqrt{2}$. Therefore, we have the system

$$c_0 + i\alpha\sqrt{2}c_1 = \exp(-\alpha h\sqrt{2}),$$
$$c_0 + c_1 i\alpha/\sqrt{2} = \exp(-\alpha h/\sqrt{2})$$

from which we obtain

$$\begin{aligned} c_0 &= [2 - \exp(-\alpha h\sqrt{2})]\exp(-\alpha h\sqrt{2}), \\ c_1 &= \frac{i\sqrt{2}}{\alpha}[1 - \exp(-\alpha h/\sqrt{2})]\exp(-\alpha h/\sqrt{2}). \end{aligned} \tag{7.15.10}$$

According to (7.15.3), we now have

$$\varphi(\lambda, h) = c_0 + c_1\lambda,$$

so that best linear predictor is [see (7.13.11)]

$$\begin{aligned} P_t\xi(t+h) &= \int_{-\infty}^{\infty} e^{i\lambda t}\varphi(\lambda, h)\eta(d\lambda) \\ &= c_0\int_{-\infty}^{\infty} e^{i\lambda t}\eta(d\lambda) + c_1\int_{-\infty}^{\infty} \lambda e^{i\lambda t}\eta(d\lambda) \\ &= c_0\xi(t) + c_1\xi'(t) \end{aligned}$$

with the coefficients c_0 and c_1 given by (7.15.10).

Remark 7.15.1. The last example is easy to generalize. If the spectral density of $\xi(t)$ is in the form

$$f(\lambda) = \left|\prod_{j=1}^{n} (\lambda - \beta_j)\right|^{-2},$$

then clearly

$$\varphi(\lambda, h) = \sum_{j=0}^{n-1} c_j\lambda^j,$$

where the c_j can be obtained as the solution of the system of linear equations

$$\sum_{j=0}^{n-1} c_j \lambda^j = e^{i\lambda h} \tag{7.15.11}$$

for $\lambda = \beta_j, j = 1, \ldots, n$. In this case, the best linear predictor of $\xi(t+h)$, given $\xi(s)$ for all $s \leq t$, is

$$P_t\xi(t+h) = c_0\xi(t) + c_1\xi'(t) + \cdots + c_{n-1}\xi^{(n-1)}(t). \tag{7.15.12}$$

When the zeros $\beta_1, \ldots, \beta_n$ in (7.15.1) are not all different, the procedure for determining the spectral characteristic $\varphi(\lambda, h)$ remains essentially the same, as the following example shows.

EXAMPLE 7.15.3. Let the spectral density be given by

$$f(\lambda) = \frac{1}{(\lambda^2 - \alpha^2)^2} = \frac{1}{|(\lambda - i\alpha)^2|^2}, \quad \alpha > 0.$$

In this case, we clearly have that $n = 2$, so that

$$\varphi(\lambda, h) = Q(\lambda, h) = c_0 + c_1\lambda.$$

We obtain the coefficients c_0 and c_1 from the fact that in such a case

$$e^{ih\lambda} - (c_0 + c_1\lambda) = 0 \quad \text{for } \lambda = i\alpha$$

and

$$\frac{d}{d\lambda}[e^{ih\lambda} - (c_0 + c_1\lambda)] = 0 \quad \text{for } \lambda = i\alpha.$$

This yields the equations

$$c_0 + c_1 i\lambda = e^{-\alpha h}$$
$$c_1 = ihe^{-\alpha h},$$

the solution of which is

$$c_1 = ihe^{-\alpha h}, \qquad c_0 = e^{-\alpha h}(1 + \alpha h). \tag{7.15.13}$$

The best linear predictor of $\xi(t+h)$ in this case is

$$P_t\xi(t+h) = c_0\xi(t) + c_1\xi'(t),$$

where c_0 and c_1 are given by (7.15.13).

In general, if β_1 appears l_1 times, $\ldots$, β_r appears l_r times, where

$$l_1 + \cdots + l_r = n,$$

we have

$$\frac{d^v}{d\lambda^v}\left(e^{i\lambda h}\prod_1^r (\lambda - \beta_j)^{l_j} - \sum_{j=0}^{n-1} c_j\lambda^j\right) = 0 \quad \text{for}$$
$$\lambda = \beta_j, \quad v = 0, 1, \ldots, l_p - 1, \quad p = 1, \ldots, r. \tag{7.15.14}$$

Problems and Complements

7.1. The covariance function of a wide sense stationary process $\{\xi(t); t \in R\}$ is given by $(t_0 > 0)$

$$C(t) = \begin{cases} 1 - \dfrac{|t|}{t_0} & \text{if } |t| \leq t_0 \\ 0 & \text{if } |t| > t_0. \end{cases}$$

Determine its spectral density.

7.2. Let $C(t) = e^{-|t|} \cos \alpha t$ be the covariance function of a wide sense stationary process. Find its spectral density.

7.3. Does there exist a wide sense stationary process whose covariance function is defined by

$$C(t) = \begin{cases} \sigma^2 & \text{if } 0 \leq t \leq t_0 \\ 0 & \text{if } t > t_0. \end{cases}$$

7.4. Let X be a r.v. distributed on $[0, \frac{1}{2}]$ with the probability density $f_X(\cdot)$. Let Y be another r.v. independent of X and uniformly distributed on $[-\frac{1}{2}, \frac{1}{2}]$. Show that

$$\xi(t) = \sigma \cos(tX + Y)$$

is wide sense stationary and find its spectral density.

7.5. A wide sense stationary process $\{\xi(t); t \in R\}$ has the covariance function

$$C(t) = \sigma^2 e^{-\alpha|t|}(1 + |t|), \quad \alpha > 0.$$

Find its spectral density.

7.6. Find the spectral density of a wide sense stationary process with covariance function

$$C(t) = \sigma^2 e^{-\alpha|t|} \cos \beta t, \quad \alpha > 0.$$

7.7. Is the wide sense stationary process with spectral density

(i) $$f(\lambda) = \frac{\alpha}{\lambda^2 + \beta},$$

(ii) $$f(\lambda) = K\left\{\frac{1}{\alpha^2 + (\lambda - \beta)^2} + \frac{1}{\alpha^2 + (\lambda + \beta)^2}\right\}, \quad \alpha > 0,$$

differentiable?

7.8. What are conditions for infinite differentiability of a wide sense stationary process (see Example 7.11.3).

7.9. The wide sense stationary process $\{\xi(t); t \in R\}$ has covariance function

$$C(t) = e^{-\alpha|t|}\left(1 + \alpha|t| + \frac{(\alpha t)^2}{3}\right).$$

Determine its spectral density.

7.10. Find the spectral density of the wide sense stationary process with covariance function

$$C(t) = e^{-\alpha|t|}(\cos \beta t - \alpha \sin \beta |t|).$$

7.11. Show that the orthogonal stochastic measure η has the following properties (see Section 7.3):

(i) If $A_1, A_2 \in \mathscr{S}$, $(\eta(A_1), \eta(A_2)) = m(A_1 \cap A_2)$.

(ii) If $A, B \in \mathscr{S}$ are disjoint,

$$\eta(A \cup B) = \eta(A) + \eta(B) \quad \text{(a.s.)}.$$

(iii) If $\{B\}_1^\infty \subset \mathscr{S}$ is a disjoint sequence,

$$\eta\left(\bigcup_{k=1}^{\infty} B_k\right) = \sum_{k=1}^{\infty} \eta(B_k) \quad \text{(a.s.)}.$$

7.12. Let $\eta: \mathscr{S} \times \Omega \to L\{\Omega, \mathscr{B}, P\}$ be a stochastic orthogonal measure, and let $m: \mathscr{S} \to R_+$ [see (7.3.3) for the definition] be the measure associated with η. Define

$$\eta_1(B) = \int_B h(s)\eta(ds), \qquad m_1(B) = \int_B |h(s)|^2 m(ds),$$

where $h \in L_2\{R, \mathscr{R}, m\}$. Show that η_1 is an orthogonal stochastic measure with $m_1(\cdot)$ as associated measure.

7.13. (Continuation) If $p(\cdot) \in L_2\{R, \mathscr{R}, m_1\}$ show that

$$\int_B p(s)\eta_1(ds) = \int_B p(s)h(s)\eta(ds) \quad \text{(a.s.)}, \quad B \in \mathscr{S}.$$

7.14. (Continuation) If $|h| > 0$ (a.e.) $[m]$, then

$$\eta(B) = \int_B h^{-1}(s)\eta_1(ds), \quad B \in \mathscr{S}.$$

7.15. Let $\{Z_n\}_1^\infty \subset L_2\{\Omega, \mathscr{B}, P\}$ be an orthogonal family such that $\sum_1^\infty \|Z_n\|^2 < \infty$. Let $\{S, \mathscr{S}\}$ be a measurable space with $\mathscr{S}$ containing singletons as elements, i.e., $x \in S \Rightarrow \{x\} \in \mathscr{S}$. Define $\eta: \mathscr{S} \to L_2\{\Omega, \mathscr{B}, P\}$ by

$$\eta(B) = \sum_{n=1}^{\infty} Z_n I_B(x_n), \quad \{x_n\}_1^\infty \subset S, \quad B \in \mathscr{S}.$$

Show that $\eta(\cdot)$ is an orthogonal stochastic measure with associated measure $m(\cdot)$ given by

$$m(B) = \sum_{n=1}^{\infty} \|Z_n\|^2 I_B(x_n).$$

Show also that

$$\int_S h(s)\eta(ds) = \sum_{n=1}^{\infty} h(x_n)Z_n, \quad h \in L_2\{S, \mathscr{S}, m\}.$$

7.16. Let η be an orthogonal random measure with $m(\cdot)$ as associated measure. If B_1, $B_2 \in \mathscr{S}$ show that

(i) $\|\eta(B_1) - \eta(B_2)\|^2 = m(B_1 \triangle B_2)$,
(ii) $\eta(B_1 \cup B_2) = \eta(B_1) + \eta(B_2) - \eta(B_1 \cap B_2)$ (a.s.),
(iii) $\eta(B_1 \triangle B_2) = \eta(B_1) + \eta(B_2) - 2\eta(B_1 \cap B_2)$ (a.s.).

7.17. (Continuation) If $B_1 \subset B_2$, show that
(i) $\|\eta(B_1)\| \leq \|\eta(B_2)\|$,
(ii) $\eta(B_2 - B_1) = \eta(B_2) - \eta(B_1)$ (a.s.).

7.18. Let $\{\xi(t); t \in R\}$ be a wide sense stationary process with $E\{\xi(t)\} = 0$ and spectral measure F (we use the same symbol for Lebesgue–Stieltjes measure and the function F that generates the measure). Show that

$$\underset{t\to\infty}{\text{l.i.m.}} \int_0^t \xi(s)\,ds = F(0).$$

7.19. (Continuation) Under the same conditions as in the previous problem, show that

$$\lim_{t\to\infty} \frac{1}{t} \int_0^t C(s)\,ds = F(0),$$

where C is the covariance function of $\xi(t)$.

7.20. Let $\{X(t); t \in R\}$ be a wide sense stationary process with continuous covariance function and spectral representation

$$X(t) = \int_{-\infty}^{\infty} e^{i\lambda t} \eta_X(d\lambda).$$

Let $\xi(t)$ be defined by

$$\xi(t) = \int_{-\infty}^{\infty} h(s) X(t-s)\,ds,$$

where $h \in L_2\{R, \mathcal{R}, F_X\}$. Determine the covariance function and the spectral density of $\xi(t)$, assuming that the spectral density of $X(t)$ is $\varphi_X(\cdot)$.

7.21. Let $\{\xi(t); t \geq 0\}$ be a standard Brownian motion and Z a r.v. independent of $\xi(t)$, with $E\{Z\} = 0$ and $E|Z|^2 = 1$. Suppose that the process $\{X(t); t \geq 0\}$ is observed where $X(t) = Zt + \xi(t)$. Given $X(s)$, $s \leq t$, determine the best q.m. estimate $\tilde{Z}_t$ of Z in the form

$$\tilde{Z}_t = \int_0^t \psi(t,s)\,dX(s).$$

7.22. Let the process $\{X(t); t \in R\}$ be defined by

$$X(t) = Z \cos 2\alpha t + N(t),$$

where $\{N(t); t \in R\} \subset L_2\{\Omega, \mathcal{B}, P\}$ with $E\{N(t)\} = 0$,

$$E\{N(s)\bar{N}(t+s)\} = e^{|t|}$$

and Z is a r.v. with $E\{Z\} = 0$, $E|Z|^2 = 1$, and $E\{ZN(t)\} = 0$. Determine a q.m. estimator of Y_t of Y given $X(s)$ for all $s \leq t$ in the form

$$\tilde{Y}_t = \int_0^t \psi(t,s)\,dX(s) + X(0).$$

7.23. Suppose we observe

$$X(t) = \xi(t) + \zeta(t),$$

where $\{\xi(t); t \in R\}$ is a wide sense stationary process with spectral density $f_\xi(\lambda)$ and $\{\zeta(t); t \in R\}$ a stationary noise process with spectral density $f_\zeta(\lambda)$. Assuming that $\xi(t)$ and $\xi(s)$ are orthogonal for all $s, t \in R$, show that the best q.m. linear estimate of $\xi(t)$ based on $X(s)$ is given by

$$\int_{-\infty}^{\infty} e^{i\lambda t} \frac{f_\xi(\lambda)}{f_\zeta(\lambda) + f_\xi(\lambda)} \eta_X(d\lambda).$$

7.24. (Continuation) Show that the best linear estimate of $\xi(t + h)$ given $X(s)$, $s \leq t$, is given by

$$\int_{-\infty}^{\infty} e^{i\lambda t}(\lambda, h)\eta_X(d\lambda).$$

7.25. In Problem 7.23, suppose that f_ζ and f_ξ are specified as follows:

$$f_\xi(\lambda) = \frac{K_1}{\lambda^2 + \alpha^2}, \qquad f_\zeta(\lambda) = \frac{K_2}{\lambda^2 + \beta^2},$$

$K_i > 0$, $\alpha, \beta > 0$. Show that, for $h > 0$,

$$\varphi(\lambda, h) = \frac{K_1}{K_1 + K_2} \frac{\alpha + \beta}{\alpha + \gamma} \frac{\beta + i\lambda}{\gamma + i\lambda} e^{-\alpha h},$$

where

$$\gamma = \frac{K_1 \beta^2 + K_2 \alpha^2}{K_1 + K_2}.$$

CHAPTER 8

Markov Processes I

8.1. Introduction

The concept of a Markov random process was defined in Section 1.7 of Chapter 1. Various particular cases were discussed in subsequent chapters. Without doubt, this has been the most extensively studied class of random processes.

Let $\{\xi(t); t \in T\}$ be a Markov process on a probability space $\{\Omega, \mathscr{B}, P\}$ assuming values in a set S. Denote by $\mathscr{S}$ a σ-algebra of subsets of S. As usual, the measure space $\{S, \mathscr{S}\}$ will be called the "state space" of the random process $\xi(t)$. The process $\xi(t)$ is said to be a homogeneous Markov process if its transition probability has the property

$$P\{\xi(s + t) \in B | \xi(s) = x\} = P(x, t, B). \tag{8.1.1}$$

for all $t > 0$, s, $s + t \in T$, $x \in S$, and $B \in \mathscr{S}$. In the following (unless otherwise stated), we will deal exclusively with homogeneous Markov processes.

The transition probability (8.1.1) must satisfy the Chapman–Kolmogorov equation

$$P(x, s + t, B) = \int_S P(x, s, dy)P(y, t, B). \tag{8.1.2}$$

In addition, the following also must hold. For each $t \in T$ and $x \in S$,

$$P(x, t, \cdot) \text{ is a probability measure on } \mathscr{S} \tag{8.1.3}$$

and, for every $t \in T$ and $B \in \mathscr{S}$,

$$P(\cdot, t, B) \text{ is an } \mathscr{S}\text{-measurable function on } S. \tag{8.1.4}$$

In the following, we take $T = [0, \infty)$.

The class of homogeneous Markov processes has an important feature. From the Chapman–Kolmogorov equation (8.1.2), we see that knowing the transition probability (8.1.1) for all x, B, and $s \leq \varepsilon$, where $\varepsilon > 0$ is arbitrarily small, we can determine $P(x, u, B)$ for any $u > \varepsilon$. In other words, the local behavior of a homogeneous Markov process (in a neighborhood of zero) determines its global behavior.

Every function $P(x, t, B)$ having properties (8.1.2)–(8.1.4) is said to be a transition probability. Let $\pi(\cdot)$ be a probability distribution on the state space $\{S, \mathscr{S}\}$. If $P(x, t, B)$ is a transition probability, does there exist a Markov process with $\pi(\cdot)$ as its initial distribution (see Definition 1.5.3) and transition probability $P(x, t, B)$? The answer is affirmative. This is a stochastic process $\{\xi(t); t \geq 0\}$ with marginal distributions

$$P\{\xi(t_1) \in B_1, \ldots, \xi(t_n) \in B_n\}$$
$$= \int_S \pi(dx_0) \int_{B_1} P(x_0, t_1, dx_1) \ldots \int_{B_n} P((x_{n-1}, t_n - t_{n-1}, dx_n), \quad (8.1.5)$$

where $0 < t_1 < \cdots < t_n$, $\{B_k\}_1^n \subset \mathscr{S}$, $n = 1, 2, \ldots$, and

$$\pi(B) = P\{\xi(0) \in B\}. \quad (8.1.6)$$

The family of finite-dimensional distributions (8.1.5) is a consistent one which, according to the Kolmogorov theorem 1.3.1, determines completely and uniquely a probability measure, which will be denoted by P_π, on the measurable space $\{S^{R_+}, \mathscr{S}^{R_+}\}$, where, as usual, $R_+ = [0, \infty)$.

In the sequel, the following notation will be often useful. For any $x \in S$, we will write

$$P_x(\cdot) = P_\pi\{\cdot | \xi(0) = x\}. \quad (8.1.7)$$

Because

$$\pi(B) = P_\pi\{\xi(0) \in B\},$$

we have

$$P_\pi(G) = \int_S P_x(G)\pi(dx), \quad (8.1.8)$$

where $G \in \mathscr{S}^{R_+}$. The expectation operator corresponding to P_x will be denoted by E_x.

Example 8.1.1. As an illustration, consider a homogeneous Markov process $\{\xi(t); t \geq 0\}$ with state space $S = \{-1, 1\}$, an initial distribution $\pi(\{1\}) = \frac{1}{2}$, and a transition probability satisfying

$$P(1, t, \{1\}) = P(-1, t, \{-1\}) = \alpha(t), \quad (8.1.9)$$

where $\alpha(t)$ is a continuous function on $[0, \infty)$ such that $\alpha(0) = 1$. We want to determine $P_\pi\{\xi(t) = 1\}$ and $\alpha(t)$ for all $t \geq 0$.

The first part of the problem is simple. For every $t > 0$,

$$P_\pi\{\xi(t) = 1\} = \tfrac{1}{2}P(1, t, \{1\}) + \tfrac{1}{2}P(-1, t, \{1\}).$$

From this and (8.1.9), we have

$$P_\pi\{\xi(t) = 1\} = \tfrac{1}{2}.$$

To determine $\alpha(t)$, we will use the fact that $P(1, t, \{1\})$ is the transition probability of the Markov process $\xi(t)$ and, therefore, it must satisfy the Chapman–Kolmogorov equation

$$P(1, t + s, \{1\}) = P(1, s, \{1\})P(1, t, \{1\}) + P(1, s, \{-1\})P(-1, t, \{1\}).$$

From this and (8.1.9), we readily deduce the functional equation

$$\begin{aligned}\alpha(t + s) &= \alpha(s)\alpha(t) + [1 - \alpha(s)][1 - \alpha(t)] \\ &= 2\alpha(s)\alpha(t) + 1 - \alpha(s) - \alpha(t).\end{aligned} \tag{8.1.10}$$

Now, by means of the substitution

$$\alpha(t) = \tfrac{1}{2}[1 + h(t)],$$

Equation (8.1.10) becomes

$$h(t + s) = h(s)h(t). \tag{8.1.11}$$

This is clearly the Cauchy functional equation whose only continuous solution is

$$h(t) = e^{-\lambda t}, \quad \lambda > 0.$$

From this and (8.1.11), we obtain that

$$\alpha(t) = (1 + e^{-\lambda t})/2.$$

Remark 8.1.1. A more conceptually complex concept of a Markov process was described in Dynkin's book *Markov Processes*. According to Dynkin's definition of a Markov process, its evolution takes place only up to a certain random time τ and then it terminates. More precisely, at time $\tau(\omega)$, the trajectory $\xi(t, \omega)$ vanishes. For instance, if $\{\xi(t); t \geq 0\}$ is a Markov process on a probability space $\{\Omega, \mathscr{B}, P\}$ and a random time $\tau: \Omega \to (0, \infty)$ is given, then, for any $\omega \in \Omega$, the trajectory $\xi(t, \omega)$ is defined only on $[0, \tau(\omega)]$.

Markov processes considered in this book are those for which $P\{\tau(\omega) = +\infty\} = 1$. Such processes are often called "conservative."

Remark 8.1.2. It is reasonable to suppose that no transition can take place in zero time. For this reason, we will assume from now on that

$$P(x, 0, B) = \begin{cases} 1 & \text{if } x \in B \\ 0 & \text{if } x \notin B. \end{cases} \tag{8.1.12}$$

We will also assume that $P(x, t, B)$ is continuous at zero, i.e., that

$$\lim_{t \to 0+} P(x, t, B) = \begin{cases} 1 & \text{if } x \in B \\ 0 & \text{if } x \notin B, \end{cases} \tag{8.1.13}$$

for all $x \in S$ and $B \in \mathscr{S}$.

8.2. Invariant Measures

Let $\{\xi(t); t \geq 0\}$ be a (not necessarily homogeneous) Markov process with state space $\{S, \mathscr{S}\}$ and transition probability $P(s, t, x, B)$ $(0 \leq s < t)$. It is not difficult to see that if $\xi(t)$ is strictly stationary, then it must be a homogeneous Markov process. This readily follows from

$$\begin{aligned} P(s, t, x, B) &= P\{\xi(t) \in B \mid \xi(s) = x\} \\ &= P\{\xi(t - s) \in B \mid \xi(0) = x\} = P(0, t - s, x, B) = P(x, t - s, B). \end{aligned}$$

The converse, however, does not hold, in general. In other words, a homogeneous Markov process is not necessarily strictly stationary.

Under what conditions is a homogeneous Markov process strictly stationary? To answer this question, we need the concept of an invariant measure.

Definition 8.2.1. An invariant measure of a homogeneous Markov process with a state space $\{S, \mathscr{S}\}$ and transition probability $P(x, t, B)$ is a measure μ on $\{S, \mathscr{S}\}$ satisfying the following condition: for each $B \in \mathscr{S}$ and $t \geq 0$

$$\mu(B) = \int_S P(x, t, B)\mu(dx). \tag{8.2.1}$$

An invariant measure is not necessarily a probability measure on $\{S, \mathscr{S}\}$. When $\mu(S) = 1$, it is often called a "stationary distribution" of the Markov process.

Example 8.2.1. Let $\{\xi(t); t \geq 0\}$ be a standard Brownian motion process. As we know (see Remark 3.1.1), $\xi(t)$ is a homogeneous Markov process with the state space $\{R, \mathscr{R}\}$ and transition probability

$$P(x, t, B) = (2\pi t)^{-1/2} \int_B \exp\left(-\frac{(y - x)^2}{2t}\right) dy. \tag{8.2.2}$$

From this, we readily deduce that

$$P(x, t, B) \to 0 \quad \text{as } t \to +\infty \tag{8.2.3}$$

for every compact $B \in \mathscr{R}$. Therefore, it there exists a finite invariant measure μ, we must have

$$\mu(B) = \int_{-\infty}^{\infty} P(x, t, B)\mu(dx)$$

for every $t \geq 0$ and compact $B \in \mathscr{R}$. By letting $t \to +\infty$, we would get from (8.2.3) that

$$\mu(B) = 0 \quad \text{for all compact } B \in \mathscr{R}.$$

Consequently, the only finite invariant measure of a standard Brownian motion is $\mu \equiv 0$.

On the other hand, if μ is the Lebesgue measure, we have, for all $t > 0$,

$$\int_{-\infty}^{\infty} P(x, t, B)\, dx = \int_{-\infty}^{\infty} \int_{B} P(x, t, dy)\, dx$$
$$= \int_{B} \left[(2\pi t)^{-1/2} \int_{-\infty}^{\infty} \exp\left(-\frac{(y - x)^2}{2t} \right) dx \right] dy = \int_{B} dy.$$

In other words, when μ is the Lebesgue measure, (8.2.1) holds. This clearly implies that the Lebesgue measure is an invariant measure of a standard Brownian motion process.

Let $\{\xi(t); t \geq 0\}$ be a strictly stationary Markov process. Then, as we have just seen, it must be homogeneous. Denote by $P(x, t, B)$ its transition probability and by $\pi(\cdot)$ its initial distribution. Clearly then, for all t,

$$\pi(B) = P\{\xi(t) \in B\}, \quad B \in \mathscr{S}.$$

From this, we obtain

$$\pi(B) = \int_{S} P(x, t, B)\pi(dx).$$

Consequently, every one-dimensional marginal distribution of the process is an invariant measure.

Conversely, if μ is a stationary distribution of a homogeneous Markov process, then the process is also strictly stationary under P_μ. As a matter of fact, because μ can be considered as the initial distribution of the homogeneous Markov process, say $\{\xi(t); t \geq 0\}$, with a state space $\{S, \mathscr{S}\}$, we have, for any $t > 0$ and $B \in \mathscr{S}$,

$$P\{\xi(t) \in B\} = \int_{S} P\{\xi(0) \in dx, \xi(t) \in B\}$$
$$= \int_{S} P(x, t, B)\mu(dx) = \mu(B).$$

From this, we deduce [see (8.1.5)] that

$$P\{\xi(t_1 + s) \in B_1, \xi(t_2 + s) \in B_2\}$$
$$= \int_{S} \int_{B_1} \int_{B_2} P\{\xi(s) \in dx, \xi(t_1 + s) \in dx_1, \xi(t_2 + s) \in dx_2\}$$

$$= \int_S \int_{B_1} \int_{B_2} P(x, t_1, dx_1)P(x_1, t_2 - t_1, dx_2)\mu(dx)$$

$$= \int_S P\{\xi(0) \in dx, \xi(t_1) \in B_1, \xi(t_2) \in B_2\}$$

$$= P\{\xi(t_1) \in B_1, \xi(t_2) \in B_2\},$$

and so on.

8.3. Countable State Space

A class of homogeneous Markov processes with countable state space is of considerable interest in a variety of applications, particularly in biology and engineering. In this section, we outline some fundamental properties of this particular class.

Let $\{\xi(t); t \geq 0\}$ be a homogeneous Markov process with countable state space S. Processes of this type are often called "continuous time Markov chains" or simply Markov chains. In the following, without loss of generality, we may assume that

$$S = \{0, 1, \ldots\}.$$

We will denote by $(s, t \geq 0)$

$$p_{ij}(t) = P\{\xi(s + t) = j | \xi(s) = i\} \tag{8.3.1}$$

the transition probability from a state $i \in S$ to another state $j \in S$ after a time-lapse of duration t. In accordance with (8.1.12) and (8.1.13), we have that

$$p_{ij}(0) = \begin{cases} 1 & \text{if } i = j \\ 0 & \text{if } i \neq j, \end{cases} \tag{8.3.2}$$

and that

$$\lim_{t \to 0+} p_{ij}(t) = \begin{cases} 1 & \text{if } i = j \\ 0 & \text{if } i \neq j. \end{cases} \tag{8.3.3}$$

The last condition means that $p_{ij}(t)$ is continuous at $t = 0$.

For any $t \geq 0$ and $i, j \in S$,

$$0 \leq p_{ij}(t) \leq 1 \quad \text{and} \quad \sum_{k=0}^{\infty} p_{ik}(t) = 1, \tag{8.3.4}$$

so that the transition probability matrix

$$M^t = \begin{bmatrix} p_{00}(t) & p_{01}(t) & \cdots \\ p_{10}(t) & p_{11}(t) & \cdots \\ \vdots & \vdots & \vdots \end{bmatrix} \tag{8.3.5}$$

is a stochastic matrix. Due to (8.3.3), we clearly have that

$$\lim_{t\to 0+} M^t = I,$$

where I is the unit matrix.

The Chapman–Kolmogorov equation has the form $(s, t \geq 0)$

$$p_{ij}(s+t) = \sum_{k=0}^{\infty} p_{ik}(s)p_{kj}(t). \tag{8.3.6}$$

By means of this equation, one obtains readily that

$$M^{s+t} = M^s \cdot M^t. \tag{8.3.7}$$

This clearly implies that the family of transition matrices $\{M^t; t \geq 0\}$ forms a semigroup. Next, we shall show that $p_{ij}(t)$ is continuous for all $t \geq 0$.

Proposition 8.3.1. *For every $i, j \in S$ and $s, t \geq 0$,*

$$|p_{ij}(s+t) - p_{ij}(s)| \leq 1 - p_{ii}(t). \tag{8.3.8}$$

Proof. From the Chapman–Kolmogorov equation [(8.3.6)], we obtain

$$p_{ij}(s+t) - p_{ij}(s) = \sum_{k\neq i}^{\infty} p_{ik}(t)p_{kj}(s) - p_{ij}(s) + p_{ii}(t)p_{ij}(s). \tag{8.3.9}$$

On the other hand,

$$\sum_{k\neq i}^{\infty} p_{ik}(t)p_{kj}(s) \leq \sum_{k\neq i}^{\infty} p_{ik}(t) = 1 - p_{ii}(t),$$

which yields

$$p_{ij}(s+t) - p_{ij}(s) \leq [1 - p_{ii}(t)][1 - p_{ij}(s)].$$

In a similar fashion, we deduce from (8.3.9) that

$$\begin{aligned} -[p_{ij}(s+t) - p_{ij}(s)] &= -\sum_{k\neq i}^{\infty} p_{ik}(t)p_{kj}(s) + p_{ij}(s) - p_{ii}(t)p_{ij}(s) \\ &\leq p_{ij}(s)[1 - p_{ii}(t)] \leq [1 - p_{ii}(t)], \end{aligned}$$

which proves (8.3.8). □

Corollary 8.3.1. *From* (8.3.8), *we have that, for any* $0 < t < s$,

$$|p_{ij}(s-t) - p_{ij}(s)| \leq 1 - p_{ii}(t).$$

This, (8.3.8), *and* (8.3.3) *clearly imply that $p_{ij}(t)$ is uniformly continuous on R_+ for all $i, j \in S$.*

The question of differentiability of $p_{ij}(t)$ is of central importance in applications. This stems from the fact that attempts to calculate $p_{ij}(t)$, under various regularity assumptions on the Markov chain $\xi(t)$, lead to a system of difference-differential equations whose solution may give $p_{ij}(t)$. Further discussion on this subject requires the following lemma due to Hille.

Lemma 8.3.1. *Let $H: R \to R$, where $H(t) = 0$ for all $t \leq 0$, be subadditive, i.e., for all $s, t \in R$,*

$$H(s + t) \leq H(s) + H(t), \tag{8.3.10}$$

then

$$\lim_{t \to 0+} H(t)/t = \sup_{0 \leq t < \infty} H(t)/t = c, \tag{8.3.11}$$

where $c \geq 0$ may be $+\infty$.

PROOF. From the definition of $H(\cdot)$, it follows that $H(t) \geq 0$ for all $t \in R$. Now, choose any $c' < c$. Then, from the definition of c, it follows that there exists $s > 0$ such that

$$c' < H(s)/s.$$

Next, fix s and write

$$s = nt + \delta, \quad n = 1, 2, \ldots,$$

where $t > 0$ and $0 \leq \delta < t$. Clearly then,

$$c' < \frac{1}{s} H(s) \leq \frac{nH(t) + H(\delta)}{s} = \frac{nt}{s}\frac{H(t)}{t} + \frac{H(\delta)}{s}.$$

When $t \to 0$, $\delta \to 0$ and $n \to \infty$ so that $nt/s \to 1$. Consequently, from the last inequality, we obtain

$$c' \leq \liminf_{t \to 0+} H(t)/t,$$

which proves the lemma. □

Proposition 8.3.2. *For each $i \in S$, the limit*

$$\lim_{t \to 0+} (1 - p_{ii}(t))/t = q_{ii} \tag{8.3.12}$$

exists but may be equal to $+\infty$.

PROOF. Write

$$H_i(t) = -\ln p_{ii}(t). \tag{8.3.13}$$

From (8.3.6), we clearly have

$$p_{ii}(s + t) \geq p_{ii}(s)p_{ii}(t),$$

so that

$$H_i(s + t) \leq H_i(s) + H_i(t),$$

and, due to (8.3.2), $H(0) = 0$. Consequently, all the conditions of Lemma 8.3.1 are satisfied. Now, from (8.3.13), we have, as $t \to 0$,

$$\frac{1}{t}(1 - p_{ii}(t)) = \frac{H_i(t)}{t}[1 + o(H_i(t))].$$

This and Lemma 8.3.1 prove the assertion. □

Remark 8.3.1. For all $i \in S$, $p_{ii}(t)$ is differentiable at zero.

Next, we discuss the question of differentiability of the transition probabilities $p_{ij}(t)$ at $t = 0$ for $i \neq j$. The following proposition shows that $p'_{ij}(0) = q_{ij}$ always exists and that $|q_{ij}| < \infty$.

Proposition 8.3.3. *For all $i \neq j$, from S, the following limit exists:*

$$\lim_{h \to 0+} p_{ij}(h)/h = q_{ij} < \infty. \tag{8.3.14}$$

Proof. Let $c \in (\frac{1}{2}, 1)$ be fixed. Then, due to (8.3.3), there exists an $s > 0$ sufficiently small so that $p_{ii}(s) > c$ and $p_{jj}(s) > c$. Let $0 < s \leq nh < \delta$. Consider a homogeneous Markov chain $\{\xi_n\}_0^\infty$ with state space S such that

$$P\{\xi_{n+1} = j | \xi_n = i\} = p_{ij}(h).$$

Then, for all $n > 1$, we have

$$\begin{aligned} p_{ij}(nh) = P\{\xi_n = j | \xi_0 = i\} &\geq P\{\xi_1 = j, \xi_n = j | \xi_0 = i\} \\ &\quad + \sum_{r=1}^{n-1} P\{\xi_1 \neq j, \ldots, \xi_{r-1} \neq j, \xi_r = i, \xi_{r+1} = j, \xi_n = j | \xi_0 = i\} \\ &\geq cp_{ij}(h)\left(1 + \sum_{r=1}^{n-1} P\{\xi_1 \neq j, \ldots, \xi_{r-1} \neq j, \xi_r = i | \xi_0 = i\}\right). \end{aligned}$$

On the other hand,

$$\begin{aligned} &P\{\xi_1 \neq j, \ldots, \xi_{r-1} \neq j, \xi_r = i | \xi_0 = i\} \\ &\quad = P\{\xi_1 \neq j, \ldots, \xi_{r-2} \neq j, \xi_r = i | \xi_0 = i\} \\ &\quad\quad - P\{\xi_1 \neq j, \ldots, \xi_{r-2} \neq j, \xi_{r-1} = j, \xi_r = i | \xi_0 = i\} \\ &\quad = P\{\xi_1 \neq j, \ldots, \xi_{r-3} \neq j, \xi_r = i | \xi_0 = i\} \\ &\quad\quad - P\{\xi_1 \neq j, \ldots, \xi_{r-3} \neq j, \xi_{r-2} = j, \xi_r = i | \xi_0 = i\} \\ &\quad\quad - P\{\xi_1 \neq j, \ldots, \xi_{r-2} \neq j, \xi_{r-1} = j, \xi_r = i | \xi_0 = i\}, \end{aligned}$$

and so on. Continuing in this fashion, we obtain

$$\begin{aligned} &P\{\xi_1 \neq j, \ldots, \xi_{r-1} \neq j, \xi_r = i | \xi_0 = i\} \\ &\quad = P\{\xi_r = i | \xi_0 = i\} - \sum_{k<r} P\{\xi_1 \neq j, \ldots, \xi_{k-1} \neq j, \xi_k = j\} P\{\xi_r = i | \xi_k = j\} \\ &\quad \geq c - (1 - c) \sum_{k<r} P\{\xi_1 \neq j, \ldots, \xi_{k-1} \neq j, \xi_k = j | \xi_0 = i\} \geq 2c - 1. \end{aligned}$$

Consequently,

$$p_{ij}(nh) \geq (2c - 1)np_{ij}(h).$$

Set $t < \delta$, $h < \delta$, and $n = [t/h]$, where $[x]$ is the integer part of x; then,

$$p_{ij}(h)/h \leq p_{ij}([t/h]h)/\{[t/h]hc(2c - 1)\}.$$

By letting $h \to 0+$, we obtain

$$\limsup_{h \to 0+} (p_{ij}(h)/h) \leq p_{ij}(t)/\{tc(2c - 1)\} < \infty$$

for all $t > 0$. Therefore,

$$\limsup_{h \to 0+} (p_{ij}(h)/h) \leq \liminf_{t \to 0+} (p_{ij}(t)/\{t(2c - 1)c\}),$$

which means that the limit

$$q_{ij} = \lim_{h \to 0+} (p_{ij}(h)/h) < \infty$$

exists. This proves the proposition. □

If $S_0 \subset S$ is any finite subset which does not contain the state i, we have

$$(1 - p_{ii}(h))/h \geq \left(\sum_{j \in S_0} p_{ij}(h)\right) \Big/ h.$$

By letting $h \to 0+$, it follows [see (8.3.12) and (8.3.14)) that

$$q_{ii} \geq \sum_{j \in S_0} q_{ij} \Rightarrow q_{ii} \geq \sum_{j \neq i}^{\infty} q_{ij}.$$

If $q_{ii} < \infty$ and

$$q_{ii} = \sum_{j \neq i}^{\infty} q_{ij}, \tag{8.3.15}$$

the state i is called "stable" or "regular." Otherwise, it is called "nonregular."

Remark 8.3.2. The matrix

$$A = \begin{bmatrix} -q_{00} & q_{01} & q_{02} & \cdots \\ q_{10} & -q_{11} & q_{12} & \cdots \\ q_{20} & q_{21} & -q_{22} & \cdots \\ \vdots & \vdots & \vdots & \vdots \end{bmatrix} \tag{8.3.16}$$

is called the infinitesimal generator of the Markov chain $\{\xi(t); t \geq 0\}$. Its essential role will become clear in what follows.

Remark 8.3.3. The state $i \in S$ is "instantaneous" if $q_{ii} = \infty$ and "absorbing" if $q_{ii} = 0$.

Proposition 8.3.4. *Assume that* $q_{ii} < \infty$ *for each* $i \in S$*, then the transition probabilities* $p_{ij}(t)$ *are differentiable for all* $t \geq 0$ *and* $i, j \in S$. *In addition,*

$$p'_{ij}(t) = \sum_{k \neq i}^{\infty} q_{ik} p_{kj}(t) - q_{ii} p_{ij}(t). \tag{8.3.17}$$

PROOF. Consider any two states $i, j \in S$. Then for any $h > 0$ and a finite subset $N_0 \subset S$ containing i and j, we have

$$(p_{ij}(t + h) - p_{ij}(t))/h = \left(\sum_{k \in N_0} p_{ik}(h) p_{kj}(t) - p_{ij}(t) + \sum_{k \in N_0 c} p_{ik}(h) p_{kj}(t) \right) \Big/ h. \tag{8.3.18}$$

Clearly,

$$\sum_{k \in N_0 c} p_{ik}(h) p_{kj}(h) \leq \left(1 - \sum_{k \in N_0} p_{ik}(h) \right). \tag{8.3.19}$$

Due to (8.3.15), given any arbitrary small $\varepsilon > 0$, we can find N_0 such that

$$\sum_{k \in N_0} q_{ik} \leq \varepsilon.$$

Now, choose $h_0 > 0$ so that for each $h \leq h_0$, $k \neq i$, and $k \in N_0$,

$$|p_{ik}(h)/h - q_{ik}| \leq \varepsilon / \|N_0\|$$

and

$$|(1 - p_{ii}(h))/h - q_{ii}| \leq \varepsilon / \|N_0\|,$$

where $\|N_0\| = \text{Card}\{N_0\}$ (the number of elements in N_0). Consequently, from (8.3.18) and the condition of the proposition, we have

$$\begin{aligned} \sum_{k \in N_0 - \{i\}} q_{ik} p_{kj}(t) - q_{ii} p_{ij}(t) &\leq \liminf_{h \to 0+} (p_{ij}(t + h) - p_{ij}(t))/h \\ &\leq \limsup_{h \to 0+} (p_{ij}(t + h) - p_{ij}(t))/h \\ &\leq \sum_{k \in N_0 - \{i\}} q_{ik} p_{kj}(t) - q_{ii} p_{ij}(t) + 2\varepsilon. \end{aligned}$$

By letting $\varepsilon \to 0$, (8.3.17) follows. □

Remark 8.3.4. The system (8.3.17) is called the Kolmogorov backward equations. The matrix form of it is

$$dM^t/dt = AM^t, \qquad M^0 = I. \tag{8.3.20}$$

Remark 8.3.5. The following also holds:

$$dM^t/dt = M^t \cdot A, \tag{8.3.21}$$

which is called the Kolmogorov forward equation.

8.4. Birth and Death Process

The birth and death process is an important example of the homogeneous Markov chain $\{\xi(t); t \geq 0\}$ with state space $S = \{0, 1, \dots\}$. Here, $\xi(t)$ represents the size of a population at time t, which fluctuates according to the following rules:

$$\begin{aligned} &\text{if at time } t \text{ the chain is in a state } i \in S, \\ &\text{in one transition it can go only to } i-1 \text{ or } i+1. \end{aligned} \tag{8.4.1}$$

The transition from state i to $i + 1$ indicates a "birth" in the population, whereas the transition from i to $i - 1$ indicates a "death." Two or more simultaneous births or deaths are not permitted. We also suppose that "autogenesis," i.e., transition from 0 to 1 is not excluded.

Assume that the chain $\{\xi(t); t \geq 0\}$ is stochastically continuous and that all states are stable [i.e., $q_{ii} < \infty$ and (8.3.15) holds]; then,

$$\lim_{h \to 0} (1 - p_{ii}(h))/h = q_{ii} < \infty, \quad i \in S,$$

for all $i \in S$ and, from assumption (8.4.1) we have, as $h \to 0$,

$$p_{ij}(h) = o(h) \quad \text{if } |i - j| > 1, \tag{8.4.2}$$

$$p_{i,i+1}(h) = \lambda_i h + o(h), \qquad p_{i,i-1}(h) = \mu_i h + o(h), \tag{8.4.3}$$

where [in each case $o(h)$ may depend on i]

$$\lambda_i = q_{i,i+1}, \qquad \mu_i = q_{i,i-1}, \qquad \mu_0 = 0. \tag{8.4.4}$$

From this, one can deduce the following: $\lambda_i h + o(h)$ represents the probability of a birth in $(t, t + h)$ given that $\xi(t) = i$. Similarly, $\mu_i h + o(h)$ represents the probability of a death in $(t, t + h)$ given $\xi(t) = i$. Because all the states are stable, we have that

$$q_{ii} = \lambda_i + \mu_i. \tag{8.4.5}$$

The infinitesimal generator A (see (8.3.16)] in this case is

$$A = \begin{bmatrix} -\lambda_0 & \lambda_0 & 0 & 0 & \cdots \\ \mu_1 & -(\lambda_1 + \mu_1) & \lambda_1 & 0 & \cdots \\ 0 & \mu_2 & -(\lambda_2 + \mu_2) & \lambda_2 & \cdots \\ 0 & 0 & \mu_3 & -(\lambda_3 + \mu_3) & \cdots \\ \vdots & \vdots & \vdots & \vdots & \vdots \end{bmatrix}.$$

The parameters λ_i and μ_i are called the birth and death rates, respectively. Now, from the Kolmogorov backward matrix equation

$$dM^t/dt = AM^t,$$

[see (8.3.7)] we readily deduce that

$$\begin{aligned} p_{0j}'(t) &= -\lambda_0 p_{0j}(t) + \lambda_0 p_{1j}(t), \quad j = 0, 1, \ldots, \\ p_{ij}'(t) &= \mu_i p_{i-1,j}(t) - (\lambda_i + \mu_i) p_{ij}(t) + \lambda_i p_{i+1,j}(t), \quad i \geq 1. \end{aligned} \tag{8.4.6}$$

This is the Kolmogorov backward system of difference differential equations for a birth and death process.

From the Kolmogorov forward equation (8.3.21), we obtain

$$\begin{aligned} p_{i0}'(t) &= -\lambda_0 p_{i0}(t) + \mu_1 p_{i,1}(t), \quad i \in S, \\ p_{ij}'(t) &= \lambda_{j-1} p_{i,j-1}(t) - (\lambda_j + \mu_j) p_{ij}(t) + \mu_{j+1} p_{i,j+1}(t), \quad j \geq 1. \end{aligned} \tag{8.4.7}$$

Denote by

$$P\{\xi(t) = j\} = p_j(t);$$

then, from (8.4.7), we obtain that

$$\begin{aligned} p_0'(t) &= -\lambda_0 p_0(t) + \mu_1 p_1(t), \\ p_j'(t) &= \lambda_{j-1} p_{j-1}(t) - (\lambda_j + \mu_j) p_j(t) + \mu_{j+1} p_{j+1}(t). \end{aligned} \tag{8.4.8}$$

Remark 8.4.1. If $\mu_k \equiv 0$, the process $\xi(t)$ is called a "pure birth" process. If, on the other hand, $\lambda_i \equiv 0$, the process $\xi(t)$ is called a "pure death" process.

The following is an application of the birth and death process in a telephone traffic problem.

Example 8.4.1. Consider a telephone exchange where the number of available lines is so large that for all practical purposes it can be considered infinite. Denote by $\xi(t)$ the number of lines in use at time t. Then, our physical intuition is not violated by assuming that $\{\xi(t); t \geq 0\}$ is a birth and death process. It also seems reasonable to assume that, for all $t > 0$, $h > 0$, and $i = 0, 1, \ldots,$

$$P\{\xi(t+h) = i+1 | \xi(t) = i\} = \lambda h + o(h)$$

because the probability that a call will occur in $(t, t+h)$ is independent of the number of busy lines at time t. On the other hand,

$$P\{\xi(t+h) = i-1 | \xi(t) = i\} = \mu_i h + o(h)$$

as $h \to 0$, for the obvious reasons.

Let us calculate $p_i(t)$ under the assumption that

$$\mu_i = i\mu.$$

In this case, the system of difference-differential equations (8.4.8) becomes

$$p_0'(t) = -\lambda p_0(t) + \mu p_1(t), \tag{8.4.9}$$

$$p_j'(t) = \lambda p_{j-1}(t) - (\lambda + j\mu) p_j(t) + \mu(1+j) p_{j+1}(t). \tag{8.4.10}$$

This is not a recursive system (i.e., it cannot be solved successively). For this reason we are compelled to use the method of probability generating functions.

Set

$$g(t,u) = \sum_{j=0}^{\infty} p_j(t)u^j;$$

then taking into account (8.4.9) and (8.4.1), we obtain

$$\begin{aligned}\frac{\partial g(t,u)}{\partial t} &= \sum_{j=0}^{\infty} p_j'(t)u^j \\ &= -\lambda p_0(t)(1-u) + \mu p_1(t)(1-u) - \lambda p_1(t)(1-u)u \\ &\quad + 2\mu p_2(t)u(1-u) - \cdots \\ &= -\lambda(1-u)\sum_{j=0}^{\infty} p_j(t)u^j + \mu(1-u)\sum_{j=1}^{\infty} jp_j(t)u^{j-1} \\ &= -\lambda(1-u)g(t,u) + \mu(1-u)\frac{\partial g(t,u)}{\partial u}.\end{aligned}$$

Thus, the generating function $g(t,u)$ satisfies the linear partial differential equation

$$\frac{\partial g(t,u)}{\partial t} - \mu(1-u)\frac{\partial g(t,u)}{\partial u} = -\lambda(1-u)g(t,u). \tag{8.4.11}$$

Suppose that $\xi(0) = i_0$; then,

$$g(0,u) = u^{i_0}. \tag{8.4.12}$$

Solving this equation by standard methods and using the initial condition (8.4.12), we obtain

$$g(t,u) = \{1-(1-u)e^{-\mu t}\}^{i_0}\exp\left\{-\frac{\lambda}{\mu}(1-u)(1-e^{-\mu t})\right\}. \tag{8.4.13}$$

Note that the first term on the right side of (8.4.13) is the probability generating function of the binomial distribution with $p = \exp\{-\mu t\}$, whereas the second term is the probability generating function of the Poisson distribution with mean value

$$\Lambda(t) = (1-e^{-\mu t})/\mu.$$

Therefore,

$$\xi(t) = \xi_0(t) + \xi_1(t), \tag{8.4.14}$$

where $\xi_0(t)$ is a binomial component, $\xi_1(t)$ is a Poisson component, and $\xi_0(t)$ is independent of $\xi_1(t)$ for all $t \geq 0$.

We now expand (8.4.13) to obtain $p_k(t)$. After some calculations, we obtain

$$p_j(t) = \exp\left\{-\frac{\lambda}{\mu}(1 - e^{-\mu t})\right\} \cdot \sum_{k=0}^{\min\{i_0, j\}} \binom{i_0}{k} \left(\frac{\lambda}{\mu}\right)^{j-r} \frac{e^{-\mu t k}(1 - e^{-\mu t})^{i_0+j-2k}}{(j-k)!},$$

where $j = 0, 1, \ldots$. From (8.4.14), we deduce that

$$E\{\xi(t)\} = i_0 e^{-\mu t} + \frac{\lambda}{\mu}(1 - e^{-\mu t}).$$

These formulas were first obtained by Palm.

8.5. Sample Function Properties

Let $\{\xi(t); t \geq 0\}$ be a homogeneous Markov process with the state space $S = \{0, 1, \ldots\}$ and transition probabilities $p_{ij}(t)$ which satisfy condition (8.3.3). In this section, we outline some basic properties of its sample functions, which are step functions. First, it is not difficult to establish that the process is stochastically continuous (see Definition 1.9.1) which is equivalent to

$$\lim_{h \to 0+} P\{\xi(t \pm h) = \xi(t)\} = 1, \quad h \geq 0. \tag{8.5.1}$$

To show that (8.5.1) holds, consider

$$P\{\xi(t + h) = \xi(t)\} = \sum_{j=0}^{\infty} p_{jj}(h) p_j(t).$$

From condition (8.3.3) and the bounded convergence theorem, it follows that $P\{\xi(t + h) = \xi(t)\} \to 1$ as $h \to 0+$. In a similar fashion, one can show that $P\{\xi(t - h) = \xi(t)\} \to 1$ as $h \to 0+$.

Because the process is stochastically continuous, it must be separable and measurable on every compact interval. In such a case (see Proposition 1.10.1), there exists a separable version, which is stochastically equivalent to $\xi(t)$, all of whose sample functions are continuous from the right. In the following, we will deal exclusively with this version, which we will denote by the same symbol $\xi(t)$.

Because the trajectories of $\xi(t)$ are (right continuous) step sample functions, it seems intuitively justifiable that, at least when a state $i \in S$ is regular, to assume that if for some t

$$\xi(t, \omega) = i,$$

then there exists an $h > 0$, which may depend on ω and i, such that $\xi(t + h, \omega) = i$. This gives rise to the following question. How long does the process $\xi(t)$ stay in a state after entering it? To answer this question, denote by

$$\tau(t) = \inf\{h > 0; \xi(t) \neq \xi(t + h)\}$$

the length of time the process remains in the state in which it is at time t.

Let us calculate the conditional probability

$$\varphi_i(s) = P\{\tau(t) > s | \xi(t) = i\}, \quad i \in S, \quad s \geq 0 \tag{8.5.2}$$

[this conditional probability does not depend on t because the process $\xi(t)$ is homogeneous]. Because

$$\{\tau(t) > s + u\} = \{\tau(t) > s, \tau(t + s) > u\}, \tag{8.5.3}$$

it follows that

$$\begin{aligned} \varphi_i(s + u) &= P\{\tau(t) > s, \tau(t + s) > u | \xi(t) = i\} \\ &= P\{\tau(t) > s | \xi(t) = i\} P\{\tau(t + s) > u | \xi(t) = i, \tau(t) > s\}. \end{aligned} \tag{8.5.4}$$

Now, $\{\xi(t) = i, \tau(t) > s\} \subset \{\xi(t + s) = i\}$. Consequently, due to Markov property of $\xi(t)$,

$$\begin{aligned} P\{\tau(t + s) > u | \xi(t) = i, \tau(t) > s\} \\ &= P\{\tau(t + s) > u | \xi(t) = i, \tau(t) > s, \xi(t + s) = i\} \\ &= P\{\tau(t + s) > u | \xi(t + s) = i\} = \varphi_i(u). \end{aligned}$$

From this and (8.5.4), we deduce

$$\varphi_i(s + u) = \varphi_i(s)\varphi_i(u). \tag{8.5.5}$$

This is a Cauchy functional equation. Because $\varphi_i(s)$ is continuous at $s = 0$, it is continuous at all points $s \geq 0$. In addition, $0 \leq \varphi_i(s) \leq 1$, which implies that the only solution of (8.5.5) must be

$$\varphi_i(s) = e^{-\lambda_i s} \quad \text{for all } s \geq 0, \tag{8.5.6}$$

where $\lambda_i \geq 0$, and λ_i may be $+\infty$. We now can give the following definition (see also Remark 8.3.3).

Definition 8.5.1. A state $i \in S$ is called "absorbing" if $\lambda_i = 0$, stable if $0 < \lambda_i < \infty$, and instantaneous if $\lambda_i = \infty$.

Clearly, if i is absorbing and if $\{\xi(t) = i\}$ occurs, it follows that $\xi(t + s) = i$ for all $s \geq 0$. In other words, if $\xi(t)$ enters i, then it stays there forever.

On the other hand, if i is a stable state, then

$$P\{0 < \tau(t) < \infty | \xi(t) = i\} = 1.$$

Finally, if i is an instantaneous state,

$$P\{\tau(t) = 0 | \xi(t) = i\} = 1,$$

which implies that the process exists from an instantaneous state as soon as it enters it. This, however, is not possible because, by assumption, we are dealing with a separable version with right-continuous trajectories. Consequently, the state space S of such a Markov chain does not contain instantaneous states. Conversely, if the state space S of a Markov chain does not contain instantaneous states, the process is Cadlag (i.e., continuous on the right and having limits on the left).

Before concluding this section, let us describe once more the structure of the sample functions of a homogeneous Markov chain $\{\xi(t); t \geq 0\}$ with state space $S = \{0, 1, \ldots\}$, where each element of S is a stable state. At time $t = 0$, the process is in a state, say i_0. It stays there for a random time T_1 at the end of which it jumps to a new state. It remains in the new state for a random time T_2 and then it jumps to another state, and so on. Because all the states in S are stable, each sample function of the process $\xi(t)$ is a right-continuous step function with left-hand limits.

Set $T_0 = 0$ and

$$\tau_n = \sum_{k=1}^{n} T_k, \quad n = 0, 1, \ldots. \tag{8.5.7}$$

Then we have the following definition.

Definition 8.5.2. A homogeneous Markov chain $\{\xi(t); t \geq 0\}$ with state space $S = \{0, 1, \ldots\}$ is said to be regular if its sample functions are (a.s.) continuous from the right and

$$\sup_n \tau_n = +\infty \quad \text{(a.s.)}. \tag{8.5.8}$$

Remark 8.5.1. The times $\tau_1, \tau_2, \ldots$ are the instants of transitions of the process $\xi(t)$.

8.6. Strong Markov Processes

To investigate some deeper properties of a Markov process, the concept of stopping time is required. Consider a stochastic process $\{X(t); t \in R_+\}$ defined on a probability space $\{\Omega, \mathscr{B}, P\}$, with an arbitrary state space $\{S, \mathscr{S}\}$. We will assume that the stochastic process is directly given (see Section 1.2 of Chapter 1). In other words, $\Omega = S^{R_+}$ and $\mathscr{S} = \mathscr{S}^{R_+}$. Therefore, each $\omega \in \Omega$ is a function $\omega: R_+ \to S$.

We now define the concept of a "stopping time" associated with the process $X(t)$. This random variable is also known as a "Markov time" or a "random variable independent of the future" (Dynkin).

Definition 8.6.1. A random variable

$$\tau: \Omega \to [0, \infty]$$

is called a "stopping time" if it has the property that for any two $\omega_1, \omega_2 \in \Omega$ such that

$$\omega_1(s) = \omega_2(s) \quad \text{for all } 0 \leq s \leq t$$

and $\tau(\omega_1) \leq t$, then $\tau(\omega_1) = \tau(\omega_2)$.

From the definition of a stopping time τ, it clearly follows that the occurrence or nonoccurrence of the random event $\{\omega; \tau \leq t\}$ can be established from the observation of the random process $X(s)$ on $[0, t]$ alone.

EXAMPLE 8.6.1. Clearly every constant $\tau = t > 0$ is a stopping time. The first visiting time of a certain set $A \in \mathscr{S}$ by the process $X(t)$ is also stopping time, i.e.,

$$\tau_A(\cdot) = \inf\{s; X(s, \cdot) \in A\}.$$

Clearly, in this case $0 \leq \tau_A(\cdot) \leq \infty$. The nth visiting time of A is also a stopping time. However, the time of the last visit to A is not a stopping time.

The sum $\tau_1 + \tau_2$ of two stopping times τ_1 and τ_2 is also a stopping time. The random variables

$$\inf\{\tau_1, \tau_2\} \quad \text{and} \quad \sup\{\tau_1, \tau_2\}$$

are also stopping times.

Consider now a homogeneous Markov process $\{\xi(t); t \geq 0\}$ with an arbitrary state space $\{S, \mathscr{S}\}$. These properties of $\xi(t)$ imply that, for any $0 \leq s_1 < \cdots < s_m < s$ and $0 < t_1 < \cdots < t_n$,

$$\begin{aligned} P\{\xi(s + t_1) \in B_1, \ldots, \xi(s + t_n) \in B_n \mid \xi(s_1) = x_1, \ldots, \xi(s_m) = x_m, \xi(s) = x\} \\ = P_x\{\xi(t_1) \in B_1, \ldots, \xi(t_n) \in B_n\} \end{aligned} \tag{8.6.1}$$

[see (8.1.7)] where $\{B_i\}_1^n \subset \mathscr{S}$.

It seems intuitively reasonable to expect that relation (8.6.1) will hold if s is replaced with a stopping time τ. This, however, is not true in general. Markov processes having this property are called "strong Markov processes." The following definition gives a precise formulation of the concept.

Definition 8.6.2. A measurable homogeneous Markov process with a state space S is said to have the strong Markov property if, for any stopping time τ,

$$\begin{aligned} P\{\xi(\tau + t_1) \in B_1, \ldots, \xi(\tau + t_n) \in B_n \mid \xi(s); s < \tau, \xi(\tau) = x\} \\ = P_x\{\xi(t_1) \in B_1, \ldots, \xi(t_n) \in B_n\}. \end{aligned} \tag{8.6.2}$$

When is a homogeneous Markov process endowed with the strong Markov property? The following result is due to Dynkin and Yushkevich (1956).

Proposition 8.6.1. *A homogeneous Markov process $\{\xi(t); t \geq 0\}$ with state space $\{S, \mathscr{S}\}$ and transition probability $P(x, t, B)$ has the strong Markov property if and only if for any $x \in S$, $B \in \mathscr{S}$, and $t > 0$,*

$$P_x\{\xi(\tau + t) \in B\} = \int_S P(y, t, B) P_x\{\xi(\tau) \in dy\}. \tag{8.6.3}$$

There is an extremely useful criterion for determining whether a Markov process has the strong Markov property. To formulate this criterion, some new notation is needed. Suppose that S is a metric space. Then, $\mathscr{S}$ is the Borel algebra. Denote by **B** the space of all bounded Borel measurable functions on S. With the usual supremum norm

$$\|h\| = \sup_x |h(x)|, \tag{8.6.4}$$

the space **B** is a Banach space.

For each $t \geq 0$, define an operator T^t by

$$(T^t h)(x) = \int_S h(y)P(x,t,dy) \tag{8.6.5}$$

or equivalently

$$(T^t h)(x) = E_x\{h(\xi(t))\}. \tag{8.6.6}$$

Because

$$|(T^t h)(x)| = \left|\int_S h(y)P(x,t,dy)\right| \leq \sup_y |h(y)| = \|h\|,$$

$T^t: \mathbf{B} \to \mathbf{B}$ is a contraction.

Proposition 8.6.2. *Let $\{\xi(t); t \geq 0\}$ be a homogeneous Markov process with state space $\{S, \mathscr{S}\}$ and a transition probability $P(x,t,B)$ which has the property that the operator* (8.6.5) *maps continuous bounded functions into continuous bounded functions. Then, the process $\xi(t)$ possesses the strong Markov property.*

Remark 8.6.1. Processes satisfying conditions of Proposition 8.6.2 are called "Feller processes."

Remark 8.6.2. The following result is due to Yushkevich. Let $\{\xi(t); t \geq 0\}$ be a homogeneous Markov process with countable state space S. If $\xi(t)$ is stochastically continuous, there exists a separable version which possesses the strong Markov property. In other words, if all elements of S are stable states the Markov chain is a strong Markov process.

8.7. Structure of a Markov Chain

Let us now return to Markov processes with countable state spaces and examine in more detail some questions concerning their stochastic structure. To this end, consider a homogeneous Markov chain $\{\xi(t); t \geq 0\}$ with state space $S = \{0, 1, \ldots\}$ and transition probability $p_{ij}(t)$. We will assume that condition (8.3.3) holds; in such a case, $p_{ij}(t)$ is called a "standard" transition probability. From Section 8.5, we see then that the process $\xi(t)$ must be stochastically continuous. This, on the other hand, implies that $\xi(t)$ is separable and measurable. Then (see Proposition 1.10.1) there exists a separable version all of whose sample functions are continuous from the right and have left-hand limits.

According to what was stated in Section 8.5, the state space S of $\xi(t)$ does not contain instantaneous states. In the rest of this section, we will deal exclusively with the separable version of the Markov chain, which will be denoted by the same symbol $\xi(t)$.

As before, let us denote by $\tau_1, \tau_2, \ldots$ the instants of transitions of the process $\xi(t)$, and by $T_0 = 0$, $T_n = \tau_n - \tau_{n-1}$, $n = 1, 2, \ldots$. Clearly, each τ_n is a stopping time. Because the process $\xi(t)$ is measurable, $X_0, X_1, \ldots$, defined by

$$X_0 = \xi(0), \qquad X_n = \xi(\tau_n), \quad n = 1, 2, \ldots, \tag{8.7.1}$$

are clearly random variables. Consider the bivariate sequence

$$\{(X_n, T_n)\}_0^\infty, \qquad T_0 = 0. \tag{8.7.2}$$

What is the stochastic structure of this bivariate process? This and related questions will be discussed in the rest of this section. For this purpose, the following two lemmas are needed.

Lemma 8.7.1. *For all $i \in S$ and $t \geq 0$,*

$$P_i\{T_1 > t\} = e^{-\lambda_i t}. \tag{8.7.3}$$

PROOF. Notice that

$$P_i\{T_1 > t\} = P_i\{\xi(s) = i, 0 \leq s \leq t\}. \tag{8.7.4}$$

On the other hand, for any $t \geq 0$, $u \geq 0$,

$$\begin{aligned} P_i&\{T_1 > t + u\} \\ &= P_i\{\xi(s) = i, 0 \leq s \leq t + u\} \\ &= P_i\{\xi(s) = i, 0 \leq s \leq t, \xi(v) = i, t \leq v \leq u + t\} \\ &= P_i\{\xi(s) = i, 0 \leq s \leq t\} P_i\{\xi(v) = i, t \leq v \leq t + u | \xi(s) = i, 0 \leq s \leq t\}. \end{aligned}$$

Now, because $\xi(t)$ is a homogeneous Markov process, we readily deduce that

$$P_i\{\xi(v) = i, t \leq v \leq t + u | \xi(s) = i, 0 \leq s \leq t\} = P_i\{\xi(v) = i, 0 \leq v \leq u\}.$$

Therefore, taking into account (8.7.4), we obtain

$$P_i\{T_1 > t + u\} = P_i\{T_1 > t\} P_i\{T_1 > u\}.$$

From this, in the same way as in Section 8.5, we conclude that

$$P_i\{T_1 > t\} = e^{-\lambda_i t},$$

which proves the assertion. □

Lemma 8.7.2. *Under P_i, the T_1 and X_1 are independent random variables.*

Proof. Denote by

$$t + T_0 = \inf\{s; \xi(s) \neq i \text{ and } s > t\}$$

the first exit time from the state i after time t. Then, for all $t > 0$ and $i \neq j \in S$, we have (if $P_i\{T_1 > t\} > 0$)

$$\begin{aligned} P_i\{T_1 > t, X_1 = j\} &= P_i\{T_1 > t, \xi(t) = i, \xi(T_1) = j\} \\ &= P_i\{T_1 > t, \xi(t) = i, \xi(t + T_0) = j\} \\ &= P_i\{\xi(t + T_0) = j \mid T_1 > t, \xi(t) = i\} P_i\{T_1 > t, \xi(t) = i\}. \end{aligned}$$

Because, by assumption $P_i\{T_1 > t\} > 0$,

$$\begin{aligned} P_i\{\xi(t + T_0) = j \mid T_1 > t, \xi(t) = i\} &= P_i\{\xi(t + T_0) = j \mid \xi(t) = i\} \\ &= P_i\{\xi(T_1) = j\} \end{aligned}$$

(due to homogeneity of the Markov process). This gives

$$P_i\{T_1 > t, X_1 = j\} = P_i\{\xi(T_1) = j\} P_i\{T_1 > t\},$$

which proves the assertion. □

The following proposition is the central result of this section.

Proposition 8.7.1. *The bivariate sequence $\{(T_i, X_i\}_0^\infty$ represents a Markov renewal process. In other words, for any $n = 0, 1, \ldots, t > 0, j \in S$,*

$$P\{T_{n+1} > t, X_{n+1} = j \mid (T_i, X_i)_0^n\} = P_i\{T_1 > t, X_1 = i\} \tag{8.7.5}$$

on the set $\{X_n = i\}$.

Proof. Let us begin with some facts. First, $\xi(t)$ has the strong Markov property. Second, knowing $\{(T_i, X_i)\}_0^n$ is equivalent to knowing $\xi(t)$ for all $0 \leq t \leq \tau_n$. Therefore,

$$\begin{aligned} P\{T_{n+1} > t, X_{n+1} = j \mid (T_i, X_i)_1^n\} &= P\{T_{n+1} > t, X_{n+1} = j \mid \xi(t); t \leq \tau_n\} \\ &= \int_t^\infty P\{T_{n+1} \in ds, \xi(\tau_n + s) = j \mid \xi(\tau_n)\} \\ &= \int_t^\infty P_{X_n}\{T_{n+1} \in ds, \xi(s) = j\} \\ &= P_{X_n}\{T_1 > t, \xi(T_1) = j\} \\ &= P_{X_n}\{T_1 > t, X_1 = j\}. \qquad \text{(a.s.)} \end{aligned}$$

This proves the proposition. □

Corollary 8.7.1. *Write*

$$P_{ij} = P_i\{X_1 = j\}. \tag{8.7.6}$$

Then, from the last proposition and Lemmas 8.7.1 *and* 8.7.2, *it follows that on* $\{X_n = i\}$, *for all* $n = 0, 1, \ldots,$

$$P\{T_{n+1} > t, X_{n+1} = j | (T_i, X_i)_0^n\} = P_{ij} e^{-\lambda_i t}. \tag{8.7.7}$$

Corollary 8.7.2. *From* (8.7.7), *we easily deduce that* $\{X_n\}_0^\infty$ *is a homogeneous Markov chain with state space S and the transition probabilities*

$$P_{ij} = P\{X_{n+1} = j | X_n = i\}. \tag{8.7.8}$$

In addition,

$$P\{T_{n+1} > t | X_n = i\} = e^{-\lambda_i t}. \tag{8.7.9}$$

Corollary 8.7.3. *From Proposition* 8.7.1, *it also follows that*

$$\begin{aligned} P\{T_1 > t_1, \ldots, T_n > t_n | X_0 = i_0, \ldots, X_n = i_n\} \\ = P\{T_1 > t_1 | X_0 = i_0\} \ldots P\{T_n > t_n | X_{n-1} = i_{n-1}\} \\ = \prod_{k=0}^{n-1} e^{-\lambda_k t_{k+1}}. \end{aligned} \tag{8.7.10}$$

Corollary 8.7.4. *It is not difficult to show that the bivariate sequence of random variables* $\{(\tau_n, X_n)\}_0^\infty$ *represents a homogeneous Markov chain with transition probabilities* ($s < t$)

$$P\{\tau_{n+1} > t, X_{n+1} = j \| \tau_n = s, X_n = i\} = P_{ij} e^{\lambda_i (t-s)}. \tag{8.7.11}$$

The next topic to be discussed is the question of regularity of a Markov chain (see Definition 8.1.5). Under what conditions on $\{\lambda_i\}_0^\infty$ does (8.5.8) hold? If the sequence $\{\lambda_i\}_0^\infty$ is bounded,

$$\sup_i \lambda_i \leq \beta < \infty, \tag{8.7.12}$$

then the process is certainly regular. This can be proved as follows. If (8.7.12) holds, then $e^{-\lambda_i t} \geq e^{-\beta t}$ for all $t \geq 0$. From this and (8.7.10), we have

$$P\{T_1 \leq t_1, \ldots, T_n \leq t_n\} \leq \prod_1^n (1 - e^{-\beta t_j}).$$

From this, we deduce that, for all $n = 1, 2, \ldots,$

$$\tau_n \overset{\text{d.}}{\geq} \sum_{k=1}^n Z_k \quad (\overset{\text{d.}}{\geq} \text{ means in distribution}),$$

where $\{Z_k\}_0^\infty$ is an i.i.d. sequence of random variables with common distribution function $1 - e^{-\beta t}$. Hence,

$$P\{\tau_n \leq t\} \leq P\left\{\sum_{k=1}^{n} Z_k \leq t\right\} = e^{-\beta t} \sum_{k=n}^{\infty} \frac{(\beta t)^k}{k!}$$

for all $0 < t < \infty$. By letting $n \to \infty$, we deduce that

$$P\{\tau \leq t\} \leq e^{-\beta t} \lim_{n\to\infty} \sum_{k=n}^{\infty} \frac{(\beta t)^k}{k!} = 0.$$

The following proposition provides a more general criterion for regularity of $\xi(t)$.

Proposition 8.7.2. *A necessary and sufficient condition for a homogeneous Markov chain to be regular is that*

$$P\left\{\sum_{k=1}^{\infty} \lambda_{X_k}^{-1} = \infty\right\} = 1.$$

PROOF. According to (8.7.9), for any $i \in S$ and $\alpha > 0$,

$$E\{e^{-\alpha T_n} | X_{n-1} = i\} = \int_0^\infty e^{-\alpha s} P\{T_n \in ds | X_{n-1} = i\}$$

$$= \lambda_i \int_0^\infty e^{-\alpha s} e^{-\lambda_i s}\, ds = \lambda_i/(\alpha + \lambda_i).$$

From this and (8.7.10), we deduce that, for every $n = 1, 2, \ldots$,

$$E\{e^{-\alpha\tau_n} | X_0 = i_0, \ldots, X_n = i_n\} = \prod_{k=0}^{n-1} \lambda_{i_k}/(\alpha + \lambda_{i_k})$$

or

$$E\{e^{-\alpha\tau_n}\} = E\left\{\prod_{k=0}^{n-1} \lambda_{X_k}/(\alpha + \lambda_{X_k})\right\}.$$

Set $\tau = \sum_{k=1}^{\infty} T_k$; then

$$E\{e^{-\alpha\tau}\} = E\left\{\prod_{k=0}^{\infty} \lambda_{X_k}/(\alpha + \lambda_{X_k})\right\}.$$

Because

$$\lim_{\alpha \to 0+} E\{e^{-\alpha\tau}\} = E\{I_{\{\tau<\infty\}}\},$$

we have

$$\lim_{\alpha \to 0} E\{I_{\{\tau<\infty\}}\} = E\left\{\lim_{\alpha\to 0} \prod_{k=0}^{\infty} \lambda_{X_k}/(\alpha + \lambda_{X_k})\right\}.$$

Finally, it is well known that

$$\lim_{\alpha \to 0} \prod_{k=0}^{\infty} \lambda_{X_k}/(\alpha + \lambda_{X_k}) = \begin{cases} 1 & \text{if } \sum_{k=0}^{\infty} \lambda_{X_k}^{-1} < \infty \\ 0 & \text{if } \sum_{k=0}^{\infty} \lambda_{X_k}^{-1} = \infty. \end{cases}$$

This yields

$$P\{\tau < \infty\} = P\left\{\sum_{k=0}^{\infty} \lambda_{X_k}^{-1} < \infty\right\}$$

and the assertion holds. □

8.8. Homogeneous Diffusion

Let $\{\xi(t); t \geq 0\}$ be a real homogeneous Markov process (i.e., $S = R$, the real line, and $\mathscr{S} = \mathscr{R}$, the σ-algebra of Borel subsets of R) with (a.s.) continuous sample functions and a transition probability $P(x, t, B)$. The stochastic process $\xi(t)$ is termed a "homogeneous diffusion" if the transition probability satisfies the following conditions: for each $\delta > 0$ and $x \in R$,

$$\text{(i)} \qquad \lim_{t \to 0+} \frac{1}{t} \int_{|y-x|>\delta} P(x, t, dy) = 0,$$

$$\text{(ii)} \qquad \lim_{t \to 0+} \frac{1}{t} \int_{|y-x|\leq\delta} (y - x)P(x, t, dy) = b(x), \tag{8.8.1}$$

$$\text{(iii)} \qquad \lim_{t \to 0+} \frac{1}{t} \int_{|y-x|\leq\delta} (y - x)^2 P(x, t, dy) = \sigma^2(t).$$

The first condition seems justified in view of (a.s.) continuity of the sample paths of the process. The function $b(\cdot)$ characterizes the "average trend" of evolution of the process over a short period of time, given that $\xi(0) = x$ and is called the "drift" coefficient. Finally, the non-negative quantity $\sigma^2(\cdot)$ determines the mean square deviation of the process $\xi(t)$ from its mathematical expectation. The function $\sigma^2(\cdot)$ is called the "diffusion" coefficient. We will assume that the functions $b(\cdot)$ and $\sigma^2(\cdot)$ are finite.

The three conditions (8.8.1) can be written in slightly different form as follows:

$$\lim_{t \to 0+} \frac{1}{t}(1 - P_x\{\xi(t) \in [x - \delta, x + \delta]\}) = 0,$$

$$\lim_{t \to 0+} \frac{1}{t} E_x[\xi(t) - \xi(0)] = b(x),$$

$$\lim_{t \to 0+} \frac{1}{t} E_x[\xi(t) - \xi(0)]^2 = \sigma^2(x).$$

The central point in the theory of diffusion processes is that, for given functions $b(\cdot)$ and $\sigma^2(\cdot)$, there is a unique and completely determined transition probability $P(x, t, B)$ which yields a homogeneous Markov process with (a.s.) continuous sample functions. There are various methods for determining $P(x, t, B)$, ranging from purely analytical to purely probabilistic. The method presented here was developed by Kolmogorov in 1931.

Consider

$$u(t, x) = \int_{-\infty}^{\infty} \varphi(y) P(x, t, dy), \tag{8.8.2}$$

where $\varphi(\cdot)$ is bounded continuous function. We shall attempt to derive the partial differential equation which $u(t, x)$ satisfies, subject to the initial condition

$$u(0+, x) = \varphi(x). \tag{8.8.3}$$

The equation that we will derive is known as Kolmogorov's backward diffusion equation, which is satisfied by the function $u(t, x)$.

Proposition 8.8.1. *Assume that conditions* (8.8.1) *hold and that the function* $u(t, x)$ *defined by Equation* (8.8.2) *has a continuous second partial derivative in* x *for all* $t > 0$. *Then,*

$$\frac{\partial u}{\partial t} = b(x)\frac{\partial u}{\partial x} + \frac{1}{2}\sigma^2(x)\frac{\partial^2 u}{\partial x^2}, \quad t > 0. \tag{8.8.4}$$

PROOF. Starting with the Chapman–Kolmogorov equation (8.1.2), we readily obtain that, for each $h > 0$,

$$\begin{aligned} u(t + h, x) &= \int_{-\infty}^{\infty} \varphi(y) P(x, t + h, dy) \\ &= \int_{-\infty}^{\infty} P(x, h, dz) \int_{-\infty}^{\infty} \varphi(y) P(z, t, dy) \\ &= \int_{-\infty}^{\infty} u(t, z) P(x, h, dz). \end{aligned}$$

Consequently,

$$\frac{u(t + h, x) - u(t, x)}{h} = \frac{1}{h}\int_{-\infty}^{\infty} P(x, h, dy)\{u(t, y) - u(t, x)\}. \tag{8.8.5}$$

Now, according to condition (8.8.1.i) and the fact that $\varphi(\cdot)$ is a bounded function, Equation (8.8.5) can be written as follows: for each $\delta > 0$ (as $h \to 0$),

$$\frac{u(t + h, x) - u(t, x)}{h} = \frac{1}{h}\int_{x-\delta}^{x+\delta} P(x, h, dy)\{u(t, y) - u(t, x)\} + o(1). \tag{8.8.6}$$

Next, we expand $u(t, x)$ in Taylor series to obtain

$$u(t, y) = u(t, x) + (y - x)\frac{\partial u(t, x)}{\partial x} + \frac{(y - x)^2}{2!}\cdot\frac{\partial^2 u(t, x)}{\partial x^2} + (y - x)^2 R(t, y)$$

with $R(t, y) \to 0$ as $y \to x$. Substituting this expansion into the right-hand side of Equation (8.8.6) and then using conditions (8.8.1.ii) and (8.8.1.iii), we arrive at

$$\lim_{h \to 0} \frac{u(t + h, x) - u(t, x)}{h} = b(x)\frac{\partial u(t, x)}{\partial x} + \frac{1}{2}\sigma^2(x)\frac{\partial^2 u(t, x)}{\partial x^2}.$$

To complete the proof, we must consider the case $h < 0$. This is handled in essentially the same way beginning with

$$\frac{u(t + h, x) - u(t, x)}{h} = \frac{1}{h}\int_{-\infty}^{\infty} P(x, |h|, dy)\{u(t + h, y) - u(t + h, x)\}.$$

The analysis is the same as before (we use the Taylor expansion) except that, in obtaining the limit as $h \to 0$, it is necessary to invoke the joint continuity of $\partial u/\partial x$ and $\partial^2 u/\partial x^2$ in t and x. In particular, we need the fact that $R(t, y) \to 0$ as $y \to x$. This proves the assertion. □

Remark 8.8.1. The error term in the Taylor expansion leads to an integral bounded by the quantity

$$\sup_{x-\delta \le y \le x+\delta} |R(t, y)|\frac{1}{h}\int_{x-\delta}^{x+\delta} P(x, h, dy)(y - x)^2.$$

Its lim sup as $h \to 0$ is not larger, then $\max|R(t, y)|\sigma^2(x) \to 0$ as $\delta \to 0$.

Remark 8.8.2. To obtain the transition probability $P(x, t, dy)$, assume that the transition probability density exists, i.e.,

$$P(x, t, B) = \int_B p(x, t, y)\, dy, \tag{8.8.7}$$

and that the derivatives

$$\frac{\partial p(x, t, y)}{\partial x}, \quad \frac{\partial^2 p(x, t, y)}{\partial x^2}$$

exist. Then, taking into account Equation (8.8.2), we can write Equation (8.8.4) as

$$\int_{-\infty}^{\infty} \varphi(y)\left\{\frac{\partial p(x, t, y)}{\partial t} - b(x)\frac{\partial p(x, t, y)}{\partial x} - \frac{1}{2}\sigma^2(x)\frac{\partial^2 p(x, t, y)}{\partial x^2}\right\} dy = 0,$$

where $\varphi(\cdot)$ is an arbitrary continuous bounded function. This, then, implies the equation

$$\frac{\partial p(x,t,y)}{\partial t} - b(x)\frac{\partial p(x,t,y)}{\partial x} - \frac{1}{2}\sigma^2(x)\frac{\partial^2 p(x,t,y)}{\partial x^2} dy = 0. \tag{8.8.8}$$

Under appropriate conditions on $b(\cdot)$ and $\sigma(\cdot)$, a solution of this equation exists and is unique. It represents a transition probability of the homogeneous diffusion process if, for instance, the coefficients $b(x)$ and $\sigma^2(x)$ are bounded and satisfy the Lipschitz conditions

$$|b(y) - b(x)| \leq C|y - x|,$$
$$|\sigma^2(y) - \sigma^2(x)| \leq C|y - x|$$

and if $\sigma^2(x) \geq \sigma^2 > 0$.

Assume again that conditions (8.8.1) hold. If the transition probability density exists, if it has a derivative

$$\frac{\partial p(x,t,y)}{\partial t}$$

which is continuous with respect to t and y, and if the function $b(y)p(x,t,y)$ is twice continuously differentiable with respect to y, the equation satisfied by $p(x,t,y)$ for fixed x turns out to be

$$\frac{\partial p(x,t,y)}{\partial t} = -\frac{\partial}{\partial y}[b(y)p(x,t,y)] + \frac{1}{2}\frac{\partial^2}{\partial y^2}[\sigma(y)p(x,t,y)]. \tag{8.8.9}$$

This is the Fokker–Planck equation or the forward Kolmogorov equation. We will not derive this equation here.

The study of diffusion processes via the backward partial differential equation has been reasonably successful in many cases. The basic problem with this method is that partial differential equations of this type are difficult to solve. In addition, the probabilistic input is very small.

An alternative approach, more probabilistically appealing, was proposed by P. Lévy and carried out by K. Itô in 1951. The idea here is to approach the problem in a fashion similar to Langevin's approach (see Section 3.7 of Chapter 3). Roughly speaking, what Itô has shown is that a diffusion process is governed by a stochastic differential equation of the form

$$d\xi(t) = b(\xi(t))\,dt + \sigma^2(\xi(t))\,dW(t), \tag{8.8.10}$$

where $b(\cdot)$ and $\sigma^2(\cdot)$ are those from (8.3.1) and $W(t)$ is a standard Brownian motion. Processes satisfying (8.8.10) are called locally Brownian.

Example 8.8.1. Find the solution of the stochastic differential equation

$$d\xi(t) = \alpha_0\xi(t)\,dt + \alpha_1\,dW(t),$$

where α_0 and $\alpha_1 > 0$ are constant and $W(t)$ is a standard Brownian motion. Here, clearly, $b(x) = \alpha_0 x$ and $\sigma^2(\cdot) \equiv \alpha_1$.

As we have pointed out [see Equation (3.7.5)], a solution of this equation over an interval containing a point t_0 is defined as a process having continuous sample functions which satisfies the equation

$$\int_{t_0}^{t} d\xi(s) = \alpha_0 \int_{t_0}^{t} \xi(s)\,ds + \alpha_1 \int_{t_0}^{t} dW(s)$$

or

$$\xi(t) - \xi(t_0) = \alpha_0 \int_{t_0}^{t} \xi(s)\,ds + \alpha_1[W(t) - W(t_0)].$$

To solve this equation, multiply both sides by $e^{-\alpha_0 t}$ to obtain

$$e^{-\alpha_0 t}\left[\xi(t) - \alpha_0 \int_{t_0}^{t} \xi(s)\,ds\right] = [\xi(t_0) - \alpha_1 W(t_0)]e^{-\alpha_0 t} + \alpha_1 e^{-\alpha_0 t} W(t).$$

This, on the other hand, can be written as

$$d\left(e^{-\alpha_0 t} \int_{t_0}^{t} \xi(s)\,ds\right) = [\xi(t_0) - \alpha_1 W(t_0)]e^{-\alpha_0 t}\,dt + \alpha_1 e^{-\alpha_0 t} W(t)\,dt.$$

Integrating both sides from t_0 to t, we have

$$e^{-\alpha_0 t} \int_{t_0}^{t} \xi(s)\,ds = \frac{\xi(t_0) - \alpha_1 W(t_0)}{\alpha_0}(e^{-\alpha_0 t_0} - e^{-\alpha_0 t_1}) + \alpha_1 \int_{t_0}^{t} e^{-\alpha_0 s} W(s)\,ds$$

or, equivalently,

$$\int_{t_0}^{t} \xi(s)\,ds = \frac{\xi(t_0) - \alpha_1 W(t_0)}{\alpha_0}(e^{\alpha_0(t-t_0)} - 1) + \alpha_1 \int_{t_0}^{t} e^{\alpha_0(t-s)} W(s)\,ds.$$

Finally, by differentiating with respect to t, we see that

$$\xi(t) = (\xi(t_0) - \alpha_1 W(t_0))e^{\alpha_0(t-t_0)} + \alpha_1\left(W(t) - \alpha_0 \int_{t_0}^{t} e^{\alpha_0(t-s)} W(s)\,ds\right).$$

Take $t_0 = 0$ and assume that $\xi(0) = 0$; then, $\xi(t)$ becomes

$$\xi(t) = \alpha_1 W(t) - \alpha_0 \alpha_1 \int_{0}^{t} e^{\alpha_0(t-s)} W(s)\,ds.$$

This is clearly a Gaussian process with

$$E\{\xi(t)\} = 0 \quad \text{and} \quad E\{\xi(t)\xi(u)\} = \alpha_1(e^{\alpha_0(t+u)} - e^{\alpha_0|t-u|})/\alpha_0.$$

Problems and Complements

8.1. Let $\{\xi(t); t \geq 0\}$ be a real homogeneous Markov process with transition probability $P(x, t, B)$. Show that, for any Borel function $f: R^n \to R$ and every $t_1 < \cdots < t_n$,

$$E_{x_0}\{f(\xi(t_1), \ldots, \xi(t_n))\} = \int_{-\infty}^{\infty} P(x_0, t_1, dx_1) \cdots \int_{-\infty}^{\infty} f(x_1, \ldots, x_n) P(x_{n-1}, t_n - t_{n-1}, dx_n).$$

8.2. Show that a stochastic process $\{\xi(t); t \geq 0\}$ with a state space $\{S, \mathscr{S}\}$ is a Markov process if and only if, for any $0 \leq s_1 < \cdots < s_k < t < t_1 < \cdots < t_n$,

$$P\{\xi(s_1) \leq x_1, \ldots, \xi(x_k) \leq x_k, \xi(t_1) \leq y_1, \ldots, \xi(t_n) \leq y_n | \xi(t) = x\}$$
$$= P\left(\bigcap_{i=1}^{k} \{\xi_i \leq x_i\} | \xi(t) = x\right) P\left(\bigcap_{j=1}^{n} \{\xi_j \leq y_j\} | \xi(t) = x\right).$$

8.3. (Continuation) Show that $\{\xi(t); t \geq 0\}$ is a Markov process if and only if

$$E\left\{\prod_{i=1}^{k} f_i(\xi(s_i)) \prod_{j=1}^{n} h_j(\xi(t_j)) | \xi(t) = x\right\}$$
$$= E\left\{\prod_{i=1}^{k} f_i(\xi(s_i)) | \xi(t) = x\right\} E\left\{\prod_{j=1}^{n} h_j(\xi(t_j)) | \xi(t) = x\right\},$$

where f_i and h_j are real Borel functions on $\{S, \mathscr{S}\}$.

8.4. Let $\{\xi(t); t \geq 0\}$ be a Markov process. Show that the sequence $\{\xi(k)\}_0^\infty$ has the Markov property.

8.5. (Continuation) Set $Z_k = [\xi(k)]$, where $[x]$ is the integer part of x. Is $\{Z_k\}_0^\infty$ a Markov process?

8.6. Let $\{\xi(t); t \geq 0\}$ be a real Markov process and $f: R \to R$ a Borel function. Show by a counterexample that $\{f(\xi(t)); t \geq 0\}$ is not necessarily a Markov process. However, if $f(\cdot)$ is one-to-one, the Markov property is preserved.

8.7. Let $\{\xi(t); t \geq 0\}$ with $E\{\xi(t)\} = 0$ for all $t \geq 0$ be a Gaussian random process. Show that

$$E\{\xi(t_n) | \xi(t_1), \ldots, \xi(t_{n-1})\} = E\{\xi(t_n) | \xi(t_{n-1})\}$$

is a necessary and sufficient condition for $\xi(t)$ to be a Markov process, where $0 \leq t_1 < \cdots < t_n$, $n = 2, 3, \ldots$.

8.8. Let $\{\xi_i(t); t \geq 0\}$, $i = 1, 2$, be two independent standard Brownian motions. Show that

$$X(t) = (\xi_1(t) + \xi_2(t))^2$$

is a Markov process.

8.9. Assume that $\{\xi_i(t); t \geq 0\}$, $i = 1, 2$, are two zero mean, independent, strictly stationary, Markov processes. Under what conditions is $Y(t) = \xi_1(t) + \xi_2(t)$ a Markov process?

8.10. Let $\{\xi(t); t \geq 0\}$ be a standard Brownian motion and

$$\tau_x = \inf\{t; \xi(t) = x\}.$$

Show that $\{\tau_x; x > 0\}$ is a process with independent increments and, hence, Markovian.

8.11. Let $\{\xi(t); t \geq 0\}$ be a homogeneous Markov process with state space $S = \{-1, 1\}$ such that

$$P\{\xi(s + t) = -1 | \xi(t) = -1\} = (2 + e^{-3t})/3,$$
$$P\{\xi(s + t) = -1 | \xi(t) = 1\} = (2 - e^{-3t})/3.$$

Find all invariant measures of the process.

8.12. Let $\{\xi(t); t \geq 0\}$ be a standard Brownian motion. Show that the process

$$Z(t) = |\xi(t) + x|, \quad x > 0,$$

(so-called "Brownian motion with reflecting barrier") is a homogeneous Markov process.

8.13. Let $\{\xi(t); t \geq 0\}$ be a standard Brownian motion. A Brownian bridge $\{X(t); t \in [0, t_0]\}$ is defined by

$$X(t) = \xi(t) + x - \frac{t}{t_0}(\xi(t) + y - x).$$

This is clearly a Brownian process that starts at point x at time $t = 0$ and passes through point y at time t_0. Show that $X(t)$ is a Markov process and determine its mean and covariance function.

8.14. (Continuation) Show that $X(t)$ and $X(t_0 - t)$ have the same distribution.

8.15. Let $\{\xi(t); t \geq 0\}$ be a standard Brownian motion. Show that $X(t) = e^{-t}\xi(e^{2t})$ is a strictly stationary Markov process. Find its mean and covariance function.

8.16. Let $\{X(t); t \geq 0\}$ be a stationary Gaussian process. If $X(t)$ has the Markov property, show that its covariance function has the form $ce^{-\alpha|t|}$, $c > 0$, $\alpha > 0$.

8.17. Let $\{\xi(t); t \geq 0\}$ be a homogeneous Markov process with transition probability $P(x, t, B)$. If

$$P(x + h, t, B + h) = P(x, t, B)$$

(spatial homogeneity) where $B + h = \{x + h; x \in B\}$, then show that $\xi(t)$ has stationary independent increments.

8.18. Let $\{\xi(t); t \geq 0\}$ be a homogeneous Markov process with state space $S = \{0, 1, \ldots\}$. Determine

$$p_{ij}(t) = P\{\xi(s + t) = j | \xi(s) = i\}$$

assuming that

$$p_{ij}(t) = 0 \quad \text{if } i > j; \qquad p_{i,i+1}(h) = \lambda h + o(h) \quad \text{as } h \to 0+;$$

$$p_{ij}(h) = o(h) \quad \text{if } h \to 0+ \text{ when } j - i \geq 2.$$

8.19. Assume that a state $j \in S = \{0, 1, \ldots\}$ is stable (see Definition 8.3.15). Show that

$$p'_{ij}(t) = -p_{ij}(t)q_i + \sum_{k \neq j}^{\infty} p_{ik}(t)q_{kj}$$

(this is the forward Kolmogorov equation).

8.20. Let $\{\xi(t); t \geq 0\}$ be a homogeneous Markov chain with state space $S = \{0, 1, \ldots\}$ and transition probability $p_{ij}(t)$. Assume that

$$p_{i,i+1}(h) = \lambda h + o(h), \qquad p_{i,i-1}(h) = i\mu h + o(h)$$

as $h \to 0+$. Determine $P\{\xi(t) = k\}$ and show that

$$\xi(t) = \xi_0(t) + \xi_1(t),$$

where $\xi_0(t)$ is binomial, $\xi_1(t)$ is Poisson, and $\xi_0(t)$ is independent of $\xi_1(t)$.

8.21. Let $\{\xi(t); t \geq 0\}$ be a "pure birth" process (see Remark 8.4.1). Denote by $0 < \tau_1 < \tau_2 < \cdots$ its discontinuity points. Show that $\{Z_n\}_1^\infty$ is a sequence of independent random variables where $Z_1 = \tau_1$, $Z_n = \tau_n - \tau_{n-1}$, $n \geq 2$.

8.22. (Continuation) Show that

$$P\{Z_{n+1} \leq t\} = 1 - e^{-\lambda_n t},$$

where λ_n is defined by (8.4.4).

8.23. (Continuation) Show that $(i \leq j)$

$$p_{ij}(t) = P\{\xi(s + t) = j | \tau_i = s\}.$$

8.24. (Continuation) Show that

$$p_{ij}(t) = P\left\{\sum_{k=i+1}^{j} Z_k \leq t\right\} - P\left\{\sum_{k=i+1}^{j+1} Z_k \leq t\right\}.$$

8.25. (Continuation) Show that $(\lambda_i \neq \lambda_j \text{ if } i \neq j)$

$$P\{\xi(t) = n\} = \prod_{j=0}^{n-1} \lambda_j \sum_{k=0}^{n} \left(\exp^{-\lambda_k t} \Big/ \prod_{j \neq k}^{n} (\lambda_j - \lambda_k)\right), \quad n \geq 1,$$

$$P\{\xi(t) = 0\} = e^{-\lambda_0 t}.$$

8.26. (Continuation) Show that

$$p_{kn}(t) = \prod_{j=k}^{n-1} \lambda_j \left(\sum_{i=k}^{n} e^{-\lambda_i t} \Big/ \prod_{j \neq k}^{n} (\lambda_j - \lambda_k)\right), \quad n > k,$$

$$p_{kk}(t) = e^{-\lambda_k t}.$$

8.27. Let $\{\xi(t); t \geq 0\}$ be a pure birth process. In order that

$$\sum_{n=0}^{\infty} P\{\xi(t) = n\} = 1 \quad \text{for all } t,$$

show that it is necessary and sufficient that the series

$$\sum_{k=0}^{\infty} \lambda_k^{-1}$$

diverges.

8.28. Let $\{\xi(t); t \geq 0\}$ represent the size of a bacterial population which may grow (each bacterium can split into two) or decline (by dying). In a small time interval $(t, t + \Delta t)$, each bacterium independently of others has a probability

$$\lambda \Delta t + o(\Delta t) \quad \text{as } \Delta t \to 0, \quad \lambda > 0,$$

of splitting in two and a probability

$$\mu \Delta t + o(\Delta t) \quad \text{as } \Delta t \to 0, \quad \mu > 0,$$

of dying. Form a system of differential equations determining

$$p_k(t) = P\{\xi(t) = k\}, \quad n = 1, \ldots,$$

and solve it assuming that $\xi(0) = 1$. Verify

$$\sum_{k=0}^{\infty} p_k(t) = 1.$$

8.29. (Continuation) Show that $\xi(t)$ is a homogeneous Markov process and determine $p_{ij}(t) = P\{\xi(s+t) = j \mid \xi(s) = i\}$.

8.30. The linear birth and death process with immigration. Let $\{\xi(t); t \geq 0\}$ be the size of a population at time t. Members of the population behave independently of one another. In $(t, t + \Delta t)$, a member will give a birth to another member with probability $\lambda\Delta t + o(\Delta t)$ as $t \to 0$. With probability $\mu\Delta t + o(\Delta t)$, the member will die. In the same time interval, an immigrant will join the population with probability $\alpha\Delta t + o(\Delta t)$ as $\Delta t \to 0$. If $\xi(0) = i$, determine the probability generating function for $\xi(t)$.

8.31. (Continuation) Show that

$$E\{\xi(t)\} = \alpha/(\mu - \lambda) + [i - \alpha/(\mu - \lambda)]e^{-(\mu-\lambda)t}$$

and determine $E\{\xi(t)\}^2$.

CHAPTER 9

Markov Processes II: Application of Semigroup Theory

9.1. Introduction and Preliminaries

Let $\{\xi(t); t \geq 0\}$ be a real homogeneous Markov process with transition probability $P(x, t, B)$. In applications the following situation is typical. The transition probability is known for all t in a neighborhood of the origin. Then, $P(x, t, B)$ can be determined for all $t > 0$ by means of the Chapman–Kolmogorov equation (8.1.2). As in the case of a countable state space, we will show that, under assumption (8.1.13), the transition probability $P(x, t, B)$ is completely determined by the value of

$$\frac{\partial P(x, t, B)}{\partial t} \quad \text{at } t = 0. \tag{9.1.1}$$

Our goal is to deduce everything about the behavior of the Markov process $\xi(t)$ from (9.1.1). Naturally, when we say everything about the behavior of $\xi(t)$, we mean everything which does not depend on the initial distribution because we only make use of the transition probability.

The basic tool to achieve this goal is semigroup theory. The modern theory of homogeneous Markov process is basically semigroup theory, whose elements will be discussed in the rest of this chapter. This approach not only elucidates various aspects of this important class of Markov processes, but also provides a unified treatment of the theory which is not attainable by other methods.

We begin with a brief review of some concepts that were discussed in some detail in Chapter 5. Denote by **B** the set of all real bounded Borel functions defined on R. With the supremum norm

$$\|h\| = \sup_{x \in R} |h(x)|, \tag{9.1.2}$$

the set **B** becomes a Banach space. A mapping

$$T: \mathbf{B} \to \mathbf{B}$$

is called a "linear operator" (see Definition 5.7.1) if, for any two $h_1, h_2 \in \mathbf{B}$,

$$T(\alpha h_1 + \beta h_2) = \alpha T h_1 + \beta T h_2, \tag{9.1.3}$$

where α and β are two fixed numbers.

We say that T is bounded if there exists a positive constant $M < \infty$ such that

$$\|Th\| \leq M \cdot \|h\| \quad \text{for all } h \in \mathbf{B}. \tag{9.1.4}$$

The smallest M for which (9.1.4) holds is called the "norm of the operator" T and is denoted by $\|T\|$. Thus,

$$\sup_{\substack{h \in \mathbf{B} \\ h \neq 0}} \frac{\|Th\|}{\|h\|} = \|T\|. \tag{9.1.5}$$

From (9.1.5), we clearly have

$$\|Th\| \leq \|h\| \cdot \|T\|.$$

If $\|T\| \leq 1$, the operator T is said to be a "contraction."

EXAMPLE 9.1.1. Let $\mathbf{B} = C[0, 1]$ be the set of continuous functions on $[0, 1]$. Define

$$T: \mathbf{B} \to \mathbf{B}$$

by

$$(Th)(x) = xh(x) \quad \text{for each } h \in \mathbf{B} \text{ and } x \in [0, 1].$$

Clearly,

$$\begin{aligned} T(h_1 + h_2)(x) = x(h_1(x) + h_2(x)) &= xh_1(x) + xh_2(x) \\ &= (Th_1)(x) + (Th_2)(x) \end{aligned}$$

so that T is linear. In addition,

$$\|T\| = \sup_{0 \leq x \leq 1} x|h(x)| \leq \|h\|,$$

which implies that $\|T\| \leq 1$ (in fact, $\|T\| = 1$).

Definition 9.1.1. A one-parameter family $\{T^t; t \geq 0\}$ of bounded linear operators on a Banach space **B** is called a "contraction semigroup" if

(i) $\|T^t u\| \leq \|u\|$ for all $u \in \mathbf{B}$,
(ii) $T^{t+s} = T^t \cdot T^s = T^s \cdot T^t$ for all $s, t \geq 0$.

In the following we will say, for short, that $\{T^t; t \geq 0\}$ is a semigroup. A semigroup is called "strongly continuous" if

$$T^0 = I \quad \text{and} \quad \|T^t - I\| \to 0 \quad \text{as } t \to 0+, \tag{9.1.6}$$

where I is the identity operator. Since, for any $s, t \geq 0$,

$$\|T^{t+s} - T^t\| = \|T^t(T^s - I)\| \leq \|T^t\| \cdot \|T^s - I\|,$$

because T^t is a contraction, this implies uniform continuity of T^t.

In the theory of homogeneous Markov processes, of particular interest is the semigroup induced by the transition probability. Let $\{\xi(t); t \geq 0\}$ be a real homogeneous Markov process with the transition probability $P(x, t, B)$. Let $\mathbf{B}$ be the set of all real bounded Borel functions on R. For each $f \in \mathbf{B}$ and $t \geq 0$, define

$$(T^t f)(x) = \int_{-\infty}^{\infty} f(y) P(x, t, dy). \tag{9.1.7}$$

The family of operators T^t defined by (9.1.7) is clearly a contraction. On the other hand, from the Chapman—Kolmogorov equation, we have

$$\begin{aligned}
(T^{t+s}f)(x) &= \int_{-\infty}^{\infty} f(y) P(x, t + s, dy) \\
&= \int_{-\infty}^{\infty} f(y) \int_{-\infty}^{\infty} P(x, t, dz) P(z, s, dy) \\
&= \int_{-\infty}^{\infty} P(x, t, dz) \int_{-\infty}^{\infty} f(y) P(z, s, dy) \\
&= \int_{-\infty}^{\infty} P(x, t, dz)(T^s f)(z) = (T^t T^s f)(x).
\end{aligned}$$

In other words,

$$T^{t+s} = T^t \cdot T^s = T^s \cdot T^t$$

and this is the semigroup property.

9.2. Generator of a Semigroup

Let $\{T^t; t \geq 0\}$ be a strongly continuous semigroup on a Banach space $\mathbf{B}$. If a sequence $\{h_n\}_1^\infty \subset \mathbf{B}$ is such that $\|h_n - h\| \to 0$ as $n \to \infty$, we say that it strongly converges to h and write

$$h = s \lim_{n \to \infty} h_n.$$

Clearly, the strong convergence is nothing but pointwise uniform convergence.

Definition 9.2.1. The (infinitesimal) generator A of the semigroup is defined by

$$Af = s \lim_{t \to 0+} \frac{T^t f - f}{t} \tag{9.2.1}$$

at those $f \in \mathbf{B}$ for which this limit exists.

Denote by $\mathbf{D}_A \subset \mathbf{B}$ the subset of elements where the limit (9.2.1) exists. Clearly, if $f_1, f_2 \subset \mathbf{D}_A$, then

$$\alpha f_1 + \beta f_2 \in \mathbf{D}_A.$$

Consequently, A is linear operator on $\mathbf{D}_A$. We now prove the following proposition.

Proposition 9.2.1. *If $f \in \mathbf{D}_A$, the function $T^t f$ is differentiable in t and*

(a) $$T^t f \in \mathbf{D}_A,$$

(b) $$\frac{dT^t f}{dt} = AT^t f = T^t A f, \tag{9.2.2}$$

(c) $$T^t f - f = \int_0^t T^s A f \, ds.$$

PROOF. (a) We have to show that the limit

$$s \lim_{h \to 0+} \frac{T^h T^t f - T^t f}{h}$$

exists. This, however, follows from

$$\frac{T^h T^t f - T^t f}{h} = T^t \frac{T^h f - f}{h} \tag{9.2.3}$$

and the fact that the limit on the right-hand side of the last equation exists as $h \to 0$.

(b) From (9.2.3), we have that, for all $t \geq 0$, there exists

$$s \lim_{h \to 0+} \frac{T^{t+h} f - T^t f}{h} = \frac{d^+ T^t f}{dt}$$

and that

$$\frac{d^+ T^t f}{dt} = T^t A f = A T^t f. \tag{9.2.4}$$

To complete the proof we have to show that

$$\frac{d^+ T^t f}{dt} = \frac{d^- T^t f}{dt} \quad \text{for each } f \in \mathbf{D}_A.$$

Consider $0 < s < t$ and write

$$\left\| \frac{T^t f - T^{t-s} f}{s} - T^t A f \right\| = \left\| \frac{T^t f - T^{t-s} f}{s} - T^{t-s} A f + T^{t-s} A f - T^t A f \right\|$$
$$\leq \left\| T^{t-s} \left(\frac{T^s f - f}{s} - A f \right) \right\| + \| T^{t-s} (T^s A f - A f) \|$$
$$\leq \| T^{t-s} \| \left(\left\| \frac{T^s f - f}{s} - A f \right\| + \| T^s A f - A f \| \right) \to 0$$

as $s \to 0$ which completes the proof of (b).

(c) The last part follows from

$$\int_0^t \frac{d}{ds} T^s f \, ds = T^t f - f = \int_0^t T^s A f \, ds,$$

which proves the proposition. □

In the following we will define the concept of a strongly integrable function. Let W_t be a mapping

$$W: [a, b] \to \mathbf{B}.$$

If there exists the limit

$$s \lim_{\delta \to 0} \sum_{k=0}^{n-1} (t_{k+1} - t_k) W_{t_{k+1}}, \tag{9.2.5}$$

where $a = t_0 < t_1 < \cdots < t_n = b$ and $\delta = \max_{0 \leq k \leq n-1} (t_{k+1} - t_k)$, the mapping W_t is called "strongly integrable" and the limit (9.2.5) is denoted by

$$\int_a^b W_t \, dt.$$

The following properties of the integral are simple to verify. If W_t is strongly integrable, then TW_t is strongly integrable on $[a, b]$ and

$$\int_a^b T W_t \, dt = T \left(\int_a^b W_t \, dt \right), \tag{9.2.6}$$

where $T: \mathbf{B} \to \mathbf{B}$ is a linear operator. If W_t is strongly continuous and integrable, then

$$\left\| \int_a^b W_t \, dt \right\| \leq \int_a^b \| W_t \| \, dt. \tag{9.2.7}$$

In addition, if W_t is strongly integrable and strongly continuous from the right,

$$s \lim_{h \to 0+} \frac{1}{h} \int_a^{a+h} W_t \, dt = W_a. \tag{9.2.8}$$

Finally, if dW_t/dt is strongly continuous on $[a,b]$, then

$$\int_a^b \frac{dW_t}{dt}\,dt = W_b - W_a. \tag{9.2.9}$$

The next proposition holds for strongly continuous semigroups.

Proposition 9.2.2. *Let* $\{T^t; t \geq 0\}$ *be a strongly continuous semigroup; then, for all* $t \geq 0$,

$$\int_0^t T^s f\,ds \in \mathbf{D}_A. \tag{9.2.10}$$

Proof. Observe that for all $0 < u < t$, we have [see (9.2.6)]

$$\frac{1}{u}(T^u - I)\int_0^t T^s f\,ds = \frac{1}{u}\left\{\int_0^t T^{s+u} f\,ds - \int_0^t T^s f\,ds\right\}$$

(after the substitution $\tau = u + s$ and $[0,t] = [0,u] \cup [u,t]$)

$$= \frac{1}{u}\left\{\int_u^{u+t} T^\tau f\,d\tau - \int_0^t T^s f\,ds\right\}$$

$$= \frac{1}{u}\left\{\int_t^{t+u} T^\tau f\,d\tau - \int_0^u T^s f\,ds\right\}.$$

By letting $u \to 0$ and invoking (9.2.8), we have

$$s\lim_{u\to 0+} \frac{1}{u}(T^u - I)\int_0^t T^s f\,ds = T^t f - f.$$

Thus, the limit exists and the proposition follows. □

How big is the set $\mathbf{D}_A$? Does it contain enough elements to make the concept of a generator useful? The following corollary answers these questions.

Corollary 9.2.1. *If* A *is the generator of a strongly continuous semigroup* $\{T^t; t \geq 0\}$ *on* $\mathbf{B}$, *then* $\mathbf{D}_A$ *is dense in* $\mathbf{B}$. *As a matter of fact, because* $T^t f$ *is strongly continuous, it follows from* (9.2.10) *that, for every* $f \in \mathbf{B}$,

$$\frac{1}{t}\int_0^t T^s f\,ds \in \mathbf{D}_A.$$

On the other hand, due to (9.2.8),

$$s\lim_{t\to 0+} \frac{1}{t}\int_0^t T^s f\,ds = f \in \mathbf{B}.$$

In other words, any element $f \in \mathbf{B}$ *is a strong limit of a family of elements from* $\mathbf{D}_A$.

Proposition 9.2.3. *The operator A is closed, i.e., if $\{f_n\} \subset \mathbf{D}_A$ and*

$$s\lim_{n\to\infty} f_n = f, \qquad s\lim_{n\to\infty} Af_n = h,$$

then $f \in \mathbf{D}_A$ and $Af = h$.

PROOF. From (9.2.2.c) we have, for every $n = 1, 2, \ldots$ and $t \geq 0$,

$$T^t f_n - f_n = \int_0^t T^s A f_n \, ds.$$

By letting $n \to \infty$, we obtain

$$T^t f - f = \int_0^t T^s A f \, ds = \int_0^t T^s h \, ds \in \mathbf{D}_A$$

due to (9.2.10). Dividing both sides of the last equation by t and letting $t \to 0+$, we conclude that $Af = h$, which proves the proposition. □

9.3. The Resolvent

This section is concerned with solutions of the differential equation (9.2.2.b). In the numerical case, the semigroup property leads to the classical Cauchy functional equation

$$f(s+t) = f(s)f(t), \qquad f(0) = 1$$

whose only possible continuous, in fact, the only measurable solutions, are $f(t) = e^{\alpha t}$ and $f(t) \equiv 0$.

In the case of the differential equation,

$$\frac{dT^t}{dt} = AT^t, \qquad T^0 = I,$$

to be also exponential, namely,

$$T^t = e^{tA},$$

would require that $\mathbf{B} = \mathbf{D}_A$ so that A is a bounded operator. This, however, is not always the case, although A is always the limit of bounded operators

$$(T^t - I)/t.$$

Therefore, a new method to solve this differential equation is required. To this end, we need the concept of a resolvent.

Definition 9.3.1. Let $\{T^t; t \geq 0\}$ be a contraction semigroup on a Banach space $\mathbf{B}$, and let

$$R_\lambda f = \int_0^\infty e^{-\lambda t} T^t f \, dt, \quad f \in \mathbf{B}, \quad \lambda > 0. \tag{9.3.1}$$

The family of operators $\{R_\lambda; \lambda > 0\}$ is called the "resolvent" of the semigroup.

From (9.3.1), we see that $R_\lambda f$ is the Laplace transform of the continuous function $T^t f$. Because $\lambda > 0$, the domain of definition of R_λ is the whole $\mathbf{B}$. Consequently, R_λ is a bounded linear operator. In fact,

$$\|R_\lambda f\| \le \int_0^\infty e^{-\lambda t}\|T^t f\|\,dt \le \|f\|/\lambda, \tag{9.3.2}$$

and linearity is apparent. Note that the integral (9.3.1) is defined in the Riemann sense.

Proposition 9.3.1. *Let $\{T^t; t \ge 0\}$ be a strongly continuous semigroup on a Banach space $\mathbf{B}$ and A its generator; then, for each $\lambda > 0$,*

$$R_\lambda: \mathbf{B} \to \mathbf{D}_A. \tag{9.3.3}$$

In addition, the mapping

$$(\lambda_I - A): \mathbf{D}_A \to \mathbf{B} \tag{9.3.4}$$

is one-to-one and

$$R_\lambda = (\lambda_I - A)^{-1}. \tag{9.3.5}$$

PROOF. Let us first prove that (9.3.3) holds. To this end, write

$$\frac{1}{t}(T^t - I)R_\lambda h = \frac{1}{t}(T^t - I)\int_0^\infty e^{-\lambda s}T^s f\,ds$$

[by invoking (9.2.8)]

$$= \frac{1}{t}\int_0^\infty e^{-\lambda s}\{T^{s+t}h - T^s h\}\,ds$$

$$= \frac{1}{t}\left\{e^{\lambda t}\int_t^\infty e^{-\lambda u}T^u h\,du - \int_0^\infty e^{-\lambda s}T^s h\,ds\right\}$$

$$= -\frac{1}{t}e^{\lambda t}\int_0^t e^{-\lambda u}T^u h\,du + \frac{1}{t}(e^{\lambda t} - 1)\int_0^\infty e^{-\lambda s}T^s h\,ds$$

$$= -\frac{1}{t}e^{\lambda t}\int_0^t e^{-\lambda u}T^u h\,du + \frac{1}{t}(e^{\lambda t} - 1)R_\lambda h.$$

By letting $t \to 0+$, we see that

$$AR_\lambda h = -h + \lambda R_\lambda h. \tag{9.3.6}$$

Consequently, the limit

$$s\lim_{t\to 0+}(T^t - I)R_\lambda h/t$$

exists, which proves (9.3.3).

From Equation (9.3.6), we obtain

$$\lambda R_\lambda h - AR_\lambda h = h \tag{9.3.7}$$

so that $f = R_\lambda h$ is a solution of the equation

$$\lambda f - Af = h. \tag{9.3.8}$$

To complete the proof we have to show that (9.3.8) has a unique solution f for each $\lambda > 0$. Assume that f_1 and f_2 are two solutions of (9.3.8); then, $\varphi = f_1 - f_2 \in \mathbf{D}_A$ and

$$\lambda\varphi - A\varphi = 0. \tag{9.3.9}$$

On the other hand, from Proposition 9.2.1, we see that $T^t\varphi \in \mathbf{D}_A$ and satisfies

$$\frac{dT^t\varphi}{dt} = T^tA\varphi = \lambda T^t\varphi.$$

Thus,

$$\frac{d}{dt}(e^{-\lambda t}T^t\varphi) = 0.$$

Consequently, for all $t \geq 0$,

$$e^{-\lambda t}T^t\varphi = C \quad \text{(a constant)}.$$

For $t = 0$, we obtain that $C = \varphi$, which gives

$$0 \leq \|\varphi\| \leq e^{-\lambda t}\|T^t\varphi\| \leq e^{-\lambda t}\|\varphi\| \to 0$$

as $t \to +\infty$, so that $\|\varphi\| = 0$, and uniqueness is proved. This shows that $\lambda_I - A$ is one-to-one on $\mathbf{D}_A$. Finally, from (9.3.7), we obtain that

$$R_\lambda = (\lambda_I - A)^{-1}$$

and that

$$A = \lambda_I - R_\lambda^{-1}. \qquad \square$$

Remark 9.3.1. Proposition 9.3.1 shows that the mapping (9.3.3) is one-to-one and onto.

Corollary 9.3.1. *For any $f \in \mathbf{B}$, we have*

$$s\lim_{\lambda \to +\infty} \lambda R_\lambda f = f. \tag{9.3.10}$$

In fact,

$$\begin{aligned}
\|f - \lambda R_\lambda f\| &= \left\| f - \lambda \int_0^\infty e^{-\lambda t}T^t f\,dt \right\| \\
&= \lambda \left\| \int_0^\infty e^{-\lambda t}[f - T^t f]\,dt \right\| = \left\| \int_0^\infty e^{-u}[f - T^{u\lambda^{-1}}f]\,du \right\| \\
&\leq \int_0^\infty e^{-u}\|f - T^{u\lambda^{-1}}f\|\,du.
\end{aligned}$$

Now, for each $u \geq 0$, $\|f - T^{u\lambda^{-1}}f\| \to 0$ as $\lambda \to \infty$ and the integrand is bounded by $2e^{-u}\|f\|$. Now invoking the dominated convergence theorem, (9.3.10) follows.

9.4. Uniqueness Theorem

In this section, we show that, under certain regularity conditions, semigroups are uniquely determined by their generators. Thus, given a generator, we should be able, at least in some cases, to recover the corresponding semigroup. This appears simple enough if the generator is bounded. If not, there are considerable difficulties that must be resolved to achieve this goal.

Let **B** be a Banach space. The space of all bounded, continuous linear functionals l: $\mathbf{B} \to R$ with the norm

$$\|l\| = \sup_{\substack{f \neq 0 \\ f \in \mathbf{B}}} \frac{|l(f)|}{\|f\|} \tag{9.4.1}$$

is also a Banach space, which is called the "conjugate" space of **B** and denoted by $\mathbf{B}^*$. We now need the following result.

Lemma 9.4.1. *Let the mapping*

$$w: R_+ \to \mathbf{B}$$

be continuous and bounded. Then,

$$\int_0^\infty e^{-\lambda t} w(t)\, dt = 0 \tag{9.4.2}$$

for all $\lambda > 0$ implies that $w(t) \equiv 0$.

PROOF. Let $l \in \mathbf{B}^*$; then $l(w(t))$ is a bounded and continuous real function. Due to linearity of l, we have, for all $\lambda > 0$,

$$\int_0^\infty e^{-\lambda t} l(w(t))\, dt = l\left(\int_0^\infty e^{-\lambda t} w(t)\, dt\right) = 0.$$

This clearly implies that $l(w(t)) = 0$ for all $t \geq 0$ because of the uniqueness theorem for the Laplace transforms of real functions. Because this holds for any $l \in \mathbf{B}^*$, the assertion follows. □

We are now ready to prove the following proposition.

Proposition 9.4.1. *Let $\{T_0^t; t \geq 0\}$ and $\{T_1^t; t \geq 0\}$ be two strongly continuous semigroups on **B** with common generator A; then, $T_0^t = T_1^t$ for all $t \geq 0$.*

PROOF. From Proposition 9.3.1 we deduce that T_0^t and T_1^t have the same resolvent,

$$R_\lambda = (\lambda_I - A)^{-1}.$$

Consequently, for every $f \in \mathbf{B}$ and $\lambda > 0$,

$$\int_0^\infty e^{-\lambda t}(T_0^t f - T_1^t f)\,dt = R_\lambda f - R_\lambda f = 0.$$

From this and the previous lemma, the assertion follows. □

Next, we shall attempt to construct a semigroup from its generator A when A is bounded. To this end, the following notation is needed. Define $\exp\{A\}$ as

$$e^A = \sum_{k=0}^\infty A^k/k!, \qquad A^0 = I, \tag{9.4.3}$$

where A^n is the n-iterate of A, i.e., $A^n = A \cdot A^{n-1}$, $n = 1, 2, \ldots$. Because

$$\|A^n\| \le \|A\|^n$$

and A is bounded, it follows that the series (9.4.3) converges absolutely.

Proposition 9.4.2. *If the generator A of a semigroup $\{T^t; t \ge 0\}$ is bounded, then for all $t \ge 0$,*

$$T^t = e^{tA}. \tag{9.4.4}$$

PROOF. Let us show that

$$\left\| \frac{e^{tA} - I}{t} - A \right\| \to 0 \quad \text{as } t \to 0. \tag{9.4.5}$$

Clearly,

$$\|e^{tA} - I - tA\| \le \sum_{k=2}^\infty \|A\|^k t^k/k! = e^{t\|A\|} - 1 - t\|A\|,$$

so that (9.4.5) holds. Because $\exp\{tA\}$ is a semigroup, the assertion now follows from Proposition 9.4.1. □

From this proposition we see that a semigroup T^t must be of the form (9.4.4) if its generator A is bounded. In addition, $\mathbf{D}_A = \mathbf{B}$. Therefore, every bounded operator generates a semigroup.

If the generator A is not bounded, the situation is considerably more complicated. However, from Proposition 9.4.1, we know that A determines uniquely its semigroup. The next proposition shows that, under certain conditions, T^t is the unique solution of the differential equation (9.2.1.b).

Proposition 9.4.3. *Let A be the generator of a semigroup $\{T^t; t \ge 0\}$. If $f \in \mathbf{D}_A$, the function $W(t) = T^t f$ is the unique solution of the differential equation*

$$\frac{dW(t)}{dt} = AW(t) \tag{9.4.6}$$

subject to the following conditions:

(i) $W(t)$ *is strongly differentiable for all* $t > 0$;
(ii) $\|W(t)\| \leq Ke^{\alpha t}$, *where* $K, \alpha \in (0, \infty)$ *are some constants;*
(iii) $W(t) \to f$ *strongly as* $t \to 0+$.

PROOF. According to (9.2.1.b), the function $W(t)$ satisfies Equation (9.4.6) and condition (i). To show uniqueness, assume that $W_1(t)$ and $W_2(t)$ are two different solutions of (9.4.6) satisfying (i)–(iii) and set $V(t) = W_1(t) - W_2(t)$. Then $V(t)$ is also a solution of (9.4.6) which satisfies conditions (i) and (ii) and $V(t) \to 0$ (strongly) if $t \to 0+$. Set $U(t) = e^{-\lambda t}V(t)$; then, from (9.4.6), we have

$$\begin{aligned}\frac{d}{dt}U(t) &= -\lambda e^{-\lambda t}V(t) + e^{-\lambda t}\frac{d}{dt}V(t)\\ &= -\lambda U(t) + e^{-\lambda t}AV(t)\\ &= -\lambda U(t) + AU(t) = -R_\lambda^{-1}U(t).\end{aligned}$$

Hence,

$$U(t) = -R_\lambda \frac{d}{dt}U(t).$$

Integrating both sides of this equation, we get

$$\int_0^t U(s)\,ds = -R_\lambda \int_0^t \frac{d}{ds}U(s)\,ds = -R_\lambda U(t).$$

When $t \to \infty$, the left-hand side tends to the Laplace transform of $V(t)$, while the right-hand side converges to 0. Therefore, for a suitable $\lambda > 0$, the Laplace transform of $V(t)$ vanishes. This and Lemma 9.4.1 prove the assertion. □

9.5. The Hille–Yosida Theorem

In the previous section, we have established that every bounded operator generates a semigroup specified by (9.4.4). If an operator is unbounded, it may not be the generator of a semigroup. When is such an operator the generator of a semigroup? This question is answered by the following proposition due to Hille and Yosida, which characterizes the operators that are generators of semigroups.

Proposition 9.5.1. *Let* **B** *be a Banach space and* A *a linear operator with domain* $\mathbf{D}_A \subset \mathbf{B}$. *In order to be the generator of a strongly continuous semigroup on* **B**, *it is necessary and sufficient that the following conditions hold:*

(i) $\mathbf{D}_A$ *is dense in* $\mathbf{B}$;
(ii) *for every* $\lambda > 0$ *and* $h \in \mathbf{B}$, *the equation*

$$\lambda f - Af = h \tag{9.5.1}$$

has a unique solution $f \in \mathbf{D}_A$;
(iii) *the solution* f *satisfies* $\|f\| \leq \|h\|/\lambda$.

PROOF. The necessity of these conditions was already proved in the previous section. To prove their sufficiency, assume that they hold. Then, from (ii), it clearly follows that the operator $\lambda I - A$ is invertible. Set

$$R_\lambda = (\lambda I - A)^{-1}, \quad \lambda > 0$$

(we will see later that R_λ is indeed the resolvent). Clearly, $R_\lambda\colon \mathbf{B} \to \mathbf{D}_A$ (onto) and by (iii) we have that $\|R_\lambda\| \leq \lambda^{-1}$.

Define $A_\lambda = \lambda A R_\lambda$ and write

$$A_\lambda = \lambda\{-(\lambda I - A)R_\lambda + \lambda R_\lambda\} = \lambda(\lambda R_\lambda - I). \tag{9.5.2}$$

From this, we deduce that A_λ is bounded, i.e.,

$$\|A_\lambda\| \leq \lambda\|\lambda R_\lambda - I\| \leq \lambda(1 + 1) = 2\lambda.$$

Let us prove that, as $\lambda \to \infty$,

$$A_\lambda f \to Af \quad \text{strongly for all } f \in \mathbf{D}_A. \tag{9.5.3}$$

To this end, we will show first that, as $\lambda \to \infty$,

$$\lambda R_\lambda f \to f \quad \text{strongly for all } f \in \mathbf{D}_A. \tag{9.5.4}$$

Because [see (9.5.2)]

$$\lambda R_\lambda f - f = \frac{1}{\lambda} A_\lambda f = A R_\lambda f = R_\lambda A f,$$

we have

$$\|\lambda R_\lambda f - f\| \leq \|R_\lambda\|\,\|Af\| \leq \|Af\|/\lambda \to 0$$

as $\lambda \to \infty$, which proves (9.5.4).

Let us now show that (9.5.4) holds for every $f \in \mathbf{B}$. Due to condition (i), for any $f \in \mathbf{B}$ and $\varepsilon > 0$ arbitrarily small, there exists $f_0 \in \mathbf{D}_A$ such that $\|f - f_0\| \leq \varepsilon$. Then

$$\begin{aligned}\|\lambda R_\lambda f - f\| &\leq \|\lambda R_\lambda f_0 - f_0\| + \|\lambda R_\lambda(f - f_0)\| + \|f - f_0\| \\ &\leq \|\lambda R_\lambda f_0 - f_0\| + 2\varepsilon.\end{aligned}$$

Consequently, due to (9.5.4)

$$\limsup_{\lambda \to \infty} \|\lambda R_\lambda f - f\| \leq 2\varepsilon,$$

which shows that (9.5.4) holds for all $f \in \mathbf{B}$.

To prove (9.5.3) consider an $f \in \mathbf{D}_A$; then, due to (9.5.4),

$$A_\lambda f = \lambda R_\lambda A f \to Af \quad \text{strongly as } \lambda \to \infty.$$

The last result implies, roughly speaking, that the bounded operator A_λ approximates A for large λ.

Define

$$T_\lambda^t = \exp\{tA_\lambda\}. \tag{9.5.5}$$

This is clearly a (contraction) semigroup for each $\lambda > 0$. Taking into account (9.5.2), we can write

$$T_\lambda^t = \exp\{t\lambda^2 R_\lambda - t\lambda I\} = \exp(-t\lambda)\exp\{t\lambda^2 R_\lambda\}.$$

Therefore,

$$\|T_\lambda^t f\| \le e^{-t\lambda} \sum_{n=0}^{\infty} \frac{(t\lambda)^n}{n!} \|(\lambda R_\lambda)^n f\| \le e^{-\lambda t} e^{\lambda t} \|f\|.$$

Next, we have to prove that $T_\lambda^t f$ has a limit as $\lambda \to \infty$. To do this, we have to show that, for $f \in \mathbf{D}_A$,

$$\|T_\lambda^t f - T_\eta^t f\| \to 0 \quad \text{as } \lambda, \eta \to \infty. \tag{9.5.6}$$

Let λ and η be fixed: Set

$$\varphi(t) = T_\lambda^t f - T_\eta^t f, \quad f \in \mathbf{D}_A; \tag{9.5.7}$$

then

$$\varphi'(t) = A_\lambda T_\lambda^t f - A_\eta T_\eta^t f = A_\lambda \varphi(t) + g(t),$$

where

$$g(t) = (A_\lambda - A_\mu) T_\mu^t f.$$

It is not difficult to show that

$$\frac{d}{dt}[\varphi(t)\exp\{-tA_\lambda\}] = g(t)\exp\{-tA_\lambda\}.$$

Therefore, because $\varphi(0) = 0$ [see (9.5.7)],

$$\varphi(t)\exp\{-tA_\lambda\} = \int_0^t \exp\{-sA_\lambda\} g(s)\, ds.$$

Noting that T_λ^t and A_η commute, we can write

$$\begin{aligned}\varphi(t) &= \int_0^t \exp\{(t-s)A_\lambda\} g(s)\, ds \\ &= \int_0^t T_\lambda^{t-s} T_\eta^s (A_\lambda - A_\eta) f\, ds.\end{aligned}$$

Hence, because T_λ^t is a contraction for each $\lambda > 0$,

$$\begin{aligned}\|\varphi(t)\| &\le \int_0^t \|T_\lambda^{t-s} T_\eta^s (A_\lambda - A_\eta) f\| \, ds \\ &\le t \|(A_\lambda - A_\eta) f\|. \end{aligned} \tag{9.5.8}$$

But, by (9.5.3), the last term tends to zero as $\lambda, \eta \to \infty$, which proves (9.5.6). Notice that the convergence is uniform in t in any compact interval.

Now define the family of operators $\{T^t; t \ge 0\}$ by setting

$$T^t f = s \lim_{\lambda \to \infty} T_\lambda^t f, \quad f \in \mathbf{D}_A,$$

in the strong sense. Because the convergence is uniform in t, $T^t f$ is continuous in t and $T^t f \to f$ strongly if $t \to 0+$. In addition, for each $f \in \mathbf{D}_A$,

$$\|T^t f\| \le \|T_\lambda^t f\| + \|T^t f - T_\lambda^t f\| \le \|f\| + \varepsilon,$$

where $\varepsilon \to 0$ as $\lambda \to \infty$. Consequently, $\|T^t f\| \le \|f\|$. It is now easy to extend T to all $f \in \mathbf{B}$.

Finally, we have to show that the generator of the semigroup that we have just constructed is A. In any case, T^t has a generator, say A_0. Let us compute $A_0 f$ for $f \in \mathbf{D}_A$. Because

$$\frac{d}{dt} T_\lambda^t f = T_\lambda^t A_\lambda f,$$

it follows that

$$T_\lambda^t f - f = \int_0^t T_\lambda^s A_\lambda f \, ds. \tag{9.5.9}$$

Now

$$\|T_\lambda^t A_\lambda f - T^t A f\| \le \|T^t A f - T_\lambda^t A f\| + \|T_\lambda^t A f - T_\lambda^t A_\lambda f\|.$$

The first term on the right side tends to 0 uniformly as $\lambda \to \infty$, for t bounded. The second term is bounded by $\|A_\lambda f - Af\|$ because T_λ^t is a contraction, and this bound tends to zero as $\lambda \to \infty$. Therefore,

$$T_\lambda^t A_\lambda f \to T^t A f \quad \text{strongly as } \lambda \to \infty$$

for each fixed $f \in \mathbf{D}$, and the convergence is uniform in t over compact intervals. As a result, we can pass to the limit in (9.5.9) to obtain

$$T^t f - f = \int_0^t T^s A f \, ds. \tag{9.5.10}$$

Dividing by t and letting $t \to 0+$, we have

$$A_0 f = s \lim_{t \to 0+} \frac{T^t f - f}{t} = s \lim_{t \to 0+} \frac{1}{t} \int_0^t T^u A f \, du = Af.$$

Consequently, $\mathbf{D}_{A_0} \supset \mathbf{D}_A$ and $A_0 f = Af$ on $\mathbf{D}_A$.

Because A_0 is the generator of $\{T^t; t \ge 0\}$, $\lambda I - A_0$ maps $\mathbf{D}_{A_0}$ onto $\mathbf{B}$ by

Proposition 9.3.1. But $\lambda I - A$ agrees with $\lambda I - A_0$ on $\mathbf{D}_A$ and maps $\mathbf{D}_A$ onto $\mathbf{B}$ by condition (ii). It follows that $\mathbf{D}_{A_0} = \mathbf{D}_A$ and, thus, $A_0 = A$. This proves the proposition. □

9.6. Examples

The first five sections of this chapter are concerned with some elements of semigroup theory. The question, of course, is where does all this lead? Our principal interest here is homogeneous Markov processes and their transition probabilities. Semigroup theory is a tool which is supposed to help us to investigate what are possible systems of transition probabilities $\{P(x,t,B)\}$. The development so far clearly implies that we should study $P(x,t,B)$ through its generator. In this respect, it is of some interest to find out which operators on a suitable Banach space can be generators of semigroups. If an operator is bounded, then it is certainly the generator of a semigroup (see Proposition 9.4.2). An unbounded operator is not necessarily the generator of a semigroup. The next several examples illustrate this particular situation.

EXAMPLE 9.6.1. Let $\{\xi(t); t \geq 0\}$ be a homogeneous Markov chain with a finite state space, say $S = \{x_1, \ldots, x_N\}$ and the transition probability

$$p_{ij}(t) = P\{\xi(s+t) = x_j | \xi(s) = x_i\}.$$

We will assume that all states in S are stable (see 8.3.15). Denote by M^t the transition matrix

$$M^t = \{p_{ij}(t)\};$$

then due to condition (8.3.2), $M^0 = I$ where I is the unit matrix. In addition, condition (9.3.3) implies that

$$\lim_{t \to 0+} M^t = I.$$

From the Chapman–Kolmogorov equation, it readily follows [see (8.3.7)] that, for all $s, t \geq 0$,

$$M^{s+t} = M^s \cdot M^t = M^t \cdot M^s,$$

so that the family of transition matrices $\{M^t; t \geq 0\}$ represents a semigroup.

Denote by $\mathbf{B}$ the set of all real bounded functions on S (as a matter of fact, the Banach space $\mathbf{B}$ is the set of N-dimensional column vectors). For every $x_i \in S$ and $f \in \mathbf{B}$, we define

$$(M^t f)(x_i) = \sum_{j=1}^{N} p_{ij}(t) f(x_j). \tag{9.6.1}$$

We have

$$|(M^t f)(x_i)| \leq \|f\| \quad \text{for all } x_i \in S,$$

so that

$$\sup_{x_i} |(M^t f)(x_i)| = \|M^t f\| \leq \|f\|,$$

which clearly shows that $\{M^t; t \geq 0\}$ is a contraction semigroup.

Consider

$$\frac{1}{t}\{(M^t f)(x_i) - f(x_i)\} = \frac{1}{t}\left\{\sum_{j=0}^{N} p_{ij}(t) f(x_j) - f(x_i)\right\}$$

$$= \frac{1}{t}\left\{\sum_{j \neq i}^{N} p_{ij}(t) f(x_j) - [1 - p_{ii}(t)] f(x_i)\right\}.$$

From Propositions 9.3.2 and 9.3.3, it readily follows that the limit

$$\lim_{t \to 0+} \frac{1}{t}\{(M^t f)(x_i) - f(x_i)\} = \sum_{j \neq i}^{N} q_{ij} f(j) - q_{ii} f(x_i) \tag{9.6.2}$$

$$= (Af)(x_i)$$

exists for all $f \in \mathbf{B}$ and $x_i \in S$, where

$$A = \begin{bmatrix} -q_{11} & q_{12} & \cdots & q_{1N} \\ q_{21} & -q_{22} & \cdots & q_{2N} \\ \vdots & \vdots & \cdots & \vdots \\ q_{N1} & q_{N2} & \cdots & -q_{NN} \end{bmatrix}.$$

Because all states of S are stable, $|q_{ii}| < \infty$.

It is clear that A is a bounded operator because

$$\|Af\| = \sup_{x_i} |(Af)(x_i)| = \sup_{x_i} \left| \sum_{j \neq i}^{N} q_{ij} f(x_j) - q_{ii} f(x_i) \right|$$

$$\leq \sup_{i} \sum_{j=i}^{N} |q_{ij}| \cdot \|f\|.$$

Finally, to show that A is the generator, consider

$$\left\| \frac{M^t f - f}{t} - Af \right\| = \sup_{x_i} \left| \frac{1}{t}\{(M^t f)(x_i) - f(x_i)\} - (Af)(x_i) \right| \to 0$$

due to (9.6.2). Because the generator is bounded, we have

$$M^t = e^{tA} = I + \sum_{k=1}^{n} \frac{t^k}{k!} A^k.$$

EXAMPLE 9.6.2. Let $\{\xi(t); -\infty < t < \infty\}$ be a strictly stationary homogeneous chain with the same state space and transition probabilities as in the previous example. Denote by

$$p_k = P\{\xi(t) = x_k\};$$

then, for all $s, t \geq 0$,

$$p_j = P\{\xi(s+t) = j\} = \sum_{i=1}^{n} p_{ij}(t) p_i.$$

This implies that

$$(M^t)' Q = Q, \tag{9.6.3}$$

where $Q' = (p_1, \ldots, p_n)$ and Q' means the transpose of Q.

Consider now the case $N = 2$ and take

$$P\{\xi(t) = x_1\} = \tfrac{1}{2}, \qquad P\{\xi(t) = x_2\} = \tfrac{1}{2}.$$

Our goal is to determine M^t. First, we have

$$p_{11}(t) + p_{12}(t) = 1$$

and, from (9.6.3), we deduce

$$p_{11}(t) + p_{21}(t) = 1.$$

Set $p_{11}(t) = \alpha(t)$; from the last two equations, we obtain that

$$M^t = \begin{bmatrix} \alpha(t) & 1 - \alpha(t) \\ 1 - \alpha(t) & \alpha(t) \end{bmatrix}.$$

Consequently,

$$\left.\frac{dM^t}{dt}\right|_{t=0} = A = \alpha'(0) \begin{bmatrix} 1 & -1 \\ -1 & 1 \end{bmatrix},$$

where $0 > \alpha'(0) = -\lambda$. Set $A = -\lambda A_0$; then it is readily seen that, for all $n = 1, 2, \ldots,$

$$A^n = \tfrac{1}{2}(-2\lambda)^n A_0.$$

Because A is bounded, we have

$$\begin{aligned} M^t &= I + \frac{1}{2} \sum_{k=1}^{\infty} \frac{(-2\lambda t)^k}{k!} A_0 \\ &= I + \tfrac{1}{2} A_0 (e^{-2\lambda t} - 1) \\ &= \begin{bmatrix} 1 & 0 \\ 0 & 1 \end{bmatrix} + \begin{bmatrix} \frac{1}{2}(e^{-2\lambda t} - 1) & -\frac{1}{2}(e^{-2\lambda t} - 1) \\ -\frac{1}{2}(e^{-2\lambda t} - 1) & \frac{1}{2}(e^{-2\lambda t} - 1) \end{bmatrix} \\ &= \begin{bmatrix} \frac{1}{2}(e^{-2\lambda t} + 1) & \frac{1}{2}(1 - e^{-2\lambda t}) \\ \frac{1}{2}(1 - e^{-2\lambda t}) & \frac{1}{2}(e^{-2\lambda t} + 1) \end{bmatrix}. \end{aligned}$$

The two examples discussed above deal with generators which are bounded linear operators defined on the whole of **B**. In this respect, it is of some interest to point out that if A is a generator whose domain is the entire space **B**, then A must be bounded.

If a generator A is not bounded the solution (9.4.3), then breaks down because the series on the right-hand side does not have to converge even on $\mathbf{D}_A$. This is unfortunate because many of the most interesting examples (like the Brownian motion process) do not have bounded generators.

EXAMPLE 9.6.3. Let $\{\xi(t); t \geq 0\}$ be one-dimensional standard Brownian motion. As we have seen, this is a homogeneous Markov process with state space $\{R, \mathscr{R}\}$ and transition probability

$$P(x, t, B) = (2\pi t)^{-1/2} \int_B \exp\left\{-\frac{(y-x)^2}{2t}\right\} dy. \tag{9.6.4}$$

Let $\mathbf{B}$ be the Banach space of real bounded Borel measurable functions on R with the usual supremum norm. Simple calculation yields that, for any $f \in \mathbf{B}$,

$$(T^t f)(x) - f(x) = (2\pi)^{-1/2} \int_{-\infty}^{\infty} \exp\left\{-\frac{u^2}{2}\right\} f(u\sqrt{t} + x)\,du - f(x). \tag{9.6.5}$$

From this, it seems clear that the semigroup $\{T^t; t \geq 0\}$ induced by (9.6.4) is not strongly continuous on $\mathbf{B}$.

Denote by $C_0 \subset \mathbf{B}$ the subset of continuous functions and by $\mathbf{B}_0 \subset \mathbf{B}$ the subset on which the semigroup is strongly continuous. Let us show that $\mathbf{B}_0 \subset C_0$. To this end, consider the resolvent of the semigroup ($h \in \mathbf{B}$)

$$\begin{aligned}(R_\lambda h)(x) &= \int_0^\infty e^{-\lambda t}(T^t g)(x)\,dt \qquad \lambda > 0\\ &= \int_0^\infty e^{-\lambda t}\,dt(2\pi t)^{-1/2} \int_{-\infty}^{\infty} \exp\left\{-\frac{(y-x)^2}{2t}\right\} h(y)\,dy\\ &= (2\lambda)^{-1/2} \int_{-\infty}^{\infty} \exp\{|y-x|\sqrt{2\lambda}\} h(y)\,dy. \end{aligned} \tag{9.6.6}$$

From this we see that for every $h \in \mathbf{B}$, $(R_\lambda h)(x)$ is a bounded continuous function on R, so that

$$R_\lambda\colon \mathbf{B} \to C_0.$$

On the other hand,

$$\mathbf{D}_A = R_\lambda \mathbf{B} \subset C_0 \tag{9.6.7}$$

[see (9.3.3)], where $R_\lambda \mathbf{B} = \{R_\lambda h; h \in \mathbf{B}\}$. But according to Corollary 9.2.1, the set $\mathbf{D}_A$ is dense in $\mathbf{B}_0$ (as a matter of fact, the closure $\bar{\mathbf{D}}_A = \mathbf{B}_0$). Because C_0 is also a Banach space, it follows that $\mathbf{B}_0 \subset C_0$.

For $h \in C_0$, write

$$\begin{aligned}(R_\lambda h)(x) = (2\lambda)^{-1/2} \Big\{&\exp(-x\sqrt{2\lambda}) \int_{-\infty}^{x} \exp(y\sqrt{2\lambda}) h(y)\,dy\\ &+ \exp(x\sqrt{2\lambda}) \int_x^\infty \exp(-y\sqrt{2\lambda}) h(y)\,dy\Big\}\end{aligned}$$

and set $f(x) = (R_\lambda h)(x)$, then, after some simple calculations, we have

$$f'(x) = -2\exp(-x\sqrt{2\lambda}) \int_{-\infty}^{x} \exp(y\sqrt{2\lambda})h(y)\,dy + \sqrt{2\lambda}f(x), \tag{9.6.8}$$

$$f''(x) = 2\lambda f(x) - h(x). \tag{9.6.9}$$

The last equation shows that $f''(\cdot)$ is a bounded and continuous function on R. Consequently, $f'(\cdot)$ is uniformly continuous.

From (9.3.7), we have

$$\lambda f(x) - (Af)(x) = h(x).$$

This and (9.6.9) yield

$$(Af)(x) = \tfrac{1}{2} f''(x). \tag{9.6.10}$$

In other words, every $f \in \mathbf{D}_A$ satisfies (9.6.10).

Now, denote by $C_0^* \subset C_0$ the set of uniformly continuous functions on R. From Equation (9.6.8), it follows that $f' \in C_0$. This then implies that $f \in C_0^*$. It is also not difficult to see that $Af \in C_0^*$. From this and (9.6.10), we have that $f'' \in C_0^*$.

Let us now show that the converse also holds, i.e., if

$$f \in C_0^* \Rightarrow f'' \in C_0^* \Rightarrow f \in \mathbf{D}_A.$$

To this end, consider

$$\begin{aligned}
&\frac{1}{t}\{(T^t f)(x) - f(x)\} - \frac{1}{2} f''(x) \\
&= \frac{1}{t\sqrt{2\pi t}} \int_{-\infty}^{\infty} \exp\left\{-\frac{(y-x)^2}{2t}\right\}[f(y) - f(x)]\,dy - \frac{1}{2} f''(x)
\end{aligned}$$

(using the Taylor formula)

$$f(y) = f(x) + (y-x)f'(x) + \tfrac{1}{2}(y-x)^2 f(x + \theta(y-x)) \quad (0 < \theta < 1)$$

$$\begin{aligned}
&= \frac{1}{t\sqrt{2\pi t}} \int_{-\infty}^{\infty} \exp\left\{-\frac{(y-x)^2}{2t}\right\}\left\{(y-x)f'(x) \right. \\
&\qquad \left. + \frac{1}{2}(y-x)^2 f''(x + \theta(y-x))\right\} dy - \frac{1}{2} f''(x) \\
&= (2\pi)^{-1/2} \int_{-\infty}^{\infty} e^{-u^2/2}[\tfrac{1}{2}u^2 f''(x + \theta u\sqrt{t})]\,du - \tfrac{1}{2} f''(x) \\
&= \tfrac{1}{2}(2\pi)^{-1/2} \int_{-\infty}^{\infty} e^{-u^2/2}[f''(x + \theta u\sqrt{t}) - f''(x)]u^2\,du.
\end{aligned}$$

This implies that

$$\left\| \frac{1}{t}(T^t f - f) - \frac{1}{2} f'' \right\| \to 0$$

as $t \to 0+$.

Consequently, $\mathbf{D}_A = \{f; f \in C_0^* \text{ and } f'' \in C_0^*\}$ and $\mathbf{B}_0 = C_0^*$.
Next, assume that $P(x,t,B)$ is such that

$$P(x,t,B) = (T^t I_B)(x) \in \mathbf{D}_A;$$

then

$$\frac{\partial P(x,t,B)}{\partial t} = A(P(\cdot,t,B))(x)$$

or

$$\frac{\partial P(x,t,B)}{\partial t} = \frac{1}{2}\frac{\partial^2 P(x,t,B)}{\partial x^2}, \tag{9.6.11}$$

which is the backward Kolmogorov equation [see (8.8.8)]. From the general theory of the equation of heat conduction, we know that the unique solution of (9.6.11) is (9.6.4).

9.7. Some Refinements and Extensions

Let $\{\xi(t); t \geq 0\}$ be a homogeneous Markov process with state space $\{R, \mathscr{R}\}$ and transition probability $P(x,t,B)$. In this section, we discuss some more-refined properties of this function and of the semigroup which it induces. The first question that naturally arises is one concerning the continuity of $P(x,t,B)$ with respect to t at $t = 0$. As we have seen, this property was repeatedly used in Chapter 8 to obtain many useful results.

Which of the Markov processes have this property? In other words, when does condition (8.1.13) hold? To answer this and some related questions, denote by $U(x,\varepsilon)$ the open ε-neighborhood of $x \in R$, i.e.,

$$U(x,\varepsilon) = \{y \in R; |y - x| < \varepsilon\}.$$

Definition 9.7.1. A transition probability $P(x,t,B)$ on $\{R, \mathscr{R}\}$ is said to be "stochastically continuous" if

$$\lim_{t \to 0+} P(x,t,U(x,\varepsilon)) = 1 \tag{9.7.1}$$

for all $\varepsilon > 0$ and $x \in R$ fixed. If the limit (9.7.1) holds uniformly in x for each $\varepsilon > 0$, the transition probability is said to be "uniformly stochastically continuous."

The following proposition specifies conditions on the sample functions of the Markov process $\xi(t)$ under which (9.7.1) holds.

Proposition 9.7.1. *Assume that the sample functions of the homogeneous Markov process* $\{\xi(t); t \geq 0\}$ *are continuous from the right; then* (9.7.1) *holds.*

Proof. Let $\{t_n\} \subset R_+$ be a sequence decreasing to zero and set $B_n = \{\xi(t_n) \in U(x,\varepsilon)\}$. Because $\xi(t)$ is continuous from the right, we have

$$\{\xi(0) \in U(x,\varepsilon)\} \subseteq \bigcap_{n=1}^{\infty} \bigcup_{k=n}^{\infty} B_k = \liminf_{n\to\infty} B_n.$$

Therefore, for all $x \in R$ and $\varepsilon > 0$,

$$\liminf_{n\to\infty} P_x(B_n) \geq P_x\left\{\liminf_{n\to\infty} B_n\right\} \geq P_x\{\xi(0) \in U(x,\varepsilon)\} = 1.$$

This implies that

$$\liminf_{n\to\infty} P(x, t_n, U(x,\varepsilon)) \geq 1,$$

which proves the assertion. □

In the following, we will always assume, unless otherwise stated, that Markov processes have right-continuous sample functions. Denote as usual by **B** the Banach space of bounded measurable functions on R with supremum norm $\|f\| = \sup_x |f(x)|$, $f \in \mathbf{B}$. On **B**, we have defined a one-parameter family of bounded linear operators $\{T^t; t \geq 0\}$ induced by a transition probability $P(x,t,B)$ by

$$(T^t f)(x) = \int_{-\infty}^{\infty} f(y) P(x,t,dy) = E_x\{f(\xi(t))\}.$$

Clearly, for each $t \geq 0$,

$$T^t: \mathbf{B} \to \mathbf{B},$$

$T^{s+t} = T^t \cdot T^s$, and $\|T^t\| \leq 1$.

As we have seen in the previous sections, the theory of semigroups of bounded linear operators in a Banach space deals with the exponential functions in infinite-dimensional functional spaces. These problems were investigated independently by Hille and Yosida (1948). They introduced the concept of the infinitesimal generator A of a semigroup T^t and discussed the problem of generation of T^t in terms of A. Let us now prove the following result.

Proposition 9.7.2. *Let $f \in \mathbf{B}$ be continuous; then, for each $x \in R$,*

$$\lim_{t\to 0+} (T^t f)(x) = f(x).$$

Proof. Due to continuity of f, for each $x \in R$ and any $\delta > 0$, there exists $U(x,\varepsilon)$ such that when $y \in U(x,\varepsilon)$, $|f(y) - f(x)| < \delta$. Therefore,

$$(T^t f)(x) - f(x) = \int_{U(x,\varepsilon)} (f(y) - f(x)) P(x,t,dy) + \int_{U^c(x,\varepsilon)} (f(y) - f(x)) P(x,t,dy).$$

From this, it readily follows that

$$|(T^t f)(x) - f(x)| \leq \delta P(x, t, U(x, \varepsilon)) + 2\|f\| P(x, t, U^c(x, \varepsilon)).$$

This and the previous lemma prove the assertion. □

Corollary 9.7.1. *From the last inequality, we deduce that the semigroup is strongly continuous on the subset of continuous functions* $C \subset \mathbf{B}$ *if* $P(x, t, B)$ *is uniformly stochastically continuous.*

The following proposition is due to Hille.

Proposition 9.7.3. *Denote by* $\alpha(t) = \ln \|T^t\|$; *then,*

$$\lim_{t\to\infty} \frac{1}{t}\alpha(t) = \inf_{t>0} \frac{1}{t}\alpha(t).$$

Proof. From the semigroup property, we deduce that

$$\|T^{s+t}\| = \|T^s T^t\| \leq \|T^s\| \cdot \|T^t\|,$$

so that

$$\alpha(s + t) \leq \alpha(s) + \alpha(t).$$

Now, denote by

$$\beta = \inf_{t>0} \frac{1}{t}\alpha(t);$$

then β is either $-\infty$ or finite. Assume that β is finite. We choose, for any $\varepsilon > 0$, a number $x > 0$ such that $\alpha(x) \leq (\beta + \varepsilon)x$. Let $t > x$ and n be such that $nx \leq t \leq (n + 1)x$. Then

$$\beta \leq \frac{1}{t}\alpha(t) \leq \frac{1}{t}\alpha(nx) + \frac{1}{t}\alpha(t - nx)$$

$$\leq \frac{nx}{t}\frac{1}{x}\alpha(x) + \frac{1}{t}\alpha(t - nx)$$

$$\leq \frac{nx}{t}(\beta + \varepsilon) + \frac{1}{t}\alpha(t - nx).$$

Thus, letting $t \to \infty$ in the above inequality, we obtain

$$\lim_{t\to\infty} \frac{1}{t}\alpha(t) = \beta.$$

The case $\beta = -\infty$ is treated similarly. □

Remark 9.7.1. Lemma 8.3.1 is a version of this proposition.

The concept of the resolvent R_λ associated with a semigroup $\{T^t; t \geq 0\}$ was discussed in Section 9.3 of this chapter. According to Definition 9.3.1, this is a family of bounded linear operators $\{R_\lambda; \lambda > 0\}$ on the Banach space of bounded Borel-measurable functions on R defined by

$$(R_\lambda f)(x) = \int_0^\infty e^{-\lambda t}(T^t f)(x)\,dt, \quad \lambda > 0, \quad f \in \mathbf{B}.$$

From this and Fubini's theorem, it follows that

$$(R_\lambda f)(x) = E_x\left\{\int_0^\infty e^{-\lambda t} f(\xi(t))\,dt\right\}. \tag{9.7.2}$$

It is easy to see that $\|R_\lambda\| = \lambda^{-1}$, $(R_\lambda 1)(x) \equiv \lambda^{-1}$.

Proposition 9.7.4. *For all* $\lambda_1, \lambda_2 > 0$,

$$R_{\lambda_1} - R_{\lambda_2} + (\lambda_1 - \lambda_2)R_{\lambda_2}R_{\lambda_1} = 0. \tag{9.7.3}$$

In addition, if $f \in \mathbf{B}$ *is continuous at* x_0, *then*

$$\lim_{\lambda \to \infty} \lambda(R_\lambda f)(x_0) = f(x_0).$$

PROOF. Consider

$$\begin{aligned} T^s(R_{\lambda_1} f)(x) &= \int_{-\infty}^\infty P(x, s, dy) \int_0^\infty e^{-\lambda_1 t}(T^t f)(y)\,dt \\ &= \int_0^\infty e^{-\lambda_1 t}\,dt \int_{-\infty}^\infty \int_{-\infty}^\infty f(u) P(x, s, dy) P(y, t, du) \\ &= \int_0^\infty e^{-\lambda_1 t}\,dt \int_{-\infty}^\infty f(u) P(x, s + t, du) \\ &= \int_0^\infty e^{-\lambda_1 t}(T^{s+t} f)(x)\,dt \\ &= e^{-\lambda_1 s} \int_s^\infty e^{-\lambda_1 \tau}(T^\tau f)(x)\,d\tau. \end{aligned}$$

From this, we now have

$$\begin{aligned} \int_0^\infty e^{-\lambda_2 s}(T^s R_{\lambda_1} f)(x)\,ds &= (R_{\lambda_2} R_{\lambda_1} f)(x) \\ &= \int_0^\infty e^{-\lambda_2 s} e^{-\lambda_1 s}\,ds \int_s^\infty e^{-\lambda_1 \tau}(T^\tau f)(x)\,d\tau \\ &= \int_0^\infty e^{-\lambda_1 \tau}(T^\tau f)(x)\,d\tau \int_0^\tau e^{(\lambda_1 - \lambda_2)s}\,ds \\ &= (\lambda_1 - \lambda_2)^{-1}\{(R_{\lambda_1} f)(x) - (R_{\lambda_2} f)(x)\}, \end{aligned}$$

which proves (9.7.3). Next, from

$$\lambda(R_\lambda f)(x_0) = \int_0^\infty \lambda e^{-\lambda t}(T^t f)(x_0)\,dt$$
$$= \int_0^\infty e^{-t}(T^{t\lambda^{-1}} f)(x_0)\,dt$$

and Proposition 9.7.2, the second part of the proposition follows. □

Problems and Complements

9.1. Let $\{T^t; t \geq 0\}$ be a semigroup on a Banach space $\mathbf{B}$ and $\mathbf{B}_0 \subset \mathbf{B}$ the subset on which T^t is strongly continuous. Show that $\mathbf{B}_0$ is a closed linear manifold.

9.2. Let A be a bounded linear operator on a Banach space $\mathbf{B}$. Show that
(i) $\|e^A\| \leq \exp\{\|A\|\}$;
(ii) $e^{tI} = e^t I$;
(iii) $e^{A+B} = e^A \cdot e^B$ if $A \cdot B = B \cdot A$.

9.3. If the bounded operators A and B are such that $\|e^{tA}\| \leq 1$ and $\|e^{tB}\| \leq 1$ for all $t \geq 0$, show that $\|e^{tB}f - e^{tA}f\| \leq t\|Bf - Af\|$.

9.4. If A is a bounded operator on $\mathbf{B}$, show that $T^t = e^{tA}$ is a semigroup.

9.5. Let $C_0[-\infty, \infty]$ be the space of bounded continuous functions on $[-\infty, \infty]$. Define on C_0 the operator T^t, $t \geq 0$, by

$$(T^t f)(x) = e^{-\lambda t} \sum_{k=0}^{\infty} (\lambda t)^k f(x - k\mu)/k!$$

where $\lambda > 0$ and $\mu > 0$. Show that: (i) $T^t\colon C_0 \to C_0$; (ii) $\{T^t; t \geq 0\}$ is a semigroup. Is the semigroup strongly continuous?

9.6. (Continuation) Define T^t on C_0 by

$$(T^t f)(x) = \int_{-\infty}^{\infty} K(x - u, t) f(u)\,du,$$

where

$$K(x, t) = (2\pi t)^{-1/2} e^{-x^2/2t}, \quad -\infty < x < \infty, \quad t > 0.$$

Show that $\{T^t; t \geq 0\}$ is a strongly continuous semigroup.

9.7. Let $\{T^t; t \geq 0\}$ be a semigroup on a Banach space $\mathbf{B}$ and $\mathbf{B}_0 \subset \mathbf{B}$ the subset on which it is strongly continuous. Show that $T^t\mathbf{B}_0 = \mathbf{B}_0$ for all $t \geq 0$.

9.8. (Continuation) Show that there exists a constant $M > 0$ and $\lambda > 0$ such that, for all $t \geq 0$,

$$\|T^t\| \leq M e^{\lambda t}.$$

9.9. (Continuation) Show that for each $h \in \mathbf{B}_0$, $T^t h$ is a continuous mapping from $[0, \infty)$ into $\mathbf{B}_0$ ($\{T^t f; t \geq 0\}$ represents a curve in $\mathbf{B}_0$).

9.10. Let $\{\xi(t); t \geq 0\}$ be a homogeneous Markov process with state space $\{R, \mathscr{R}\}$ and transition probability $P(x, t, B)$ satisfying condition (8.1.13). Let **B** be the set of bounded Borel functions $f\colon R \to R$. For each $t \geq 0$, define $T^t\mathbf{B} \to \mathbf{B}$ by

$$(T^t f)(x) = \int_{-\infty}^{\infty} f(y)P(x, t, dy) = E_x\{f(\xi(t))\}.$$

Show that $\{T^t; t \geq 0\}$ is a strongly continuous contraction semigroup on **B**.

9.11. (Continuation) Show that, for each $t \geq 0$, T^t is a continuous linear operator on **B**.

9.12. (Continuation) If $h \in \mathbf{B}$ is continuous at $x_0 \in R$, show that

$$\lim_{t \to 0+} (T^t h)(x_0) = h(x_0).$$

9.13. Show that the generator A of a semigroup is a linear operator on $\mathbf{D}_A$.

9.14. Show that for every $f \in \mathbf{D}_A$, $Af \in \mathbf{B}_0$.

9.15. Show that the generator of the semigroup in Problem 9.4 is A.

9.16. Show that the generator of the semigroup in Problem 9.5 is

$$(Af)(x) = \lambda\{f(x - \mu) - f(x)\}.$$

9.17. Determine the generator of the semigroup in Problem 9.6.

9.18. Show that $R_\lambda \mathbf{B} \subseteq \mathbf{B}$.

9.19. Let $\{T^t; t \geq 0\}$ be a contraction semigroup on a Banach space **B** with generator A. Show that, for each $\lambda > 0$,

$$(\lambda I - A)\colon \mathbf{D}_A \to \mathbf{B}$$

is one-to-one, and that the inverse mapping taking **B** into $\mathbf{D}_A$ is the resolvent R_λ (see Proposition 9.3.1).

CHAPTER 10

Discrete Parameter Martingales

10.1. Conditional Expectation

The concept of a martingale introduced in Section 1.5 of Chapter 1 [see (1.5.19)] was defined in terms of the conditional expectation with respect to a σ-algebra. In this section, we will explore briefly some basic properties of this conditional expectation, which are needed in this chapter. We begin with some definitions.

Let $\{\Omega, \mathscr{B}, P\}$ be a probability space and denote by $L_2 = L_2\{\Omega, \mathscr{B}, P\}$ the Hilbert space of random variables (complex or real) defined on $\{\Omega, \mathscr{B}, P\}$, having finite variances. On L_2, the inner product is defined by

$$(Z_1, Z_2) = EZ_1\bar{Z}_2$$

(see Definition 5.1.3). The norm of an element $Z \in L_2$ is then

$$\|Z\| = (Z, Z)^{1/2} = (E|Z|^2)^{1/2}.$$

Let $\mathscr{S} \subset \mathscr{B}$ be a sub-σ-algebra; the subset of all those random variables from L_2, measurable with respect to $\mathscr{S}$, is a subspace $L_2^* = L_2\{\Omega, \mathscr{S}, P\}$ of L_2. Denote by P^* the projection operator defined on $L_2\{\Omega, \mathscr{B}, P\}$ which projects elements of this space perpendicularly onto L_2^*. According to Proposition 5.4.1., for every $Z \in L_2$, $P^*Z \in L_2^*$ so that P^*Z is $\mathscr{S}$-measurable. In the following, we will write

$$P^*Z = E\{Z|\mathscr{S}\} \tag{10.1.1}$$

and call it the "conditional expectation of Z with respect to (or given) the σ-algebra $\mathscr{S}$."

It is clear that, due to Proposition 5.3.1, the $\mathscr{S}$-measurable random variable $E\{Z|\mathscr{S}\}$ is uniquely defined up to a P-null set. If $Y \in L_2^*$, then, because $P^*Y = Y$, it follows that

$$E\{Y|\mathscr{S}\} = Y \quad \text{(a.s.)}. \tag{10.1.2}$$

Next, for any $A \in \mathscr{S}$, $I_A \in L_2^*$; consequently [see (5.2.8) and (5.2.9)], we clearly have

$$\int_A E\{Z|\mathscr{S}\}\,dP = \int P^*Z \cdot I_A\,dP = (P^*Z, I_A) = (Z, P^*I_A)$$

$$= (Z, I_A) = \int_A Z\,dP. \tag{10.1.3}$$

From this, it follows that

$$E(E\{Z|\mathscr{S}\}) = E(Z). \tag{10.1.4}$$

If $Z = I_B$, where $B \in \mathscr{B}$, the conditional expectation $E\{Z|\mathscr{S}\}$ is called the "conditional probability of B given $\mathscr{S}$" and written $P(B|\mathscr{S})$, i.e.,

$$P(B|\mathscr{S}) = E\{I_B|\mathscr{S}\}. \tag{10.1.5}$$

From this and (10.1.2), we obtain

$$P(A|\mathscr{S}) = I_A \quad \text{(a.s.)} \tag{10.1.6}$$

when $A \in \mathscr{S}$. In addition, due to (10.1.3),

$$\int_A P(B|\mathscr{S})\,dP = P(A \cap B). \tag{10.1.7}$$

We now list some basic properties of the conditional expectation which are not difficult to prove. First,

$$E\{\alpha Z_1 + \beta Z_2|\mathscr{S}\} = \alpha E\{Z_1|\mathscr{S}\} + \beta E\{Z_2|\mathscr{S}\} \quad \text{(a.s.)}, \tag{10.1.8}$$

where α and β are constants.

To show (10.1.8), we invoke (10.1.1), (5.8.4), and (5.8.5) to obtain

$$E\{\alpha Z_1 + \beta Z_2|\mathscr{S}\} = P^*(\alpha Z_1 + \beta Z_2)$$
$$= \alpha P^*Z_1 + \beta P^*Z_2 = \alpha E\{Z_1|\mathscr{S}\} + \beta E\{Z_2|\mathscr{S}\},$$

which proves (10.1.8).

Let $\mathscr{S}_1 \subset \mathscr{S}_2 \subset \mathscr{B}$ be two sub-σ-algebras, then

$$\begin{aligned} &\text{(i)} \quad E(E\{Z|\mathscr{S}_2\}|\mathscr{S}_1) = E\{Z|\mathscr{S}_1\} \quad \text{(a.s.)},\\ &\text{(ii)} \quad E(E\{Z|\mathscr{S}_1\}|\mathscr{S}_2) = E\{Z|\mathscr{S}_1\} \quad \text{(a.s.)}. \end{aligned} \tag{10.1.9}$$

To show the first equation, note that $L_2\{\Omega, \mathscr{S}_1, P\} \subset L_2\{\Omega, \mathscr{S}_2, P\}$. If we denote by P_1^*Z the orthogonal projection of $Z \in L_2$ on the first subspace and by P_2^*Z on the second one, we have

$$E(E\{Z|\mathscr{S}_2\}|\mathscr{S}_1) = E(P_2^*Z|\mathscr{S}_1) = P_1^*(P_2^*Z) = P_1^*Z$$

which proves (10.9.i). The second relation is a direct consequence of (10.1.2).

If the random variable Z is independent of $\mathscr{S}$, then

$$E\{Z|\mathscr{S}\} = E(Z) \quad \text{(a.s.)}. \tag{10.1.10}$$

As a matter of fact, for any $A \in \mathscr{S}$,

$$\int_A E\{Z|\mathscr{S}\}\,dP = \int_A Z\,dP = E(Z \cdot I_A)$$

$$= P(A)E(Z) = \int_A E\{Z\}\,dP.$$

Now consider $X, Z \in L_2$ such that $X \cdot Z \in L_2$ and assume that X is $\mathscr{S}$-measurable; then

$$E\{XZ|\mathscr{S}\} = XE\{Z|\mathscr{S}\} \quad \text{(a.s.)}. \tag{10.1.11}$$

The proof of this begins by considering the case $X = I_A$. For any $C \in \mathscr{S}$, we have

$$\int_C E\{XZ|\mathscr{S}\}\,dP = \int_C ZI_A\,dP = \int_{A\cap C} Z\,dP$$

$$= \int_{A\cap C} E\{Z|\mathscr{S}\}\,dP = \int_C XE\{Z|\mathscr{S}\}\,dP.$$

Extension of this result to simple functions, to non-negative functions, and then to a general case follows in the usual way.

Denote by

$$\mathscr{F}_n = \sigma\{X_1, \ldots, X_n\}$$

the σ-algebra generated by the random variables $X_1, \ldots, X_n$; then, we write

$$E\{Z|\mathscr{F}_n\} = E\{Z|X_1, \ldots, X_n\}. \tag{10.1.12}$$

Remark 10.1.1. For the conditional expectation of a random variable Y, given a σ-algebra $\mathscr{S}$, to exist, it suffices that $E|Y| < \infty$ (the second moment is not really necessary). In such a case, $E\{Y|\mathscr{S}\}$ is defined as any $\mathscr{S}$-measurable random variable satisfying the condition

$$\int_A E\{Y|\mathscr{S}\}\,dP = \int_A Y\,dP$$

for all $A \in \mathscr{S}$. Of course, $E\{Y|\mathscr{S}\}$ so defined has all the properties (10.1.2)–(10.1.11). In the sequel, we will assume only that the random variables we are dealing with are from $L_1 = L_1\{\Omega, \mathscr{B}, P\}$.

10.2. Discrete Parameter Martingales

Let $\{\xi_n\}_0^\infty$ be a sequence of real random variables defined on a probability space $\{\Omega, \mathscr{B}, P\}$. Denote by

$$\mathscr{F}_n = \sigma\{\xi_0, \ldots, \xi_n\}, \quad n = 0, 1, \ldots,$$

the sub-σ-algebra of $\mathscr{B}$ generated by $\xi_0, \ldots, \xi_n$. Clearly, $\{\mathscr{F}_n\}_0^\infty$ is an increasing sequence, i.e., $\mathscr{F}_0 \subset \mathscr{F}_1 \subset \cdots$, often called the "internal history" of $\{\xi_n\}_0^\infty$. The system

$$\{(\xi_n, \mathscr{F}_n)\}_0^\infty \tag{10.2.1}$$

is called a "stochastic sequence."

Definition 10.2.1. Assume that

$$\{\xi_n\}_0^\infty \subset L_1\{\Omega, \mathscr{B}, P\}; \tag{10.2.2}$$

then the stochastic sequence $\{(\xi_n, \mathscr{F}_n)\}_0^\infty$ represents a discrete parameter

$$\begin{aligned} &\text{(i) martingale if } E\{\xi_{n+1} | \mathscr{F}_n\} = \xi_n \quad \text{(a.s.)}, \\ &\text{(ii) submartingale if } E\{\xi_{n+1} | \mathscr{F}_n\} \geq \xi_n \quad \text{(a.s.)}, \\ &\text{(iii) supermartingale if } E\{\xi_{n+1} | \mathscr{F}_n\} \leq \xi_n \quad \text{(a.s.)} \end{aligned} \tag{10.2.3}$$

for all $n = 0, 1, \ldots$.

From (10.1.4) and (10.2.3.i), it follows that

$$E\{\xi_{n+1}\} = E\{\xi_n\}$$

for all $n = 0, 1, \ldots$. In other words, if $\{(\xi_n, \mathscr{F}_n)\}_0^\infty$ is a martingale $E\{\xi_n\} = E\{\xi_0\}$ for all $n \geq 1$. On the other hand, if the stochastic sequence is a submartingale, we have

$$E\{\xi_{n+1}\} \geq E\{\xi_n\}, \quad n = 0, 1, \ldots.$$

Finally, we have, for all $n = 0, 1, \ldots$,

$$E\{\xi_{n+1}\} \leq E\{\xi_n\}$$

if the stochastic sequence is supermartingale.

From the properties of conditional expectations, one easily deduces that (10.2.3) is equivalent to the following: If, for each $A \in \mathscr{F}_n$, $n = 0, 1, \ldots$,

$$\int_A \xi_{n+1}\, dP = \int_A \xi_n\, dP, \quad \{(\xi_n, \mathscr{F}_n)\}_0^\infty \text{ is a martingale;}$$

$$\int_A \xi_{n+1}\, dP \geq \int_A \xi_n\, dP, \quad \{(\xi_n, \mathscr{F}_n)\}_0^\infty \text{ is a submartingale;}$$

$$\int_A \xi_{n+1}\, dP \leq \int_A \xi_n\, dP; \quad \{(\xi_n, \mathscr{F}_n)\}_0^\infty \text{ is a supermartingale.}$$

It is said that a martingale $\{(\xi_n, \mathscr{F}_n)\}_0^\infty$ is closed from the right if there exists a random variable $\xi_\infty \in L_1\{\Omega, \mathscr{B}, P\}$ such that

$$\xi_n = E\{\xi_\infty | \mathscr{F}_n\} \quad \text{(a.s.)}$$

for all $n = 0, 1, \ldots$. On the other hand, if $\{(\xi_n, \mathscr{F}_n)\}_0^\infty$ is a submartingale and

$$\xi_n \le E\{\xi_\infty | \mathscr{F}_n\} \quad \text{(a.s.)}$$

for all $n = 0, 1, \ldots$, we say that the submartingale is closed from the right.

In the modern literature a somewhat more general definition of a martingale is formulated. Let $\{\mathscr{S}_n\}_0^\infty$ be an increasing sequence of sub-σ-algebras of $\mathscr{B}$, i.e.,

$$\mathscr{S}_0 \subset \mathscr{S}_1 \subset \cdots.$$

Such a family is usually called a "history" or "filtration." A sequence $\{\xi_n\}_0^\infty$ of random variables on $\{\Omega, \mathscr{B}, P\}$ is said to be "adapted" to the history $\{\mathscr{S}_n\}_0^\infty$ if ξ_n is $\mathscr{S}_n$-measurable for each $n = 0, 1, \ldots$.

A sequence $\{\xi_n\}_0^\infty \subset L_1\{\Omega, \mathscr{B}, P\}$ of real random variables adapted to a history $\{\mathscr{S}_n\}_0^\infty$ is called a "martingale" with respect to $\{\mathscr{S}_n\}_0^\infty$ if

$$E\{\xi_{n+1} | \mathscr{S}_n\} = \xi_n \quad \text{(a.s.)} \tag{10.2.4}$$

for every $n = 0, 1, \ldots$. In a similar fashion, one may define the concept of a submartingale and supermartingale with respect to $\{\mathscr{S}_n\}_0^\infty$.

Remark 10.2.1. It is not difficult to see that (10.2.4) leads to the following more general relation. For every $1 \le k \le n$,

$$E\{\xi_n | \mathscr{S}_k\} = \xi_k \quad \text{(a.s.)}. \tag{10.2.5}$$

This follows from (10.1.9.i) by induction because

$$\begin{aligned} E\{\xi_n | \mathscr{S}_k\} &= E(E\{\xi_n | \mathscr{S}_{n-1}\} | \mathscr{S}_k) \\ &= E\{\xi_{n-1} | \mathscr{S}_k\}, \end{aligned}$$

and so on.

Remark 10.2.2. In the following, unless otherwise stated, when we say that a sequence of random variables is a martingale (submartingale or supermartingale), it means it is so with respect to its internal history.

Proposition 10.2.1. *Let $\{\xi_n\}_0^\infty$ be a martingale and $h: R \to R$ a convex function, then $\{h(\xi_n)\}_0^\infty$ is a submartingale if $\{h(\xi_n)\}_0^\infty \subset L_1\{\Omega, \mathscr{B}, P\}$.*

PROOF.

$$E\{h(\xi_{n+1}) | h(\xi_0), \ldots, h(\xi_n)\} = E(E\{h(\xi_{n+1}) | \xi_0, \ldots, \xi_n\} | h(\xi_0), \ldots, h(\xi_n)) \quad \text{(a.s.)}.$$

From Jensen's inequality, we have

$$\begin{aligned} E\{h(\xi_{n+1}) | h(\xi_0), \ldots, h(\xi_n)\} &\ge h(E\{\xi_{n+1} | \xi_0, \ldots, \xi_n\}) \\ &= h(\xi_n) \quad \text{(a.s.)}. \end{aligned}$$

Therefore,

$$E\{h(\xi_{n+1}) | h(\xi_0), \ldots, h(\xi_n)\} \ge h(\xi_n) \quad \text{(a.s.)}. \qquad \square$$

Corollary 10.2.1. *If* $\{\xi_n\}_0^\infty$ *is a martingale,* $\{|\xi_n|\}_0^\infty$ *is a submartingale because* $|x|$ *is a convex function. The function* $x^+ = \max\{0, x\}$ *is a nondecreasing convex function. Indeed, because, for any two real numbers* x_1 *and* x_2, $(x_1 + x_2)^+ \leq x_1^+ + x_2^+$, *it follows that*

$$(px_1 + qx_2)^+ \leq px_1^+ + qx_2^+, \quad p + q = 1, \quad 0 \leq p \leq 1,$$

which proves the assertion. Therefore, $\{\xi_n^+\}_0^\infty$ *is a submartingale.* □

10.3. Examples

In this section a series of motivating examples will be presented.

EXAMPLE 10.3.1. Let $X \in L_1\{\Omega, \mathscr{B}, P\}$ be an arbitrary random variable and $\{\mathscr{F}_n\}_0^\infty$ a filtration, such that $\mathscr{F}_n \subset \mathscr{B}$ for all $n = 0, 1, \ldots$. Set

$$\xi_n = E\{X|\mathscr{F}_n\}.$$

Clearly, the sequence of random variable $\{\xi_n\}_0^\infty$ is adapted to $\{\mathscr{F}_n\}_0^\infty$. Next, due to (10.1.9.i), we have

$$\begin{aligned} E\{\xi_{n+1}|\mathscr{F}_n\} &= E\{E\{X|\mathscr{F}_{n+1}\}|\mathscr{F}_n\} \\ &= E\{X|\mathscr{F}_n\} = \xi_n, \end{aligned}$$

which implies that the sequence $\{\xi_n\}_0^\infty$ is a martingale. Many (but not all) martingales can be obtained in this way.

The next example gives a gambling interpretation of a martingale.

EXAMPLE 10.3.2. A gambler plays a sequence of independent games. Assume that X_0 is his initial fortune (before play commences), $X_0 + X_1$ is the gambler's fortune after the first play, $X_0 + X_1 + X_2$ at the end of second, and so on. In this way we obtain a sequence of partial sums $\{\xi_n\}_0^\infty$

$$\xi_n = X_0 + X_1 + \cdots + X_n$$

of independent random variables. Note that

$$\sigma\{\xi_0, \ldots, \xi_n\} = \sigma\{X_0, \ldots, X_n\}, \quad n = 0, 1, \ldots. \tag{10.3.1}$$

In fact, because $X_0 = \xi_0$ and $X_k = \xi_k - \xi_{k-1}$, $k = 1, \ldots, n$, each $X_0, X_1, \ldots, X_n$ is $\sigma\{\xi_0, \ldots, \xi_n\}$-measurable so that

$$\sigma\{X_0, \ldots, X_n\} \subset \sigma\{\xi_0, \ldots, \xi_n\}.$$

On the other hand, each ξ_k is clearly $\sigma\{X_0, \ldots, X_n\}$-measurable for every $k = 0, \ldots, n$, so that

$$\sigma\{\xi_0, \ldots, \xi_n\} \subset \sigma\{X_0, \ldots, X_n\},$$

which proves (10.3.1).

Next, invoking (10.1.10) and (10.3.1), we have

$$E\{\xi_{n+1}|\xi_0,\ldots,\xi_n\} = E\{\xi_n + X_{n+1}|X_0,\ldots,X_n\}$$
$$= \xi_n + E\{X_{n+1}\}.$$

From this, we deduce the following. The sequence of random variables $\{\xi_n\}_0^\infty$ is a martingale if $E\{X_n\} = 0$ for all $n = 0, 1, \ldots$. If $E\{X_n\} > 0$ for every $n = 0, 1, \ldots, \{\xi_n\}_0^\infty$ is a submartingale and a supermartingale if $E\{X_n\} < 0$, $n = 0, 1, \ldots$.

EXAMPLE 10.3.3. Let $\{Z_n\}_0^\infty$ be an i.i.d. sequence of (nondegenerate) random variables with common support on $(0, 1]$ and $E\{Z_1^i\} = \alpha_i > 0$, $i = 1, 2$. Set

$$L_n = \left(\prod_{j=1}^n Z_j\right)\Big/\alpha_1^n.$$

It is quite clear that the sequence of random variables $\{L_n\}_1^\infty$ represents a martingale such that

$$E\{L_n\} = 1 \quad \text{and} \quad E\{L_n^2\} = (\alpha_2/\alpha_1^2)^n \tag{10.3.2}$$

for all $n = 1, 2, \ldots$ (for an application see Example 1.6.2). From this we clearly have that $E\{L_n^2\} \to \infty$ as $n \to \infty$. In spite of this, however, $L_n \to 0$ (a.s.) as $n \to \infty$.

To show this, write L_n as

$$L_n = \exp\left\{\sum_{j=1}^n \ln(Z_j/\alpha_1)\right\}.$$

From Jensen's inequality, we have

$$E\{\ln(Z_j/\alpha_1)\} < \ln E\{Z_j/\alpha_1\} = 0.$$

Therefore, because $E\{\ln(Z_j/\alpha_1)\} < 0$, by the strong law of large numbers

$$\sum_{j=1}^n \ln(Z_j/\alpha_1) \to -\infty \quad \text{(a.s.)}$$

and, thus, $L_n \to 0$ (a.s.) as $n \to \infty$. This is an interesting example of a sequence of random variables converging to zero (a.s.) while its variance increases to infinity.

EXAMPLE 10.3.4. Let $\{X_n\}_1^\infty$ be a sequence of random variables such that the probability density of $(X_1,\ldots,X_n)$ is either $f_n^0(\cdot,\ldots,\cdot)$, or $f_n^1(\cdot,\ldots,\cdot)$ for each $n = 1, 2, \ldots$. Assume that $f_n^1 > 0$ on R^n for all $n = 1, 2, \ldots$; then the random variable $\rho_n(\omega)$, defined by

$$\rho_n = \frac{f_n^0(X_1,\ldots,X_n)}{f_n^1(X_1,\ldots,X_n)},$$

is a likelihood ratio. If f_n^1 is the true probability density of $(X_1,\ldots,X_n)$, then

$$E\{\rho_{n+1}|X_1 = x_1,\dots,X_n = x_n\}$$
$$= \int \rho_{n+1}(x_1,\dots,x_n,y)P\{X_{n+1} \in dy|X_1 = x_1,\dots,X_n = x_n\}$$
$$= \int \frac{f_{n+1}^0(x_1,\dots,x_n,y)}{f_n^1(x_1,\dots,x_n)}dy = \frac{f_n^0(x_1,\dots,x_n)}{f_n^1(x_1,\dots,x_n)}.$$

In other words,

$$E\{\rho_{n+1}|X_1,\dots,X_n\} = \rho_n \quad \text{(a.s.)},$$

which shows that $\{\rho_n\}_1^\infty$ is a martingale with respect to the filtration $\{\mathcal{F}_n\}_1^\infty$, where $\mathcal{F}_n = \sigma\{X_1,\dots,X_n\}$.

On the other hand, from (10.1.3), we see that

$$\sigma\{\rho_1,\dots,\rho_n\} \subset \sigma\{X_1,\dots,X_n\}.$$

From this and (10.1.9.i), we deduce that

$$E\{\rho_{n+1}|\rho_1,\dots,\rho_n\} = E(E\{\rho_{n+1}|X_1,\dots,X_n\}|\rho_1,\dots,\rho_n)$$
$$= E(\rho_n|\rho_1,\dots,\rho_n) = \rho_n \quad \text{(a.s.)},$$

so that $\{\rho_n\}_1^\infty$ is a martingale.

EXAMPLE 10.3.5. Consider an urn which contains $b \geq 1$ black balls and $w \geq 1$ white balls which are well mixed. Repeated drawings are made from the urn, in such a way that after each drawing the selected ball is returned to the urn along with $c \geq 1$ balls of the same color. Set

$$X_0 = b/(b + w)$$

and denote by X_n the proportion of black balls in the urn after the nth draw. Here we want to show that the sequence of random variables $\{X_n\}_0^\infty$ represents a martingale.

To this end, set $Y_0 = 1$ and, for $n \geq 1$, define Y_n as follows:

$$Y_n = \begin{cases} 1 & \text{if the } n\text{th ball drawn is black} \\ 0 & \text{if the } n\text{th ball drawn is white.} \end{cases}$$

Let b_n and w_n be the number of black and white balls, respectively, in the urn after the nth draw ($w_0 = w, b_0 = b$). Then, clearly,

$$X_n = b_n/(b_n + w_n), \quad n = 1,\dots,$$

and

$$b_{n+1} = b_n + cY_{n+1}, \qquad w_{n+1} = w_n + c(1 - Y_{n+1}).$$

Next, we, clearly, have that

$$P\{Y_{n+1} = 1|Y_0,\dots,Y_n\} = b_n/(b_n + w_n) = X_n,$$

so that (a.s.)

$$E\{X_{n+1}|Y_0,\ldots,Y_n\} = E\left\{\frac{b_{n+1}}{b_{n+1}+w_{n+1}}\middle| Y_0,\ldots,Y_n\right\}$$
$$= E\left\{\frac{b_n + cY_{n+1}}{b_n + w_n + c}\middle| Y_0,\ldots,Y_n\right\}$$
$$= \frac{b_n}{b_n + w_n + c} + \frac{cP\{Y_{n+1} = 1|Y_0,\ldots,Y_n\}}{b_n + w_n + c}$$
$$= \frac{b_n}{b_n + w_n + c} + \frac{c}{b_n + w_n + c}X_n$$
$$= \frac{1}{b_n + w_n + c}\left\{b_n + \frac{cb_n}{b_n + w_n}\right\} = X_n.$$

Because

$$\sigma\{X_0,\ldots,X_n\} \subset \sigma\{Y_0,\ldots,Y_n\},$$

we see that

$$E\{X_{n+1}|X_0,\ldots,X_n\} = E(E\{X_{n+1}|Y_0,\ldots,Y_n\}|X_0,\ldots,X_n)$$
$$= E\{X_n|X_0,\ldots,X_n\} = X_n \quad \text{(a.s.)},$$

which proves the assertion.

10.4. The Upcrossing Inequality

The proof of the fundamental convergence theorem for submartingales, which will be given in the next section, depends on a result which is known as the "upcrossing inequality." Its proof, somewhat combinatorial in nature, will be presented here.

We begin by briefly discussing the convergence of numerical sequences. Let $\{\alpha_n\}_1^\infty$ be a sequence of real numbers. From the definition of convergence, it is clear that $\{\alpha_n\}_1^\infty$ has a limit as $n \to \infty$ if and only if, for any two rational numbers $-\infty < r_0 < r_1 < \infty$, the sequence passes from below r_0 to above r_1 at most a finite number of times. In other words, if the sequence converges, the number of "upcrossings" of any such interval $[r_0, r_1]$ is finite. If, on the other hand, such an interval can be found where the number of upcrossings by $\{\alpha_n\}_1^\infty$ is infinite, the sequence diverges. As a matter of fact, if $\beta(r_0, r_1)$ represents the number of upcrossings of $[r_0, r]$ by $\{\alpha_n\}_1^\infty$

$$\underline{\lim}\, \alpha_n \leq r_0 < r_1 \leq \overline{\lim}\, \alpha_n \Leftrightarrow \beta(r_0, r_1) = \infty.$$

From this, we deduce that the sequence $\{\alpha_n\}_1^\infty$ converges if and only if $\beta(r_0, r_1) < \infty$ for any two rational numbers $-\infty < r_0 < r_1 < \infty$.

Similarly, a sequence of real random variables $\{\xi_n\}_1^\infty$ has a limit (finite or infinite) with probability 1 if and only if the number of upcrossings of any $[r_0, r_1]$ is finite with probability 1.

In this section we will establish an upper bound for the number of upcrossings of submartingales. Let $-\infty < a < b < +\infty$ be two numbers and $\{(\xi_n, \mathscr{F}_n)\}_1^\infty$ a stochastic sequence. Set

$$\begin{aligned} \tau_0 &= 0, \\ \tau_1 &= \inf\{n > 0; \xi_n \leq a\}, \\ \tau_2 &= \inf\{n > \tau_1; \xi_n \geq b\}, \\ &\vdots \\ \tau_{2k-1} &= \inf\{n > \tau_{2k-2}; \xi_n \leq a\}, \\ \tau_{2k} &= \inf\{n > \tau_{2k-1}; \xi_n \geq b\}, \end{aligned} \tag{10.4.1}$$

taking $\tau_\nu = 0$ if the corresponding set $\{\ \}$ is empty.

Next, for every $n \geq 1$, define

$$\beta_n(a, b) = \max\{i; \tau_{2i} \leq n\} \tag{10.4.2}$$

and

$$\beta_n(a, b) = 0 \quad \text{if } \tau_2 > n.$$

In other words, the random variable $\beta_n(a, b)$ represents the number of upcrossings of the interval $[a, b]$ by the sequence $\xi_1, \ldots, \xi_n$ (see Figure 10.1). Clearly, β: 0, 1, ..., $[n/2]$, where $[x]$ represents the integer part of x. The following proposition is due to Doob.

Proposition 10.4.1. *Let $\{(\xi_n, \mathscr{F}_n)\}_1^\infty$ be a submartingale. Then, for every $n \geq 1$,*

$$E\{\beta_n(a, b)\} \leq E(\xi_n - a)^+/(b - a). \tag{10.4.3}$$

PROOF. From Corollary 10.2.1, we know that the sequence

$$\{(\xi_n - a)^+, \mathscr{F}_n\}_1^\infty \tag{10.4.4}$$

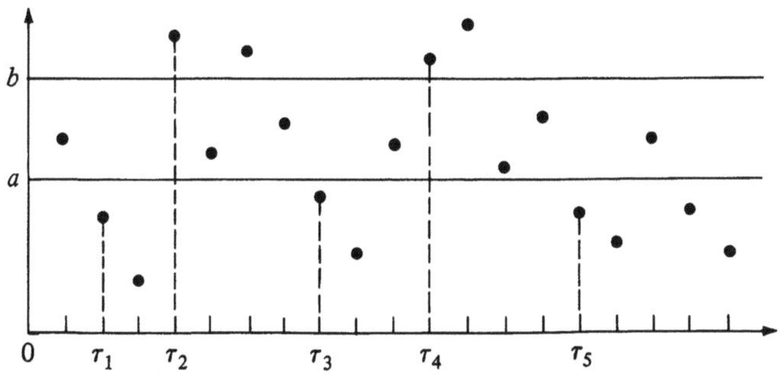

Figure 10.1

is a non-negative submartingale. In addition, the number of upcrossings of the interval $[0, b-a]$ by

$$(\xi_1 - a)^+, \ldots, (\xi_n - a)^+$$

is again $\beta_n(a, b)$. Therefore, it suffices to consider a non-negative submartingale $\{(X_n, \mathscr{F}_n)\}_1^\infty$, $X_n \geq 0$, with $a = 0$ and to show that

$$E\{\beta_n(0, b)\} \leq E\{X_n\}/b.$$

Set $X_0 = 0$, $\mathscr{F}_0 = \{\theta, \Omega\}$, and, for $i \geq 1$, write

$$\varepsilon_i = \begin{cases} 1 & \text{if } \tau_k < i \leq \tau_{k+1} \text{ for some odd } k \\ 0 & \text{if } \tau_k < i \leq \tau_{k+1} \text{ for some even } k. \end{cases}$$

In other words, [see (10.4.1)], $\varepsilon_i = 1$ if the largest τ_k left of i is a downcrossing and the τ_{k+1} is an upcrossing, and $\varepsilon_i = 0$ if the largest τ_k left of i is an upcrossing and τ_{k+1} is a downcrossing. Clearly, then, $\varepsilon_1 = 0$ because $1 \leq \tau_1$ (as a matter of fact, $\varepsilon_i = 0$ for all $i \leq \tau_i$). Now, we have

$$b\beta_n(0, b) \leq \sum_{i=1}^n (X_i - X_{i-1})\varepsilon_i$$

and

$$\begin{aligned} \{\varepsilon_i = 1\} &= \bigcup_{\text{odd } k} \{\tau_k < i \leq \tau_{k+1}\} \\ &= \bigcup_{\text{odd } k} (\{\tau_k \leq i\} - \{\tau_{k+1} \leq i\}) \in \mathscr{F}_{i-1}. \end{aligned}$$

Consequently,

$$\begin{aligned} bE\{\beta_n(0, b)\} &\leq \sum_{i=1}^n E\{(X_i - X_{i-1})\varepsilon_i\} \\ &= \sum_{i=1}^n \int_{\{\varepsilon_i = 1\}} E\{(X_i - X_{i-1})|\mathscr{F}_{i-1}\}\, dP \\ &= \sum_{n=1}^n \int_{\{\varepsilon_i = 1\}} (E\{X_i|\mathscr{F}_{i-1}\} - X_{i-1})\, dP \\ &\leq \sum_{i=1}^n \int (E\{X_i|\mathscr{F}_{i-1}\} - X_{i-1})\, dP \leq \int E\{X_n|\mathscr{F}_{n-1}\}\, dP \\ &= E\{X_n\}. \end{aligned}$$

This proves the assertion. □

10.5. Convergence of Submartingales

In this section we will prove a theorem of Doob concerning convergence of submartingales. It represents one of the fundamental results in the theory of martingales and can be considered an analog of the well-known fact in real analysis that every bounded monotone sequence has a finite limit.

Proposition 10.5.1 (Doob). *Let $\{(X_n, \mathscr{F}_n)\}_1^\infty$ be a submartingale such that*

$$\sup_n E|X_n| < \infty. \tag{10.5.1}$$

Then, there exists a random variable X_∞ such that

$$X_n \to X_\infty \quad \text{(a.s.)} \tag{10.5.2}$$

and $E|X_\infty| < \infty$.

PROOF. Assume that the submartingale does not converge; then,

$$P\{\overline{\lim}\, X_n > \underline{\lim}\, X_n\} > 0. \tag{10.5.3}$$

Because

$$\{\overline{\lim}\, X_n > \underline{\lim}\, X_n\} = \bigcup_{\alpha<\beta} \{\overline{\lim}\, X_n > \beta > \alpha > \underline{\lim}\, X_n\}$$

for all rationals $\alpha < \beta$, there exist two rational numbers $r_0 < r_1$ such that

$$P\{\overline{\lim}\, X_n > r_0 > r_1 > \underline{\lim}\, X_n\} > 0.$$

In other words, with positive probability, there is an infinite number of upcrossings of $[r_0, r_1]$.

Denote by $\beta_n(r_0, r_1)$ the number of upcrossings of $[r_0, r_1]$ by $X_1, \ldots, X_n$ and write

$$\beta(r_0, r_1) = \lim_{n\to\infty} \beta_n(r_0, r_1).$$

By (10.4.3),

$$\begin{aligned} E\{\beta_n(r_0, r_1)\} &\le E(X_n - r_0)^+/(r_1 - r_0) \\ &\le E(X_n^+ + |r_0|)/(r_1 - r_0), \end{aligned}$$

and, therefore,

$$E\{\beta(r_0, r_1)\} \le \left(\sup_n EX_n^+ + |r_0|\right)\Big/(r_1 - r_0).$$

From this (bearing in mind that $\sup_n E|X_n| < \infty \Leftrightarrow \sup_n EX_n^+ < \infty$), we deduce that

$$E\{\beta(r_0, r_1)\} < \infty,$$

which contradicts (10.5.3). Consequently, (10.5.2) holds and by Fatou's lemma

$$E|X_\infty| \le \sup_n E|X_n| < \infty,$$

so the assertion follows. □

Corollary 10.5.1. *Let $\{(X_n, \mathscr{F}_n)\}_1^\infty$ be a non-negative martingale; then, the* lim X_n *exists* (a.s.). *Indeed,*

$$\sup_n E|X_n| = \sup_n X_n = EX_1 < \infty$$

and Proposition 10.5.1 *is applicable.*

Remark 10.5.1. Proposition 10.5.1 also holds for martingales and supermartingales because every martingale is a submartingale and because $\{(-X_n, \mathscr{F}_n)\}_1^\infty$ is a submartingale if $\{(X_n, \mathscr{F}_n)\}_1^\infty$ is a supermartingale. For non-negative martingales or supermartingales, assumption (10.5.1) always holds.

Remark 10.6.2. Condition (10.5.1) does not guarantee the converges of X_n to X_∞ in the mean as the following counterexample shows. Let $\{\xi_n\}_1^\infty$ be an i.i.d. sequence of random variables with the common distribution

$$P\{\xi_i = 0\} = \tfrac{1}{2}, \qquad P\{\xi_i = 2\} = \tfrac{1}{2}.$$

Consider

$$X_n = \prod_{i=1}^{n} \xi_i, \quad n = 1, 2, \ldots;$$

then, $\{(X_n, \mathscr{F}_n)\}_1^\infty$ is clearly a martingale with $EX_n = 1$. Thus,

$$X_n \to X_\infty \quad \text{(a.s.)},$$

where $X_\infty = 0$ (a.s.). On the other hand,

$$E|X_n - X_\infty| = EX_n = 1 \quad \text{for all } n = 1, 2, \ldots.$$

However, if assumption (10.5.1) is strengthened to uniform integrability of the sequence $\{X_n\}_1^\infty$, then we have, simultaneously,

$$X_n \to X_\infty \text{ (a.s.)} \quad \text{and} \quad X_n \to X_\infty \text{ in the mean.}$$

Definition 10.5.1. A sequence of random variables $\{\xi_n\}_1^\infty$ is uniformly integrable if

$$\lim_{c \to +\infty} \left(\sup_n \int_{\{\xi_n > c\}} |\xi_n| \, dP \right) = 0. \tag{10.5.4}$$

Let us give some criteria for uniform integrability. The simplest one is the following: If there exist a random variable U such that $|\xi_n| \le U$ and $EU < \infty$, then $\{\xi_n\}_1^\infty$ is an uniformly integrable sequence. More useful is the following result.

Proposition 10.5.2. *A sequence of random variables $\{\xi_n\}_1^\infty$ is uniformly integrable if and only if:*

(i) $\sup_n E|\xi_n| < \infty$;
(ii) *for every $\varepsilon > 0$, there exists $\delta > 0$ such that for any $B \in \mathscr{B}$ with $P(B) < \delta$,* $\sup_n \int_B |\xi_n| \, dP < \varepsilon$.

PROOF. (Necessity). If the sequence is uniformly integrable, then, for any $\varepsilon > 0$, there exists $c > 0$ such that

$$\sup_n E|\xi_n| = \sup_n (E|\xi_n|I_{\{|\xi_n|\geq c\}} + E|\xi_n|I_{\{|\xi_n|<c\}})$$
$$\leq \sup_n E|\xi_n|I_{\{|\xi_n|\geq c\}} + \sup_n E|\xi_n|I_{\{|\xi_n|<c\}})$$
$$\leq \varepsilon + c,$$

which proves (i). On the other hand,

$$E|\xi_n|I_B = E|\xi_n|I_{B\cap\{|\xi_n|\geq c\}} + E|\xi_n|I_{B\cap\{|\xi_n|<c\}}$$
$$\leq E|\xi_n|I_{\{|\xi_n|\geq c\}} + cP(B).$$

Now, take c large enough so that $\sup_n E|\xi_n|I_{\{|\xi_n|\geq c\}} \leq \varepsilon/2$; then, if $P(B) < \varepsilon/2c$, we obtain

$$\sup_n E|\xi_n|I_B = \sup_n \int_B |\xi_n|\,dP < \varepsilon.$$

(Sufficiency). Take $\varepsilon > 0$ and $\delta > 0$ so that $P(B) < \delta$ implies $\sup E|\xi_n|I_B < \varepsilon$. Because

$$E|\xi_n| \geq E|\xi_n|I_{\{|\xi_n|\geq c\}} \geq cP_{\{|\xi_n|\geq c\}},$$

using Markov's inequality we have

$$\sup_n P\{|\xi_n| \geq c\} \leq \frac{1}{c}\sup_n E|\xi_n| \to 0 \quad \text{as } c \to \infty.$$

Consequently, for large c, any set $\{|\xi_n| \geq c\}$ can be taken as B. This implies that

$$\int_{\{|\xi_n|\geq c\}} |\xi_n|\,dP < \varepsilon,$$

which completes the proof of the proposition. □

Proposition 10.5.3. *Let* $\{\xi_n\}_1^\infty$ *be a uniformly integrable family; then*

$$E\{\underline{\lim}\,\xi_n\} \leq \underline{\lim}\,E\{\xi_n\} \leq \overline{\lim}\,E\{\xi_n\} \leq E\{\overline{\lim}\,\xi_n\}. \tag{10.5.5}$$

Proof. Assume $x > 0$ and write

$$E\xi_n = E\{\xi_n I_{\{\xi_n \leq -x\}}\} + E\{\xi_n I_{\{\xi_n > -x\}}\}. \tag{10.5.6}$$

Given $\varepsilon > 0$, due to uniform integrability there is x sufficiently large so that

$$\sup_n |E\{\xi_n I_{\{\xi_n < -x\}}| < \varepsilon. \tag{10.5.7}$$

On the other hand, by Fatou's lemma,

$$\underline{\lim}\,E\{\xi_n I_{\{\xi_n > -x\}}\} \geq E\{\underline{\lim}\,\xi_n I_{\{\xi_n > -x\}}\}.$$

But $\xi_n I_{\{\xi_n > -x\}}\} \geq \xi_n$, which yields

$$\underline{\lim}\,E\{\xi_n I_{\{\xi_n > -x\}}\} \geq E\{\underline{\lim}\,\xi_n\}. \tag{10.5.8}$$

Finally, (10.5.6), (10.5.7), and (10.5.8) give

$$\begin{aligned} \underline{\lim}\, E\{\xi_n\} &\geq \underline{\lim}\, E\{\xi_n I_{\{\xi_n \leq -x\}}\} - \varepsilon \\ &\geq E\{\underline{\lim}\, \xi_n\} - \varepsilon. \end{aligned}$$

Because $\varepsilon > 0$ is arbitrary, the first half of (10.2.7) follows. The second half can be proved similarly. □

Corollary 10.5.2. *If, in the last proposition, we have that $\xi_n \to \xi$ (a.s.), then ξ is integrable,*

$$E\xi_n \to E\xi \quad \text{and} \quad E|\xi_n - \xi| \to 0 \quad \text{as } n \to \infty.$$

Indeed, in such a case we have

$$E\{\xi\} \leq \underline{\lim}\, E\{\xi_n\} \leq \overline{\lim}\, E\{\xi_n\} \leq E\{\xi\}.$$

This shows that $E\{\xi\}$ is finite and that

$$\lim_{n\to\infty} E\{\xi_n\} = E\{\xi\}.$$

Finally, $|\xi_n - \xi| \to 0$ (a.s.) and, because $|\xi_n - \xi| \leq |\xi_n| + |\xi|$, it is a uniformly integrable family.

Another simple criterion for uniform integrability gives the following result.

Proposition 10.5.4. *Let $\{\xi_n\}_1^\infty \subset L_1\{\Omega, \mathscr{B}, P\}$ and $h(t) \geq 0$ be a nondecreasing function, defined for all $t \geq 0$, such that*

$$\begin{aligned} &\text{(i)} && \lim_{t\to\infty} h(t)/t = +\infty, \\ &\text{(ii)} && \sup_n E\{h(|\xi_n|)\} < \infty; \end{aligned} \tag{10.5.9}$$

then $\{\xi_n\}_1^\infty$ is uniformly integrable.

Proof. Let $\varepsilon > 0$, $Q = \sup_n E\{h(|\xi_n|)\}$, and $\alpha = Q/\varepsilon$. Take $x > 0$ so large that $h(t)/t \geq \alpha$ for $t \geq x$. Then

$$\int_{\{|\xi_n| \geq x\}} |\xi_n|\, dP \leq \frac{1}{\alpha} \int_{\{|\xi_n| \geq x\}} h(|\xi_n|)\, dP \leq Q/\alpha = \varepsilon$$

uniformly for $n \geq 1$.

10.6. Uniformly Integrable Martingales

In this section we discuss some basic features of uniformly integrable martingales.

Proposition 10.6.1. *Let $\{(X_n, \mathscr{F}_n)\}_1^\infty$ be a uniformly integrable submartingale. Then, there exists a random variable X_∞ with $E|X_\infty| < \infty$ such that*

$$\begin{aligned}&\text{(i) } X_n \to X_\infty \text{ (a.s.)},\\&\text{(ii) } E|X_n - X_\infty| \to 0 \text{ as } n \to \infty,\end{aligned} \tag{10.6.1}$$

and $\{X_1, \dots, X_\infty; \mathscr{F}_1, \dots, \mathscr{F}_\infty\}$, *where* $\mathscr{F}_\infty = \sigma\{\bigcup_1^\infty \mathscr{F}_k\}$ *is also a submartingale.*

PROOF. Because the submartingale is uniformly integrable, condition (10.5.1) holds, and, consequently, $X_n \to X_\infty$ (a.s.). From Corollary 10.5.2, we conclude that (10.6.1.ii) also holds. Finally, consider $A \in \mathscr{F}_n$ and let $N \geq n$, then

$$\int_A |X_N - X_\infty| \, dP \to 0 \quad \text{as } N \to \infty.$$

Hence,

$$\lim_{N\to\infty} \int_A X_N \, dP = \int_A X_\infty \, dP.$$

The sequence $\{EX_n I_A\}$ is nondecreasing because

$$E\{X_N | \mathscr{F}_{N-1}\} \geq X_{N-1} \Rightarrow \int_A X_N \, dP \geq \int_A X_{N-1} \, dP.$$

Therefore,

$$\int_A X_n \, dP \leq \int_A X_N \, dP \leq \int_A X_\infty \, dP,$$

which implies that

$$X_n \leq E\{X_\infty | \mathscr{F}_n\} \quad \text{(a.s.)}.$$

This completes the proof of the proposition. □

Corollary 10.6.1. *Let* $\{(X_n, \mathscr{F}_n)\}_1^\infty$ *be a submartingale such that for some* $\gamma > 1$,

$$\sup_n E|X_n|^\gamma < \infty. \tag{10.6.2}$$

Then there is an integrable random variable X_∞ *such that* (10.6.1) *holds.*

The proof follows from the fact that condition (10.6.2) guarantees the uniform integrability of $\{X_n\}_1^\infty$ (see Proposition 10.5.4).

The following result due to Lévy is concerned with the continuity property of a conditional expectation.

Proposition 10.6.2. *Let* $\{\Omega, \mathscr{B}, P\}$ *be a probability space and* $\{\mathscr{F}_n\}_1^\infty$ *a filtration, i.e.,* $\mathscr{F}_1 \subset \mathscr{F}_2 \subset \cdots \subset \mathscr{B}$. *Let* $\xi \in L_1\{\Omega, \mathscr{B}, P\}$ *and set* $\mathscr{F}_\infty = \sigma\{\bigcup_{k=1}^\infty \mathscr{F}_k\}$. *Then*

$$E\{\xi | \mathscr{F}_n\} \to E\{\xi | \mathscr{F}_\infty\} \tag{10.6.3}$$

(a.s.) and in the mean.

PROOF. Set

$$X_n = E\{\xi | \mathscr{F}_n\};$$

then, clearly, $\{X_n\}_1^\infty$ is a martingale. For any $a > 0$ and $b > 0$,

$$\begin{aligned}\int_{\{|X_n|\geq a\}} |X_n|\,dP &\leq \int_{\{|X_n|\geq a\}} E\{|\xi||\mathscr{F}_n\}\,dP \\ &= \int_{\{|X_n|\geq a\}} |\xi|\,dP \\ &= \int_{\{|X_n|\geq a, |\xi|\leq b\}} |\xi|\,dP + \int_{\{|X_n|\geq a, |\xi|>b\}} |\xi|\,dP \\ &\leq bP\{|X_n|\geq a\} + \int_{\{|\xi|>b\}} |\xi|\,dP \\ &\leq bE\{|X_n|\}/a + \int_{\{|\xi\}>b\}} |\xi|\,dP \\ &\leq bE\{|\xi|\}/a + \int_{\{|\xi|>b\}} |\xi|\,dP.\end{aligned}$$

By letting $a \to +\infty$ and then $b \to +\infty$, we obtain

$$\lim_{a\to\infty} \sup_n \int_{\{|X_n|\geq a\}} |X_n|\,dP = 0.$$

This shows that the sequence $\{X_n\}_1^\infty$ is uniformly integrable. Therefore, by Proposition 10.6.1, there exists a r.v. X_∞ such that $X_n \to X_\infty$ (a.s.) and in the mean. What remains to be shown is that

$$X_\infty = E\{\xi|\mathscr{F}_\infty\} \quad \text{(a.s.)}.$$

Let $N \geq n$ and $A \in \mathscr{F}_n$; then,

$$\int_A X_N\,dP = \int_A X_n\,dP = \int_A E\{\xi|\mathscr{F}_\infty\}\,dP = \int_A \xi\,dP.$$

Because the sequence $\{X_n\}_1^\infty$ is uniformly integrable and because

$$\int_A |X_N - X_\infty|\,dP \to 0 \quad \text{as } N \to \infty,$$

it follows that

$$\int_A X_\infty\,dP = \int_A \xi\,dP. \tag{10.6.4}$$

This equation holds for all $A \in \mathscr{F}_n$ and, therefore, for all $A \in \bigcup_1^\infty \mathscr{F}_n$. Because $E|X_\infty| < \infty$ and $E|\xi| < \infty$, the integrals in (10.6.4) are σ-additive measures agreeing on the algebra $\bigcup_1^\infty \mathscr{F}_n$. Because of the uniqueness of their extensions to $\sigma\{\bigcup_1^\infty \mathscr{F}_n\}$, Equation (10.6.4) remains valid for all $A \in \sigma\{\bigcup_1^\infty \mathscr{F}_n\} = \mathscr{F}_\infty$. Thus, for all $A \in \mathscr{F}_\infty$,

$$\int_A X_\infty \, dP = \int_A \xi \, dP = \int_A E\{\xi | \mathscr{F}_\infty\} \, dP,$$

which implies that

$$X_\infty = E\{\xi | \mathscr{F}_\infty\} \quad \text{(a.s.)}.$$

This completes the proof of the assertion. □

The following example shows that a branching process has the martingale structure.

EXAMPLE 10.6.1. Let

$$\begin{gathered} X_1^{(1)}, X_2^{(1)}, \ldots \\ X_1^{(2)}, X_2^{(2)}, \ldots \\ \vdots \\ X_1^{(n)}, X_2^{(n)}, \ldots \\ \vdots \end{gathered}$$

be a sequence of i.i.d. non-negative integer-valued random variables such that

$$E\{X_i^{(n)}\} = \mu > 0.$$

Set

$$T_0 = 1, \qquad T_1 = X_1^{(1)},$$

and, for $n \geq 2$,

$$T_n = \sum_{i=1}^{T_{n-1}} X_i^{(n)} \quad \left(\sum_{i=1}^{0} (\cdot) = 0 \right). \tag{10.6.5}$$

Note that $\{T_i\}_0^{n-1}$ is independent of $\{X_i^{(n)}\}_1^\infty$. Consequently,

$$E\{T_n\} = E\{T_{n-1}\}\mu = \mu^n.$$

Definition 10.6.1. The sequence of random variables $\{T_n\}_1^\infty$ is called a "Galton Watson branching process."

Here, T_n can be thought of as the size of a population at the nth generation, with each individual independently giving birth to a random number of offspring to produce the population in the next generation.

Set

$$U_n = T_n/\mu^n, \quad n \geq 1;$$

then, $\{U_n\}_1^\infty$ represents a martingale. To show this, consider

$$\begin{aligned}
E\{U_n|U_1,\ldots,U_{n-1}\} &= E\left\{\sum_{i=1}^{T_{n-1}} X_i^{(n)}|U_1,\ldots,U_{n-1}\right\}\Big/\mu^n \\
&= \sum_{k=1}^{\infty} E\left\{I_{\{T_{n-1}=k\}}\sum_{i=1}^{k} X_i^{(n)}|U_1,\ldots,U_{n-1}\right\}\Big/\mu^n \\
&= \sum_{k=1}^{\infty} I_{\{T_{n-1}=k\}}\sum_{i=1}^{k} E\{X_i^{(n)}|U_1,\ldots,U_{n-1}\}/\mu^n \\
&= \sum_{k=1}^{\infty} kI_{\{T_{n-1}=k\}}/\mu^{n-1} = T_{n-1}/\mu^{n-1} = U_{n-1},
\end{aligned}$$

which proves the claim.

Next, because

$$\sup_n E|U_n| = \sup_n E\{U_n\} = 1,$$

then

$$U_n \to U_\infty \text{ (a.s.)} \quad \text{and } E\{U_\infty\} < \infty.$$

In other words,

$$T_n/\mu^n \to U_\infty \quad \text{(a.s.)},$$

which implies that

$$T_n \to 0 \text{ (a.s.) if } \mu < 1 \quad \text{and} \quad T_n \to \infty \text{ (a.s.) if } \mu > 1.$$

Problems and Complements

10.1. Let $\{\Omega, \mathscr{B}, P\}$ be a probability space and $\mathscr{F}_1$ and $\mathscr{F}_2$ two independent sub-σ-algebras of $\mathscr{B}$. If $A \in \mathscr{F}_1 \cap \mathscr{F}_2$, show that $P(A) = 0$ or 1.

10.2. (Continuation) If $G \subset \Omega$ is an arbitrary subset, write

$$\mathscr{B}_G = \{G \cap B; B \in \mathscr{B}\}.$$

(i) Show that $\mathscr{B}_G$ is a σ-algebra on G (the so-called σ-algebra induced on G by $\mathscr{B}$).

(ii) Let $\mathscr{S}$ be a family of subsets of Ω. Show that

$$\sigma\{\mathscr{S}_G\} = \sigma\{\mathscr{S}\} \cap G.$$

10.3. Show that there does not exist a σ-algebra having a countably infinite number of elements.

10.4. Let $X_1, \ldots, X_n$ be n independent Poisson random variables with parameters $\lambda_1, \ldots, \lambda_n$, respectively. Determine:

(i) the conditional distribution of $(X_1,\ldots,X_{n-1})$ given that $X_1 + \cdots + X_n = N$;

(ii) $E\{X_1|X_1 + X_2\}$.

10.5. Let $X \in L_1\{\Omega, \mathscr{B}, P\}$ and $\{B_k\} \subset \mathscr{B}$ be a partition of Ω. Let $\mathscr{F}$ be a σ-algebra generated by $\{B_k\}_1^\infty$. Show that

$$E\{X|\mathscr{F}\} = \sum_{k=1}^{\infty} \alpha_k I_{B_k} \quad \text{(a.s.)},$$

where $\alpha_k = E\{X|B_k\}$.

10.6. (Continuation) If $\mathscr{F}_1$ and $\mathscr{F}_2$ are two sub-σ-algebras of $\mathscr{B}$ such that $\sigma\{X\} \vee \mathscr{F}_1$ is independent of $\mathscr{F}_2$ show that

$$E\{X|\mathscr{F}_1 \vee \mathscr{F}_2\} = E\{X|\mathscr{F}_1\} \quad \text{(a.s.)},$$

where $\mathscr{F}_1 \vee \mathscr{F}_2 = \sigma\{\mathscr{F}_1 \cup \mathscr{F}_2\}$.

10.7. Let $\{X_i\}_1^\infty \subset L_1\{\Omega, \mathscr{B}, P\}$ be an i.i.d. sequence of random variables. Set $S_n = X_1 + \cdots + X_n$ and show that

$$E\{X_1|S_n, S_{n+1}, \ldots\} = S_n/n \quad \text{(a.s.)}.$$

10.8. Let $X, Y \in L_1\{\Omega, \mathscr{B}, P\}$ and let $\mathscr{F}_1$ and $\mathscr{F}_2$ be two independent sub-σ-algebras. Show that $E\{X|\mathscr{F}_1\}$ and $E\{Y|\mathscr{F}_2\}$ are independent random variables.

10.9. (Continuation) Show that

$$E\{X|\mathscr{F}_1 \cap \mathscr{F}_2\} = E(X).$$

10.10. Let $\{X_i\}_0^\infty$ be a sequence of independent random variables such that $E(X_i) = 1$, $i = 0, 1, \ldots$. Show that the sequence $\{\xi_n\}_0^\infty$ is a martingale, where

$$\xi_n = \prod_{i=0}^{n} X_i.$$

10.11. Let $\{X_i\}_1^\infty$ be a sequence of independent random variables and $\{Y_j\}_1^\infty$ another sequence of random variables such that, for each n, the family $\{X_n, X_{n+1}, \ldots\}$ is independent of $\{Y_1, \ldots, Y_n\}$. Show that the sequence $\{Z_n\}_1^\infty$ is a martingale, where

$$Z_n = \sum_{k=1}^{n} X_k Y_k \quad \text{if } X_k Y_k \in L_1\{\Omega, \mathscr{B}, P\}, \quad k = 1, 2, \ldots.$$

10.12. Let $\{\xi_n\}_0^\infty$ be a martingale and $\varphi(\cdot)$ a convex function such that $\varphi(\xi_n) \in L_1\{\Omega, \mathscr{B}, P\}$, $n = 0, 1, \ldots$. Show that the sequence $\{(\varphi(\xi_n), \mathscr{F}_n)\}$, where $\mathscr{F}_n = \sigma\{\xi_0, \ldots, \xi_n\}$ is a submartingale.

10.13. Let $\{\mathscr{F}_k\}_1^\infty$ be a nondecreasing sequence of σ-algebras and $\mathscr{F} = \sigma\{\bigcup_{k=1}^\infty \mathscr{F}_k\}$. Let ξ be an $\mathscr{F}$-measurable random variable having finite expectation. Show that

$$E\{\xi|\mathscr{F}_n\} \to \xi \quad \text{(a.s.)}.$$

10.14. Let $\{\xi_n\}_1^\infty$ be a uniformly integrable martingale with $\xi_n \geq 0$, $\xi_n \to \xi$ (a.s.). Show that for all $n = 1, 2, \ldots$

$$\xi_n = E\{\xi|\xi_1, \ldots, \xi_n\}.$$

10.15. Let $\{X_n\}_1^\infty$ be an i.i.d. sequence of random variables with $E(X_1) = 0$, $E(X_1^2) = 1$ and $\phi(\lambda) = E(\exp\{\lambda X_1\}) < \infty$ for $|\lambda| < \varepsilon$. Set $S_n = X_1 + \cdots + X_n$. Show that as

$\lambda \to 0$

$$\phi(\lambda) = 1 + \lambda^2/2 + o(\lambda^2).$$

10.16. Let $h(t) = (2t \log\log t)^{1/2}$. Show that

$$\lim_{t \to +\infty} h(t)/t = 0.$$

10.17. Let $\lambda \in (-\varepsilon, \varepsilon)$; show that the sequence $\{Y_n\}_1^\infty$ is a martingale, where

$$Y_n = \exp\{\lambda S_n\}/(\phi(\lambda))^n.$$

10.18. Let $\{Z_k\}_1^\infty$ be a sequence of i.i.d. random variables with support in $(0, 1]$ and $\alpha_i = E(Z_1)_i < 1$, $i = 1, 2$. Show that $\{L_n\}_1^\infty$ is a martingale where $L_n = (\prod_1^n Z_i/\alpha_1^n)$. Find $\lim_{n\to\infty} E\{L_n^2\}$ and show that $L_n \to 0$ (a.s.).

10.19. (Continuation) Let $\{Y_k\}_1^\infty \subset L_2\{\Omega, \mathcal{B}, P\}$ be an i.i.d. sequence of strictly positive random variables independent of $\{Z_k\}_1^\infty$ and set

$$X_n = Y_n \prod_1^n Z_i, \qquad S_n = X_1 + \cdots + X_n.$$

Show that $S_n \to S$ (a.s.), where $E\{S\} = \alpha_1\mu_1/(1 - \alpha_1)$ and $\mu_1 = E\{Y_1\}$. Show also that $S_n \overset{d}{=} Z_1(Y_1 + S_{n-1})$.

10.20. (Continuation) Denote by $G_n(x) = P\{S_n \le x\}$, $G(x) = P\{S \le x\}$, and show that

$$G(x) = \int_0^1 h(t)\,dt \int_0^{x/t} Q\left(\frac{x}{t} - s\right) g(s)\,ds,$$

where $Q(y) = P\{Y \le y\}$ and h is the probability density of Z_1.

Bibliography

Chapter 1

Doob, J.L. (1953). *Stochastic Processes*. Wiley, New York.
Elliott, J.R. (1982). *Stochastic Calculus and Applications*. Springer-Verlag, New York.
Gihman, I.I. and Skorohod, A.V. (1969). *Introduction to the Theory of Random Processes*. Saunders, Philadelphia.
Halmos, P.R. (1950). *Measure Theory*. Van Nostrand, New York.
Kolmogorov, A.N. (1941). Über das logarithmisch normale Verteilungsgesetz der Dimensionen der Teilchen bei Zerstückelung. *Dokl. Akad. Nauk SSR*. **31**, 99–101.
Loève, M. (1977). *Probability Theory I*. Springer-Verlag, New York.
Neveu, J. (1965). *Mathematical Foundations of the Calculus of Probability*. Holden-Day, San Francisco.
Prohorov, Y.V. and Rozanov, Y.A. (1969). *Probability Theory*. Springer-Verlag, New York.
Todorovic, P. (1980). Stochastic modeling of longitudinal dispersion in a porous medium. *Math. Sci.* **5**, 45–54.
Todorovic, P. and Gani, J. (1987). Modeling of the effect of erosion on crop production. *J. Appl. Prob.* **24**, 787–797.

Chapter 2

Belayev, Yu. K. (1963). Limit theorem for dissipative flows. *Theor. Prob. Appl.* **8**, 165–173.
Daley, D.J. and Vere-Jones, D. (1988). *An Introduction to the Theory of Point Processes*. Springer Verlag, New York.
Feller, W. (1971). *An Introduction to Probability Theory and Its Applications*, Vol. 2, 2nd ed. Wiley, New York.
Grandell, J. (1976). *Doubly Stochastic Poisson Processes* (Lecture Notes Math. 529). Springer-Verlag, New York.

Kac, M. (1943). On the average number of real roots of a random algebraic equation. *Bull. Am. Math. Soc.* **49**, 314–320.
Le Cam, L. (1960). An approximation theorem for the Poisson binomial distribution. *Pacific J. Math.* **10**, 1181–1197.
Rényi, A. (1967). Remarks on the Poisson process. *Stud. Sci. Math. Hungar.* **2**, 119–123.
Rice, S.O. (1945). Mathematical analysis of random noise, *Bell Syst. Tech. J.* **24**, 46–156.
Serfling, R.J. (1975). A general Poisson approximation theorem. *Ann. Prob.* **3**, 726–731.
Todorovic, P. (1979). A probabilistic approach to analysis and prediction of floods. Proc. 42 Session $|S|$. Manila, pp. 113–124.
Westcott, M. (1976). Simple proof of a result on thinned point process. *Ann. Prob.* **4**, 89–90.

Chapter 3

Bachelier, L. (1941). Probabilités des oscillations maxima. *C.R. Acad. Sci., Paris* **212**, 836–838.
Breiman, L. (1968). *Probability*. Addison-Wesley, Reading MA.
Brown, R. (1928). A brief account of microscopical observations made in the months of June, July, and August, 1927 on the particles contained in the pollen of plants. *Philos. Mag. Ann. Philos. (New Series)*. **4**, 161–178.
Einstein, A. (1905). On the movement of small particles suspended in a stationary liquid demanded by the molecular-kinetic theory of heat. *Ann. Physik* **17**.
Freedman, L. (1983). *Brownian Motion and Diffusion*. Springer-Verlag, New York.
Hartman, P. and Wintner, A. (1941). On the law of the iterated logarithm. *Am. J. Math.* **63**, 169–176.
Hida, T. (1965). *Brownian Motion*. Springer-Verlag, New York.
Karlin, S. (1968). *A First Course in Stochastic Processes*. Academic Press, New York.
Kunita, H. and Watanabe, S. (1967). On square integrable Mortingales. *Nagoya Math. J.* **30**, 209–245.
Lévy, P. (1965). *Processus Stochastiqus et Mouvement Brownien*. Gauthier Villars, Paris.
Nelson, E. (1967). *Dynamical Theories of Brownian Motion*. Mathematical Notes, Princeton University.
Skorohod, A.V. (1964). *Random Processes with Independent Increments*. Nauka, Moscow (in Russian).
Smokuchowski, M. (1916). Drei Vortrage über Diffusion Brownche Molekulorbewegung und Koagulation von Kolloidteilchen. *Phys. Zeit.* **17**, 557–571.
Uhlenbeck, G.E. and Ornstein, L.S. (1930). On the theory of Brownian motion. *Phys. Rev.* **36**, 823–841.

Chapter 4

Anderson, T.W. (1958). *An Introduction to Multivariate Statistical Analysis*. Wiley, New York.
Doob, J.L. (1953). *Stochastic Processes*. Wiley, New York.

Feller, W. (1971). *An Introduction to Probability Theory and its Applications, Volume 2*, 2nd ed. Wiley, New York.
Ibragimov, I.A. and Rozanov, Y.A. (1978). *Gaussian Random Processes.* Springer-Verlag, New York.
Rozanov, Y.A. (1968). Gaussian infinitely dimensional distributions, *Steklov Math. Inst. Publ.* **108**, 1–136. (in Russian).

Chapter 5

Akhiezer, N.I. and Glazman, I.M. (1963). *Theory of Linear Operators in Hilbert Space*, Volumes I and II. Frederic Ungar Publishing, Co., New York.
Dudley, R.M. (1989). *Real Analysis and Probability.* Wadsworth and Brooks/Cole, Pacific Grove, CA.
Loève, M. (1978). *Probability Theory, II.* Springer-Verlag, New York.
Kolmogorov, A.N. and Fomin, S.V. (1970). *Introductory Real Analysis.* Prentice-Hall, Englewood Cliffs, NJ.
Natanson, I.P. (1960). *Theory of Functions of Real Variables*, Volume I and II. *Frederic Ungar* Publishing, New York.
Riesz, F. and Sz.-Nagy, B. (1955). *Functional Analysis.* Frederic Ungar Publishing, New York.
Robinson, E.A. (1959). *An Introduction to Infinitely Many Variates.* Hafner Publishing, New York.
Royden, H.L. (1968). *Real Analysis*, 2nd ed. The Macmillan Co., New York.
Yosida, K. (1974). *Functional Analysis*, 4th ed. Springer-Verlag, New York.
Wilansky, A. (1964). *Functional Analysis.* Blaisdell Publishing, New York.

Chapter 6

Cramér, H. (1940). On the theory of stationary random processes. *Ann. Math.* **41**, 215–230.
Cramér, H. and Leadbetter, M.R. (1967). *Stationary and Related Stochastic Processes.* Wiley, New York.
Gihman, I.I. and Skorohod, A.V. (1974). *The Theory of Stochastic Processes.* Springer-Verlag, New York.
Grenander, U. and Rosenblatt, M. (1956). *Statistical Analysis of Stationary Time Series.* Wiley, New York.
Karhunen, K. (1947). Über Lineare Methoden in der Wahrscheinlichkeitsrechnung. *Ann. Acad. Sci. Fenn.* **37**.
Khinchin, A.Y. (1938). Correlation theory of stationary random processes. *Usp. Math. Nauk.* **5**, 42–51.
Loève, M. (1946). Fonctions aléatoires du second ordre. *Rev. Sci.* **84**, 195–206.
Lovitt, W.V. (1924). *Linear Integral Equation.* McGraw-Hill, New York.
Mercer, J. (1909). Functions of positive and negative type and their connections with the theory of integral equations. *Phil. Trans.* Roy. Soc. London, Ser. A, **209**, 415–446.
Riesz, F., and Sz-Nagy, B. (1955). *Functional Analysis.* Frederic Unger Publishing, New York.
Rozanov, A.Y. (1967). *Stationary Random Processes.* Holden-Day, San Francisco.
Tricomi, F.G. (1985). *Integral Equation.* Dover Publishing, New York.

Chapter 7

Ash, R. and Gardner, M.L. (1975). *Topics in Stochastic Processes.* Academic Press, New York.

Bochner, S. (1955). *Harmonic Analysis and Theory of Probability.* University of California Press, Berkeley.

Bochner, S. (1959). *Lectures on Fourier Integrals.* (Ann. Math. Studies 42). Princeton University Press, Princeton, NJ.

Cramér, H. (1940). On the theory of stationary random processes. *Ann. Math.* **41**, 215–230.

Cramér, H. and Leadbetter, M.R. (1967). *Stationary and Related Stochastic Processes.* Wiley, New York.

Gihman, I.I. and Skorohod, A.V. (1974). *The Theory of Stochastic Processes, Volume 1.* Springer-Verlag, New York.

Hájek, J. (1958). Predicting a stationary process when the correlation function is convex. *Czech. Math. J.* **8**, 150–161.

Khinchin, A.Y. (1938). Correlation theory of stationary random processes. *Usp. Math. Nauk.* **5**, 42–51.

Kolmogorov, A.N. (1941). Interpolation and extrapolation of stationary random sequences. *Izv. Akad. Nauk. SSSR Ser. Math.* **5**, 3–14.

Krein, M.G. (1954). On the basic approximation problem in the theory of extrapolation and filtering of stationary random processes. *Dokl. Akad. Nauk. SSSR* **94**, 13–16.

Liptser, R.S. and Shiryayev, A.N. (1978). *Statistics of Random Processes, Volume 2.* Springer-Verlag, New York.

Rozanov, A.Y. (1967). *Stationary Random Processes.* Holden-Day, San Francisco.

Vong, E. and Hayek, B. (1985). *Stochastic Processes in Engineering Systems.* Springer-Verlag, New York.

Wold, H. (1954). A study in the Analysis of Stationary Time Series, 2nd ed., Almqvist and Wiksell, Stockholm.

Yaglom, A.M. (1949). On the question of linear interpolation of stationary stochastic processes. *Usp. Math. Nauk.* **4**, 173–178.

Yaglom, A.M. (1962). *An Introduction to the Theory of Stationary Random Functions.* Prentice-Hall, Englewood Cliffs, NJ.

Chapter 8

Blumenthal, R.M. and Getoor, R.K. (1968). *Markov Processes and Potential Theory.* Academic Press, New York.

Chiang, L.C. (1968). *Introduction to Stochastic Processes in Biostatistics.* Wiley, New York.

Chung, K.L. (1967). *Markov Chains with Stationary Transition Probabilities, 2nd ed.* Springer-Verlag, New York.

Doob, J.L. (1953). *Stochastic Processes.* Wiley, New York.

Dynkin, E.B. and Yushkevich A.A. (1956). Strong Markov processes. *Theory Prob. Appl.* **1**, 134–139.

Dynkin, E.B. (1961). *Foundations of the Theory of Markov Processes.* Prentice-Hall, Englewood Cliffs, NJ.

Feller, W. (1954). Diffusion processes in one dimension. *Trans. Amr. Math. Soc.* **77**, 1–31.

Feller, W. (1966). *An Introduction to Probability Theory and its Applications, Volume 2*. Wiley, New York.
Gihman, I.I. and Skorohod, A.V. (1975). *The Theory of Stochastic Processes, Volume 2*. Springer-Verlag, New York.
Gnedenko, B.V. (1976). *The Theory of Probability*. Mir Publishers, Moscow.
Hille, E. (1950). On the differentiability of semi-groups of operators. *Acta Sci. Math. Szeged* **12B**, 19–24.
Itô, K. (1963). *Random Processes II*. Izdatelstwo Inostranoy Literatury, Moscow (Russian translation from Japanese).
Karlin, S. (1968). *A First Course in Stochastic Processes*. Academic Press, New York.
Kolmogorov, A.N. (1951). On the problem of differentiability of transition probabilities of time-homogeneous Markov processes with countable number of states. *Uchenye Zapiski Moskovskovo Gos.* **148** (Matematika 4), 53–59 (in Russian).
Lamperti, J. (1977). *Stochastic Processes*. Springer-Verlag, New York.
Mandl, P. (1968). *Analytical Treatment of One-Dimensional Markov Processes*. Springer-Verlag, New York.

Chapter 9

Dynkin, E.B. (1965). *Markov Processes*. Springer-Verlag, Berlin.
Ethier, N.S. and Kurtz, T.G. (1986). *Markov Processes*. Wiley, New York.
Feller, W. (1952). The parabolic differential equations and the associated semigroups of transformations. *Ann. Math.* **55**, 468–519.
Hille, E. and Philips, R.S. (1957). *Functional Analysis and Semi-groups*. Am. Math. Soc. Colloq. Publ. 31. American Mathematical Society, Providence, RI.
Hille, E. (1948). Functional Analysis and Semi-groups. Collq. Publ. Amer. Math. Soc.
Mandl, P. (1968). *Analytical Treatment of One-Dimensional Markov Processes*. Springer-Verlag, New York.
Yosida, K. (1948). On the differentiability and representation of one-parameter semigroups of linear operators. J. Math. Soc. Japan. **1**, 15–21.
Yosida, K. (1974). *Functional Analysis*, 4th ed. Springer-Verlag, New York.

Chapter 10

Breiman, L. (1968). *Probability*. Addison-Wesley, Reading MA.
Chung, K.L. (1974). *A Course in Probability Theory*, 2nd ed. Academic Press, New York.
Doob, J.L. (1953). *Stochastic Processes*. Wiley, New York.
Loève, M. (1978). *Probability Theory II*, 4th ed. Springer-Verlag, New York.
Shiryayev, A.N. (1984). *Probability*. Springer-Verlag, New York.

Index

A

A/B/s, 18
Absorbing state, 209, 215
Adapted, 262
Adjoint operator, 123
Admissible linear filter, 175
Almost sure continuous trajectories, 24
Analytical, 189
Arc sin law, 70
Arrival times, 17, 27
Autogenesis, 211
Average hitting time, 66

B

Bachelier, 81
Backward diffusion equation, 226
Backward equation, 210, 212
Banach space, 218, 233
Basis of L_2, 118
Belayev, 52, 279
Beppo-Levi, 112
Bernoulli, 37, 39
Bernoulli random variables, 37, 39
Bessel inequality, 108
Best approximation, 171
Best linear predictor, 187
Binomial component, 213
Birth and death process, 211
Bivariate point process, 53
Bochner-Khinchin, 151
Bombardment molecular, 81
Borel cylinder, 4
Borel-Cantelli, 77
Borel-Radon measure, 35
Bounded generator, 242
Bounded linear operator, 223
Bounded random variables, 113
Branching process, 275
Brown, Robert, 15, 62
Brownian bridge, 64
Brownian motion, 15, 63
Brownian motion with drift, 64

C

Cadlag, 7, 20, 215
Cardinality, 121
Cauchy functional equation, 49, 202, 215, 238
Cauchy inequality, 109
Cauchy or fundamental sequence, 111
Cauchy theorem, 189
Cauchy sequence, 111, 202
Chapman-Kolmogorov equation, 11, 200, 202, 206, 224
Characterization of normal distribution, 96

Closed contour, 189
Closed linear manifold, 115
Closure, 19
Complete orthogonal system, 118
Completeness, 113
Conditional
  expectation, 258
  probability, 259
Conjugate space, 241
Conservative Markov process, 202
Consistency conditions, 2
Continuity (a.s.), 24
Continuity concepts, 22
Continuous functions (set), 22
Continuous in probability, 22; *see also* stochastically continuous
Continuous time Markov chain, 205
Contraction, 218, 233
Contraction semigroup, 233
Control, 162
Convergence in quadratic mean, 110
Convergence of submartingales, 268
Convex, 127, 262, 263
Countably additive, 114
Counting random function, 36
Coupling, 38
Covariance function, 129
Covariance matrix, 94
Cox, 51
Cramer, H., 79
Curve in the Hilbert space, 177

D
Death rate, 211
Death process, 212
Decomposition of Z, 116
Dense subset, 114
Deterministic process, 177
Deviation square, 223
Difference-differential equations, 212
Diffusion coefficient, 223
Diffusion process, 223
Dimensionality, 121
Dinamically neutral, 16
Dini, 141
Dirac measure, 35
Directly given, 3
Discontinuities, 23
  of the first kind, 25
Discontinuity point (fixed), 23
Discrete parameter processes, 1, 182
Dispersion (longitudinal), 16
Distance in L_2, 109
Distance (minimal), 116
Distribution (marginal), 1
Doob, J., 19, 20, 64, 79, 82, 103, 267, 268
Doubly stochastic, 51
Drift coefficient, 223
Dynkin, 202, 216, 217

E
Eigen function, 136
Eigenfunctions of integral equation, 137
Eigenvalues
  of a matrix, 93
  of integral equation, 137
Einstein, A., 62
Elementary random measure, 157
Entire function, 193
Epoch, 70
Equivalent processes, 5
Ergodicity, 145, 146
Error of estimate, 170
Essentially bounded, 113
Essential supremum, 114
Estimate, 170
Estimation, 169
Everywhere dense, 114
Everywhere dense in L_2, 114
Exceedance, 57
Excursion, 57
Extremes of Brownian motion, 67

F
Feller, 53, 75
Feller processes, 128, 218
Filtering and prediction, 169, 170
Filtration, 262, 265, 273
Finite dimensional distribution, 201
Fischer, 111
Fixed discontinuity point, 24
Flood modeling, 56

Fokker Planck, 226
Forward diffusion equation, 226
Fourier coefficients, 118
Fourier series, 118
Fredholm equation, 137
Frictional force, 81
Fubini, 28
Functional, 186, 241

G
G/G/1, 18
Galton–Watson, 275
Gambling, 263
Gaussian process, 97
Gaussian system, 93
Generalized derivative, 134
Generalized Fourier series, 118
Generating function, 213
Generator (infinitesimal), 235
Generator of a semigroup, 234
Germination process, 13
Global behavior, 201
Gram–Schmidt, 121, 136
Grandel, 51

H
Hartman, 75
Herglotz's theorem, 152
Hermitian
  form, 131
  kernel, 137
  symmetry, 130, 137
Hilbert space, 113
Hille, E., 207
Hille–Yosida, 243, 254
History, 261
Hitting times, 65
Homogeneous diffusion, 223
Homogeneous Markov process, 10, 200
Homogeneous Poisson process, 43

I
Imbibition, 13
Independent increments, 8
Indistinguishable, 6
Initial distribution, 9, 201
Inner product, 107
Innovation, 183
Instantaneous state, 209
Integral (stochastic), 86
  Riemann, 86
Interarrival times, 48
Internal history, 261
Invariant measure, 203
Inverse matrix, 92
Isometric
  isomorphism, 158
  mapping, 158
  operator, 124
Itô, 86, 226

J
Jensen's inequality, 112
Joint probability density, 94
Jump, 216

K
Kac, M., 35
Karhunen–Loève expansion, 139
Kernel of integral equation, 137
Khintchine, 75
Kolmogorov, 7, 15, 25, 75, 224, 226
Kunita, 86

L
L_2 space, 106
L_2-continuous process, 132
Langevin's equation, 81
Laplace transform, 53, 239
Law of motion (Newton), 81
Law of the iterated logarithm, 74
Le Cam, 39
Likelihood ratio, 264
l.i.m., 110
Linear estimate, 173
Linear manifold, 115, 173
Linear operator, 122
Linear predictor, 173
Linear transformation, 174
Lipschitz conditions, 226
Local behavior, 201

Loève's criterion, 113
Lévy, 226

M
M/M/1, 18
Marginal distribution, 1, 2
Marked point process, 53
Markov Gaussian process, 99
Markov inequality, 23
Markov process, 9, 200
Markov process homogeneous, 10
Markov process regular, 216
Markov property, 9
Markov renewal process, 220
Markov time, 216
Martingale, 12, 46, 258, 272
Martingale closed, 262
Mathematical modeling, 12
Matrix, 92
Maximum, 19, 56, 68
Mean rate, 43
Mean square error, 170
Measurable process, 27
Mercer, 140
Metric, 38
Moving average, 176, 183, 185

N
Newton, 12, 81
Noise, 172
Non-deterministic process, 177
Non-negative definite, 93, 130
Norm, 108, 123
Norm of an operator, 123
Normal distribution, 94
Normal operator, 124
Normally distributed, 94

O
Order statistics, 23, 44
Ornstein Uhlenbeck process, 81, 82
Orthogonal
  basis, 118
  collection, 108
  complement, 117
  matrix, 93
  projection, 115
  random measure, 157
Orthonormal collection, 108
Outer measure, 8

P
Parallelogram law, 108
Parameter set, 1
Parseval, 119
Partition, 73
Path, 2, 71
Physically realizable filter, 176
Pincherle–Goursat kernel, 138
Poincaré, H., 62
Point measure, 36
Point process on R, 34
Point process, simple, 35
Poisson process, 39, 40
Pole, 190
Pólya, 152
Porous medium, 16
Positive definite, 93, 94
Prediction (pure), 169
Process with orthogonal increments, 160
Projection operator, 124, 125
Pure birth process, 49, 212
Pure death process, 212
Purely non-deterministic process, 178

Q
Quadratic form, 93
Quadratic mean continuity, 132
Quadratic mean differentiability, 134
Quadratic variation, 73
Queues, 17

R
Rainfall, 34, 51
Random measure, 36, 175
Random process, 1
Random telegraph process, 153
Real valued, 1
Rectifiable, 74
Reflection principle, 68
Regular (stable) state, 209
Regular Markov process, 216

Renyi, 53
Resolvent, 244, 238
Rice, 35
Riemann stochastic integral, 86
Riesz–Fisher theorem, 111
Right continuous inverse, 43

S
Sample path (function), 2
Scheffé, 93
Schwarz's inequality, 107
Second order, 11, 106
Second order process, 129
Self adjoint operator, 123
Semigroup, 232, 233
Separability, 18
Separable process, 19
Separable version, 20
Serfling, 39
Shift operator, 174
Singleton, 197
Singular matrix, 92
Smoluchowski, M., 62
Soil erosion, 14
Spanned, 115
Spectral characteristic, 174
Spectral density, 121
Spectral distribution, 121, 163
Spectral representation of a process, 162
Standard Brownian motion, 63
State space, 2, 200
Stationary distribution, 203
Stationary Gaussian Markov, 102
Stationary Gaussian process, 102
Stationary stochastic process, 10, 11, 143
Stationary transition probability, 10
Stieltjes integral, 85
Stochastic
  integration, 85
  matrix, 206
  measure, 157
  process, 1
  sequence, 261
  structure, 8
Stochastically
  continuous, 22
  equivalent, 5
  equivalent (wide sense), 5
Stone, 162
Stopping time, 216
Strictly stationary, 10
Strong
  convergence, 234
  Markov process, 216
  Markov property, 217
Strongly continuous, 234
Strongly integrable, 236
Subadditive, 207
Submartingale, 12, 261
Subordinated process, 175
Subspace, 115
Supermartingale, 12, 261
Supremum norm, 232
Symmetric distribution, 153

T
Telegraph process, 144
Thinned version, 52
Thinning of a point process, 51
Todorovic, 58
Total variation distance, 38
Transition
  probability, 10
  standard, 218
  stochastically continuous, 252
Transpose matrix, 92
Triangle inequality, 109

U
Unbiased estimate, 170
Uniform integrability, 270
Uniformly integrable martingales, 272
Unitary, 115
Unitary operator, 124
Upcrossing, 266
Upcrossing inequality, 266

V
Version, 6, 9

W
Weak convergence, 52
Westcott, 52

White noise, 83, 184
Wide sense stationary, 11, 143
Wiener process, 63
Wold decomposition, 179

Y
Yaglom, 187
Yushkevich, 217, 218

Springer Series in Statistics

(continued from p. ii)

Shedler: Regeneration and Networks of Queues.
Siegmund: Sequential Analysis: Tests and Confidence Intervals.
Todorovic: An Introduction to Stochastic Processes and Their Applications.
Tong: The Multivariate Normal Distribution.
Vapnik: Estimation of Dependences Based on Empirical Data.
West/Harrison: Bayesian Forecasting and Dynamic Models.
Wolter: Introduction to Variance Estimation.
Yaglom: Correlation Theory of Stationary and Related Random Functions I: Basic Results.
Yaglom: Correlation Theory of Stationary and Related Random Functions II: Supplementary Notes and References.

MIX
Papier aus verantwortungsvollen Quellen
Paper from responsible sources
FSC® C105338
FSC
www.fsc.org

If you have any concerns about our products,
you can contact us on
ProductSafety@springernature.com

In case Publisher is established outside the EU,
the EU authorized representative is:
Springer Nature Customer Service Center GmbH
Europaplatz 3, 69115 Heidelberg, Germany

Printed by Libri Plureos GmbH
in Hamburg, Germany